AA003637

9th International Scientific Conference on Aeronautics, Automotive and Railway Engineering and Technologies (BulTrans-2017)

MATEC Web of Conferences Volume 133 (2017)

Sozopol, Bulgaria
11 - 13 September 2017

Editors:

Boyko Gigov
Nikolay Nikolov
Valeri Stoilov
Michael Todorov

ISBN: 978-1-5108-5190-0

Printed from e-media with permission by:

Curran Associates, Inc.
57 Morehouse Lane
Red Hook, NY 12571

Some format issues inherent in the e-media version may also appear in this print version.

No changes have been made to the content of these proceedings. There may be changes to pagination and minor adjustments for aesthetics.

Printed by Curran Associates, Inc. (2019)

For additional information, please contact EDP Sciences – Web of Conferences at the address below.

EDP Sciences – Web of Conferences
17, Avenue du Hoggar
Parc d'Activité de Courtabœuf
BP 112
F-91944 Les Ulis Cedex A
France

Phone: +33 (0) 1 69 18 75 75
Fax: +33 (0) 1 69 28 84 91

contact-edps@webofconferences.org

Additional copies of this publication are available from:

Curran Associates, Inc.
57 Morehouse Lane
Red Hook, NY 12571 USA
Phone: 845-758-0400
Fax: 845-758-2633
Email: curran@proceedings.com
Web: www.proceedings.com

TABLE OF CONTENTS

TECHNICAL UNIVERSITY - SOFIA
FACULTY OF TRANSPORT

INTERNATIONAL
SCIENTIFIC CONFERENCE
on Aeronautics, Automotive
and Railway Engineering and Technologies

BulTrans-2017

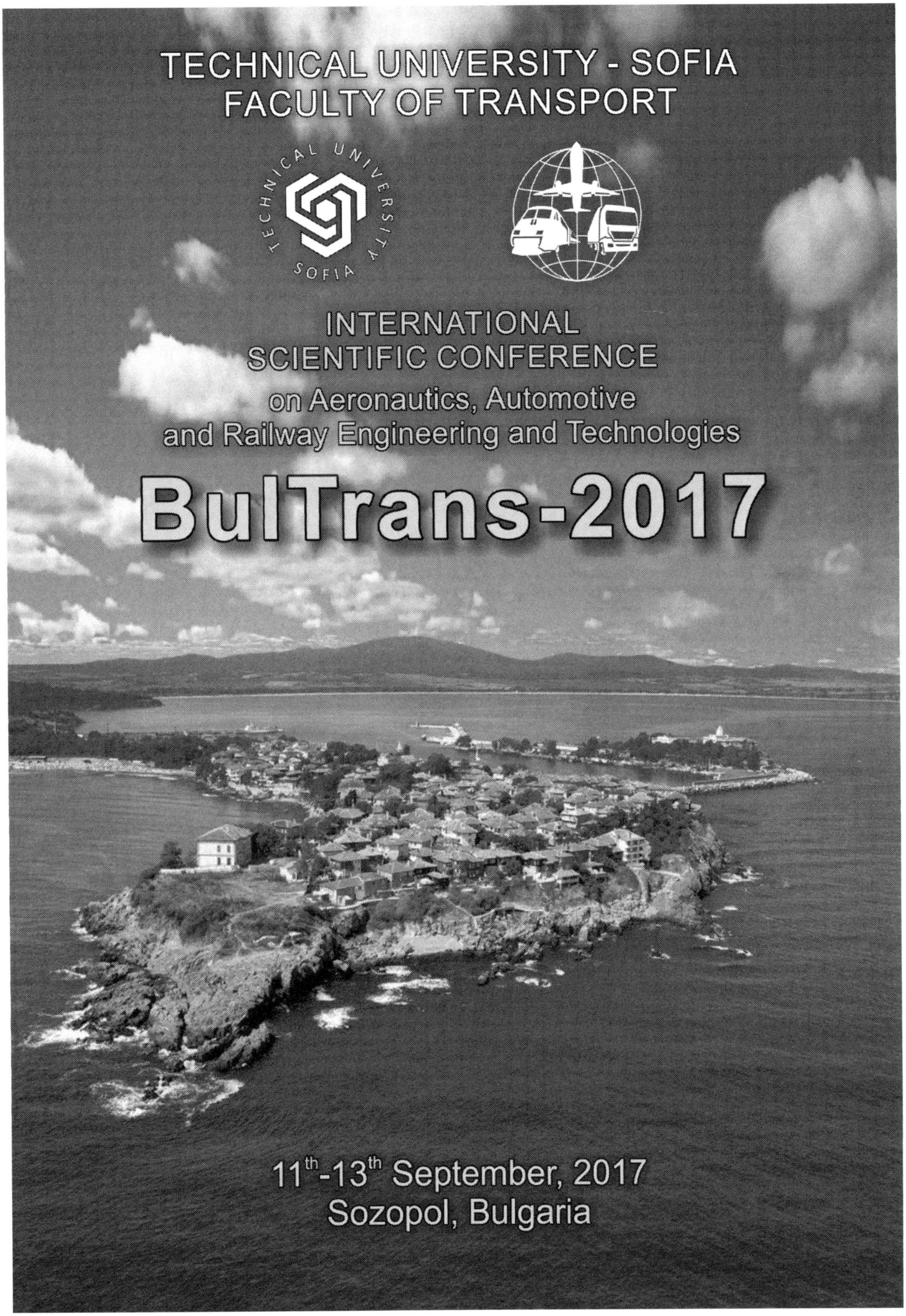

11th-13th September, 2017
Sozopol, Bulgaria

PREFACE

BulTrans International Scientific Conference on Aeronautics, Automotive and Railway Engineering and Technologies is held for the first time in 2009, as part of *The Science Days* of Technical University of Sofia. Till now, eight editions of the Conference have been organized, seven in USOB "Lazur" - Sozopol, and a jubilee forum in Sofia.

Thorough scientific achievements in the field of transport are being presented at the conference. *BulTrans* is the most prestigious scientific forum for road, railway and air transport in Bulgaria.

OPENING ADDRESS

Assoc. Prof. B. Gigov, PhD,
Chairman of
Organizing Committee

Dear colleagues and guests,

Welcome to the BulTrans-2017 Scientific Conference, organized annually by the Faculty of Transport at the Technical University - Sofia. Every year, our goal is to improve the conference by publishing valuable reports and articles in the field of transport and transport technology by Bulgarian and foreign scientists. This research is an essential base for the training of highly-qualified higher education professionals.

I wish successful and fruitful work to all participants!

Statement of Peer review

In submitting conference proceedings to MATEC Web of Conferences, I certify to the Publisher that I adhere to the **Policy on Publishing Integrity** of the journal in order to safeguard good scientific practice in publishing.

1. All articles have been subjected to peer review administered by the proceedings editors.
2. Reviews have been conducted by expert referees, who have been requested to provide unbiased and constructive comments aimed, whenever possible, at improving the work.
3. Proceedings editors have taken all reasonable steps to ensure the quality of the materials they publish and their decision to accept or reject a paper for publication has been based only on the merits of the work and the relevance to the journal.

Title, date and place of the conference

BulTrans-2017 - 9th International Scientific Conference on Aeronautics, Automotive and Railway Engineering and Technologies

11th-13th September 2017, Sozopol, Bulgaria

Proceedings editor(s):

Assoc. Prof. **Boyko Gigov**, PhD
Assoc. Prof. **Nikolay Nikolov**, PhD
Prof. **Valeri Stoilov**, PhD
Prof. **Michael Todorov**, PhD

Date and editor's signature

23 October 2017

GPS as a base for analysis of perturbations of space based and ground based applications

Tsvetelina Traykova[1,*], and *Veselin* Radkov[2]

[1]CAMO Partner Ltd., Sofia, Bulgaria
[2]Technical University – Sofia, Bulgaria

Abstract. The present work is dedicated to examining the impact of respective Navstar/GPS embarrass impact on the running of the global navigation system as well as synthesizing a computer program for their impact judgment as well. Two major groups of perturbation factors are addressed: perturbations proven impact on the orbital segment and perturbations proven impact on the running of the user's piston ring (non-orbital disturbances); it is being synthesized in the theoretical productions on the ground, plan code that yes on is modeled. The influence of those who embarrass the invoices on the accuracy of specifying the location.

1 Introduction

NAVSTAR, commonly known as Global Position System (GPS), provides worldwide three-dimensional position and velocity information to users on the Earth's surface or in the air, including precise time. In order to determine true user position, the data have to be transmitted continuously with high accuracy and perturbation effects to be minimized.

The purpose of this paper is to introduce an algorithm for computation of the errors in determining the user position, which are caused by different perturbation effects. For instance GPS satellite orbits are influenced by perturbation of factors such as gravity, oblateness of Earth, atmospheric drag, solar radiation pressure, etc. Other factors that cause an error are Satellite clock error, Relativistic Effects, Ionosphere effect, Tropospheric delay, Multipath effect affect, etc.

The presented algorithm is implemented in a computer code that allows for the analysis of the influence of the different factors mentioned above – one by one, or in a combination, on determining the user position. Numerical simulations were performed as well and the results of the analysis will be presented in a successive paper.

2 Perturbations at GPS satellite orbits

To include the effects of the perturbations, the equation of motion is written in a general form as [1]:

$$\frac{d^2r}{dt^2} = -\mu\frac{r}{r^3} + a_p, \qquad (1)$$

where r is position vector; $\mu = GM_e$, (M_e – Earth mass, $G = 6{,}6726\ (\pm 0{,}0005).10^{-11}\ m^3kg^{-1}s^{-2}$ – gravitational constant; a_p – the resultant vector of all the perturbing accelerations. The magnitude of the a_p in the Solar system for all types of satellite orbits is 10 times smaller than the central force, or $|a_p << \mu r/r^3|$. The resultant vector may consist of several types of perturbing accelerations, described in Table 1.

Perturbations at satellite orbit can be classified as conservative and non-conservative. For conservative accelerations, a_p is an explicit function of position only, and there is no net energy transfer, i.e. the mean semi-major axis of the orbit is constant. The second type, non-conservative, such as atmospheric drag and tidal friction effect, a_p is an explicit function of both position and velocity, i.e. energy transfer occurs. As a result, the mean semi-major axis of the orbit changes.

Perturbations at satellite orbit which are included in the present research are described in this work.

Table 1. Perturbation accelerations.

Gravitational:	Non-gravitational:
Third-body (sun/moon) attractions	The non-spherical Earth
Outgassing (fuel tank leaks on the spacecraft)	Atmospheric drag
	Solar-radiation pressure
	Tidal friction effect

2.1 Gravity perturbation

Each Earth satellite experiences the force of gravitational attraction from the other bodies in the system – three-body system; Sun, Moon, Earth. Forces exerted on body 1 by bodies 2 and 3 are F_{12} and F_{13}, respectively. Likewise, body 2 experiences the forces F_{21} and F_{23} whereas the forces F_{31} and F_{32} act on body 3 [2]:

$$F_{12} = -F_{21} = \frac{Gm_1m_2(R_2-R_1)}{\|R_2-R_1\|^3}; \qquad (2)$$

$$F_{13} = -F_{31} = \frac{Gm_1m_3(R_3-R_1)}{\|R_3-R_1\|^3}; \qquad (3)$$

* Corresponding author: ctraiikova@abv.bg

 DOI: 10.1051/matecconf/201713301001

$$F_{23} = -F_{32} = \frac{Gm_3 m_2 (R_3 - R_2)}{\|R_3 - R_2\|^3}. \quad (4)$$

With respect to the inertial system the acceleration of the bodies is:

$$a_i = \ddot{R}_i, i = 1, 2, 3 \dots, \quad (5)$$

where R_i is absolute position vector of body i.

The equation of motion of body 1 is:

$$F_{12} + F_{13} = m_1 a_1. \quad (6)$$

Substituting in (6) equations for F_{12} and F_{13} yields:

$$a_1 = \frac{Gm_2(R_2 - R_1)}{\|R_2 - R_1\|^3} + \frac{Gm_3(R_3 - R_1)}{\|R_3 - R_1\|^3}. \quad (7a)$$

In a similar way the equations for body 2 and 3 are:

$$a_2 = \frac{Gm_1(R_1 - R_2)}{\|R_1 - R_2\|^3} + \frac{Gm_3(R_3 - R_2)}{\|R_1 - R_2\|^3}; \quad (7b)$$

$$a_3 = \frac{Gm_1(R_1 - R_3)}{\|R_1 - R_3\|^3} + \frac{Gm_2(R_2 - R_3)}{\|R_2 - R_3\|^3}. \quad (7c)$$

The accelerations are related to the velocities by:

$$\frac{dv_i}{dt} = a_i, \ i = 1, 2, 3, \quad (8)$$

and the position vectors are likewise related to the velocities:

$$\frac{dR_i}{dt} = v_i, \ i = 1,2,3. \quad (9)$$

Equations (6-9) form a system of ordinary differential equations in the variable time t. Since there are no external forces, the acceleration on the center of the masses is zero, i.e.

$$a_G = 0, \quad (10)$$

so that

$$\frac{dv_G}{dt} = 0, \quad (11)$$

and

$$\frac{dR_G}{dt} = v_G. \quad (12)$$

Since the system above decouples with respect to the first three variables, it is easily integrable – the initial positions R_{i0} and velocities v_{i0} are given, to find v_i the first three equations have to be integrated. The resulting functions are substituted in the second three equations of the system to obtain R_i as a function of time.

The three-body system could be extended to N-body system. A geostationary satellite is related to three other bodies, but because of the lower mass of the satellite it is not taken into account in the system.

To solve the above problem the integrations are usually done numerically. For instance, writing all vectors component-wise with respect to *XYZ* basis as column vector we obtain the following presentation:

$$\{R_1\} = \begin{Bmatrix} R_{1x} \\ R_{1y} \\ R_{1z} \end{Bmatrix}; \ \{R_2\} = \begin{Bmatrix} R_{2x} \\ R_{2y} \\ R_{2z} \end{Bmatrix};$$

$$\{R_3\} = \begin{Bmatrix} R_{3x} \\ R_{3y} \\ R_{3z} \end{Bmatrix}; \{R_G\} = \begin{Bmatrix} R_{Gx} \\ R_{Gy} \\ R_{Gz} \end{Bmatrix}, \quad (13)$$

and

$$\{v_1\} = \begin{Bmatrix} v_{1x} \\ v_{1y} \\ v_{1z} \end{Bmatrix}; \{v_2\} = \begin{Bmatrix} v_{2x} \\ v_{2y} \\ v_{2z} \end{Bmatrix};$$

$$\{v_3\} = \begin{Bmatrix} v_{3x} \\ v_{3y} \\ v_{3z} \end{Bmatrix}; \{v_G\} = \begin{Bmatrix} v_{Gx} \\ v_{Gy} \\ v_{Gz} \end{Bmatrix}, \quad (14)$$

therefore

$$\{a_1\} = \begin{Bmatrix} a_{1x} \\ a_{1y} \\ a_{1z} \end{Bmatrix} = \begin{Bmatrix} \frac{Gm_2(R_{2x} - R_{1x})}{R_{12}} + \frac{Gm_3(R_{3x} - R_{1x})}{R_{13}} \\ \frac{Gm_2(R_{2y} - R_{1y})}{R_{12}} + \frac{Gm_3(R_{3y} - R_{1y})}{R_{13}} \\ \frac{Gm_2(R_{2z} - R_{1z})}{R_{12}} + \frac{Gm_3(R_{3z} - R_{1z})}{R_{13}} \end{Bmatrix}; \quad (15)$$

$$\{a_2\} = \begin{Bmatrix} a_{2x} \\ a_{2y} \\ a_{2z} \end{Bmatrix} = \begin{Bmatrix} \frac{Gm_2(R_{1x} - R_{2x})}{R_{12}} + \frac{Gm_3(R_{3x} - R_{2x})}{R_{13}} \\ \frac{Gm_2(R_{1y} - R_{2y})}{R_{12}} + \frac{Gm_3(R_{3y} - R_{2y})}{R_{13}} \\ \frac{Gm_2(R_{1z} - R_{2z})}{R_{12}} + \frac{Gm_3(R_{3z} - R_{2z})}{R_{13}} \end{Bmatrix}; \quad (16)$$

$$\{a_3\} = \begin{Bmatrix} a_{3x} \\ a_{3y} \\ a_{3z} \end{Bmatrix} = \begin{Bmatrix} \frac{Gm_2(R_{1x} - R_{3x})}{R_{12}} + \frac{Gm_3(R_{2x} - R_{3x})}{R_{13}} \\ \frac{Gm_2(R_{1y} - R_{3y})}{R_{12}} + \frac{Gm_3(R_{2y} - R_{3y})}{R_{13}} \\ \frac{Gm_2(R_{1z} - R_{3z})}{R_{12}} + \frac{Gm_3(R_{2z} - R_{3z})}{R_{13}} \end{Bmatrix}; \quad (17)$$

where

$$R_{12} = \|R_2 - R_1\|; \quad (18)$$

$$R_{13} = \|R_3 - R_1\|; \quad (19)$$

$$R_{23} = \|R_3 - R_2\|. \quad (20)$$

Next, 24-component column vector is formed

$$\{f\} = |\{R_1\} \{R_2\} \{R_3\} \{R_G\} \{v_1\} \{v_2\} \{v_3\} \{v_G\}|^T. \quad (21)$$

The derivative of this vector in the corresponding components is a column vector

$$\left\{\frac{df}{dt}\right\} = |\{v_1\} \{v_2\} \{v_3\} \{v_G\} \{a_1\} \{a_2\} \{a_3\} \{0\}|^T. \quad (22)$$

2.2 Perturbation due to Oblateness of the Earth

In order to illustrate the idea of gravitational harmonic, the concept of gravitational potentials is introduced. The gravitational field of a certain body mass may be presented by a potential function.

Earth harmonics are the result of gravitational potential using gravitational potential function. Those harmonics are the terms of a mathematical expansion through which the deviations from a sphere can be represented.

The most commonly used model of the Earth is WGS 84. The normally used harmonics are those of the lowest

 DOI: 10.1051/matecconf/201713301001

order. For example commonly used zonal harmonics are J_2, J_3, and J_4.

The second harmonic J_2 is related to the Earth's equatorial oblateness, in reference to Earth rotation and the estimated difference between the polar radius and equatorial radius – about 22 km.

If the mass of the body is uniformly distributed, the potential function is Φ is

$$\Phi = \frac{\mu}{r}, \tag{23}$$

where r is geometric range, $\mu = GM_e$, G – gravitational constant, M_e – Earth mass.

The potential function including the effect of oblateness of Earth can be presented as

$$\Phi = \frac{\mu}{r} + R, \tag{24}$$

where R is perturbing function (disturbing function), including J_2 effect, which can be given either in terms of the orbital parameters as [3]:

$$R = \frac{\mu J_2 A_e^2}{2r^3}\left[1 - \frac{3}{2}\sin^2 i\,(1 - \cos 2u)\right]. \tag{25}$$

Perturbing accelerations due to J_2 could be given in spherical coordinates as [1]:

$$F_r = -\frac{3}{2}\frac{\mu J_2 a_e^2}{r^4}\left[1 - \frac{3}{2}\sin^2 i\,(1 - \cos 2u)\right]; \tag{26}$$

$$F_s = -\frac{3}{2}\frac{\mu J_2 a_e^2}{r^4}\sin^2 i \sin 2u; \tag{27}$$

$$F_w = -\frac{3}{2}\frac{\mu J_2 a_e^2}{r^4}\sin i \cos i \sin u. \tag{28}$$

J_2 is responsible for the secular rates of the right ascension of ascending node and the argument of perigee.

2.3 Perturbation due to atmospheric drag

The motion of the satellite in high-altitude atmosphere produces drag force that changes orbital parameters. The rate of orbital decay depends on atmospheric density; this varies with time and geographic position and is not precisely known.

The drag effect on a satellite orbit is expressed through energy dissipation: when an orbit loses energy, its semi-major axis decreases. The drag effect would circularize the orbit – for orbit with large eccentricity, by gradually lowering its apogee until the satellite crashes on the Earth's surface.

For circular orbits, the orbit decay rate can be computed by the following equation:

$$\frac{da}{dt} = -\sqrt{\mu a}\,\rho\, BC, \tag{29}$$

where μ is gravitational constant, a is semi-major axis, and BC – ballistic coefficient.

When orbits are low-altitude ones, less than 1000 km, the drag effect should be considered in long-term predictions. The perturbing acceleration due to drag effect could be given as [1]:

$$a_D = -\frac{1}{2}\rho V^2 \frac{C_D A}{m} i_v, \tag{30}$$

where a_D is atmosphere drag acceleration vector; ρ – air density and parameters of the satellite: A – effective area; m – mass; V – velocity and i_v – unit vector of the velocity.

At this point the hardest obstacle is modeling the density of the atmosphere, which is function of time and altitude.

2.4 Solar radiation pressure perturbation

The solar-radiation pressure is induced by photons emitted from the sun. For the purposes of the study solar-radiation pressure at each point of the orbits is accepted to be a constant. The magnitude of solar-radiation-pressure in one astronomical unit (AU) is $4{,}65\times10^5$ dyne/cm^2.

Perturbing acceleration a_p, due to solar-radiation pressure effect, could be expressed by [1]:

$$a_p = 4{,}65.\,10^{-6}(1+\beta)\left(\frac{A}{m}\right)\left(\frac{a_\odot}{r_\odot}\right)^2, \tag{31}$$

where: β is optical reflection constant ($\beta = 1$ for total reflection, $\beta = 0$ total reception, $\beta = -1$ for total transmission); A – satellite effective area, m^2, m – satellite mass, kg; $a_\odot$ and $r_\odot$ – semi-major axis and radius of the Sun's orbit around Earth; correlation $(a_\odot/r_\odot) \approx 1$ for near Earth orbits.

3 Perturbations other than those related with satellite orbits

3.1 Satellite clock error

GPS satellites contain an atomic clock that synchronizes all onboard operations, including the generation and transmission of a signal. The acceptable deviation between satellite time and system time is 1 ms. The MCS determines and transmits clock correction parameters to the satellites for rebroadcast in the navigation message. The bias can be computed in the user/receiver by means of the following [3]:

$$\delta t_{clc} = a_{f0} + a_{f1}(t - t_{oc}) + a_{f2}(t - t_{oc})^2 + \Delta t_r, \tag{32}$$

where a_{f0} is clock bias, s; a_{f1} – clock drift; a_{f2} – frequency drift (i.e. aging), s/s^2; t_{oc} – clock data reference time, s; t – current time epoch, s; Δt_r – correction due to relativistic effects, s.

This residual clock error δt results in ranging errors that typically vary from 0,3-4 m.

3.2 Relativistic Effects

The need for relativistic correction related to special theories of relativity occurs whenever the GPS satellites and the GPS receivers:

- move with respect to the chosen isotropic light speed frame, which in the GPS system is the ECI frame;
- are located at different gravitational potentials.

In order to compensate for both of these effects, the satellite clock frequency is adjusted to 10,22999999543 MHz prior to launch. The frequency observed by the user at sea level will be 10,23 MHz; hence, the user does not have to correct for this effect.

Due to slight eccentricity of the satellite orbit the influence of other relativistic effects must be adjusted, such as periodic change in its gravitational potential. When the satellite is at perigee the velocity of the satellite is higher and gravitational potential lower and both cause delay of satellite clock. On the contrary – when the satellite is at apogee the velocity is lower and gravitational potential is higher. Both make the satellite clock to run faster. This effect could be expressed as [3]:

$$\Delta t_r = F.e.\sqrt{a}\sin E_k, \qquad (33)$$

where $F = -4{,}442807633.10^{-10}$ s/m$^{1/2}$; e – eccentricity; a – semi-major axis; E_k – eccentric anomaly of the satellite orbit.

Correcting the satellite clock will result in a more precise assessment of the time of propagation of the navigational signal.

3.3 Ionosphere effect

For the purpose of GPS applications the Earth atmosphere is divided into two sections: the ionosphere and the troposphere. The ionosphere is located between about 70 km and 1,000 km above the Earth's surface. An ionized environment causes a slow down of the group velocity of the electromagnetic wave and a rise in the carrier phase. This is known as ionospheric effect that is frequency dependent, i.e. it forms ρ_{L1} and ρ_{L2} delay. This effect can be eliminated by making ranging measurements with a dual-frequency receiver – pseudorange measurements made on both f_{L1} and f_{L2}.

To estimate the impact of ionospheric delay on measured pseudorange using f_{L1} could be expressed as follows [3]:

$$\Delta S_{iono,corr_{L1}} = \left(\frac{f_{L2}^2}{f_{L2}^2 - f_{L1}^2}\right)(\rho_{L1} - \rho_{L2}). \qquad (34)$$

The path length difference on L2 can be estimated by multiplying $\Delta S_{iono,corr_{L1}}$ by [3]:

$$(f_{L1}/f_{L2})^2 = (77/60)^2. \qquad (35)$$

These calculated adjustments could be smoothed out over time, since the errors caused by ionospheric delay do not change rapidly and are subtracted from the pseudorange measurements made by each frequency.

When a satellite is at low elevation the signal delay is three times bigger than that at the zenith. For a signal arriving at vertical incidence, the delay ranges from about 10 ns (3m) at night to as many as 50 ns (15m) during the day. At low satellite viewing angles (0° through 10°), the delay can range from 30 ns (9 m) at night to up to 150 ns (45 m) during the day.

3.4 Tropospheric delay

The troposphere is the lower part of the atmosphere that is nondispersive for frequencies up to 15 GHz. Phase and group velocity, within this media, related with C/A code and GPS navigation data for both signals with frequencies are equally delayed with respect to free-space propagation

The refractivity is often modeled as including both a dry (hydrostatic) and wet (non-hydrostatic) component. The dry component gives about 90% from tropospheric delay and can be predicted very accurately. It is much more difficult to predict the wet component, which reflects the water vapors, due to the variable nature of the condition of the atmosphere.

One accurate method for modeling the troposphere dry and wet components at zenith without meteorological sensors was developed at the University of New Brunswick (i.e. UNB3 model) [4]. This model is used in the present study.

3.5 Multipath

In the present study th emultipath is considered to be constant:

$$multipath = \mathrm{tg}0{,}25c/f_{L1}. \qquad (36)$$

Calculations are made with mask angle range of 5° - 45°, with 5° step.

4 Software tool synthesis

Computation of the errors in position determination, mentioned above, is implemented in computer code. The synthesized software tool structure is shown in Fig. 1.

Three major groups may be distinguished:

- Input module – station2.m;
- Calculation of orbital parameters including orbital perturbations module;
- User position determination module with non-orbital perturbations taken into account as well.

5 Summary

The presented algorithm, implemented in a computer code, is a tool that allows for the analysis of the impact a number of factors have when determining the user position; factors such as satellite clock error, relativistic effects, ionosphere effect, troposphere delay, multipath effect, and factors influencing the satellite orbit such as perturbation of gravity, oblateness of the Earth, atmospheric drag, solar radiation pressure, etc.

The synthesized software tool allows to research the impact of perturbation factors on the position errors – one by one, or in various combinations. Numerical simulations were performed as well and the results of the analysis will be presented in a forthcoming paper.

DOI: 10.1051/matecconf/201713301001

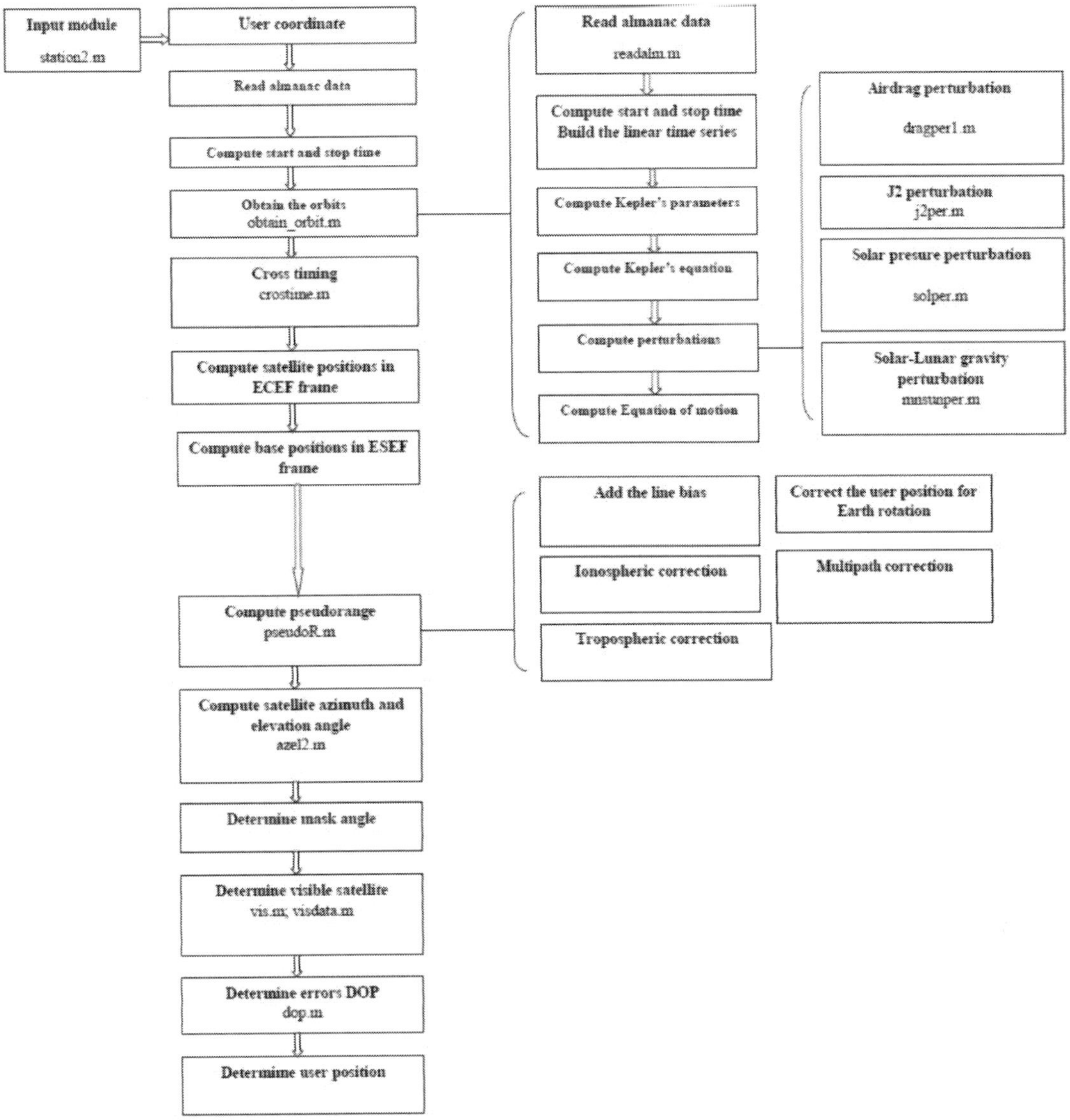

Fig. 1. The software tool structure synthesized.

References

1. V. Chobotov, *Orbital Mechanics*, Third edition; American Institute of Aeronautics & Astronautics, (2002)
2. H. Curtis, *Orbital mechanics for engineering students*, Elsevier Butterworth-Heinemann, (2005)
3. E. Kaplan, C. Hegarty, *Understanding GPS: Principles and applications*, Second edition; Artech house, (2006)
4. J. Collins, R. Langley, Nominal and extreme error performance of the UNB3 tropospheric delay model, Technical Report, University of New Brunswick, (1999)

DOI: 10.1051/matecconf/201713301002

Experimental study of canard UAV aerodynamics

Hristian Panayotov[1*], *Stanimir* Penchev[1], and *Dancho* Kolibarov[1]

[1]Technical University – Sofia, Plovdiv Branch, Department of Transport and Aviation Engineering, 4000 Plovdiv, Bulgaria

Abstract. The present paper presents the aerodynamic characteristics of a canard fixed-wing unmanned aircraft TERES-02. A wind tunnel experiment is conducted using a specially designed model of the aircraft. The model is produced through the methods of rapid prototyping using a FDM 3D printer. Aerodynamic corrections are made and thorough analysis and discussion of the results is carried out. The obtained results can be used to determine the accuracy of numerical methods for analysis of aircraft performance.

1 Main properties of TERES-02 UAV

TERES-02 (from TElesys REconnaissance and Surveillance) is a Reconnaissance and Surveillance unmanned aerial vehicle (UAV) produced by TELESYS Ltd – Plovdiv, Bulgaria and designed in collaboration with the Plovdiv Branch of TU-Sofia whose maiden flight was held successfully on July 2014 (Fig. 1).

Fig. 1. TERES-02 UAV.

The aircraft itself features canard or tail-first aerodynamic arrangement with swept wing and tip-mounted fins propelled by two-stroke, two-cylinder boxer engine with three-blade pusher propeller.

1.1 Dimensions and performance

The main dimensions and performance data are given in Table 1.

However, in order to fully analyse aircraft performance some basic aerodynamic characteristics have to be determined a priory. This could be done either theoretically (using numerical methods or analytical calculations) or experimentally. Even if theoretical calculations are used experimental validation of accuracy is also recommended. Hence an aerodynamic experiment is a must in order to properly determine aircraft performance.

Table 1. Dimensions and performance of TERES-02.

Wingspan	3840	mm
MTOW	55	kg
Payload Weight	>20	kg
Maximum Airspeed	240	km/h
Cruising Speed	110	km/h
Flight Endurance	>8	hours

1.2 Geometry of the wind tunnel model

A 1:10 scale model of the real aircraft is designed to conduct an aerodynamic experiment. This is defined by the dimensions of the test section (600×400 mm) of the wind tunnel (Fig. 6). The maximum allowed wingspan of the model is 400 mm and maximal cross-sectional area is 1800 mm^2. Also a special support of the model is designed to ensure attachment of the model itself to the internal aerodynamic balances. The main requirements towards the support are: minimum parasite drag; minimum interference with the model; minimum cross-sectional area; maximum strength and stiffness.

A semi-view of the 3D model of the assembly – model and support is shown on Fig. 2. The propeller and the lending gear are not modelled. Fins are modelled as a separate part from the aircraft and will be produced alone. The support is attached to the model with a screw mounted at the top of the fuselage. The support has mountable nose to ensure access to internal aerodynamic balance attachment screw.

* Corresponding author: hristian@tu-plovdiv.bg

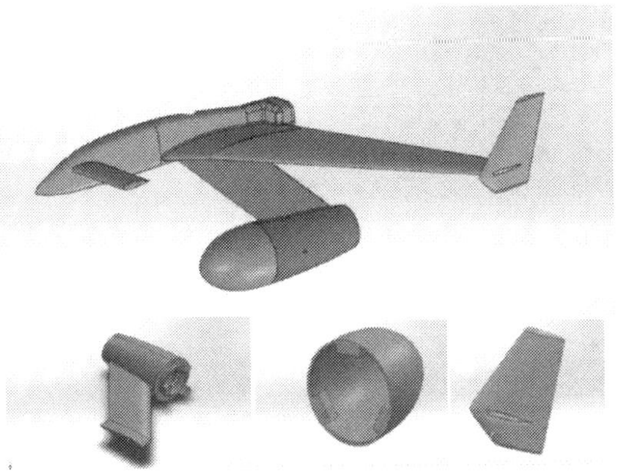

Fig. 2. Experimental wind-tunnel 3D model and its parts.

2 Production of the model

The designed wind tunnel model (Fig. 2) is to be produced using FDM (Fused Deposition Modeling or Fused Filament Fabrication) 3D printer followed by finish work. The chosen material is ColorFabb XT-CF20 carbon fiber reinforced polymer that possesses high strength and stiffness. Also the 3D printed parts surface is rough or "shagreen" so it is appropriate to be finished with filler spray and finally polished. Wanhao Duplicator 5S is the 3D printer used. It is a high precision, high resolution machine with enough build volume dimensions to create wind tunnel models.

Before printing the model it has to be "sliced" by software to produce numerical control code for the printer. Fig. 3 shows the sliced model of the fuselage and wing part. Here the operation regimes of the printer are controlled as well. The nozzle diameter is 0,4 mm, the layer resolution is 0,1 mm, the X,Y precision of head positioning is 12,5 microns. The temperature of the nozzle is 240°C and the temperature of the heated bed is 75°C to avoid temperature stress in the material and following deformations during fabrication.

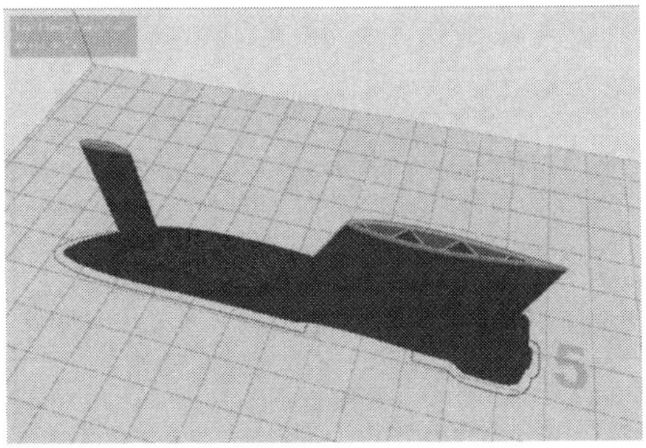

Fig. 3. 3D printer slicer software.

Fig. 4 shows the "raw" printed parts for the model without the finish work and Fig. 5 shows the finished model in the test area of the wind tunnel. The model is mounted onto internal aerodynamic balances.

Fig. 4. Printed parts for the wind tunnel model.

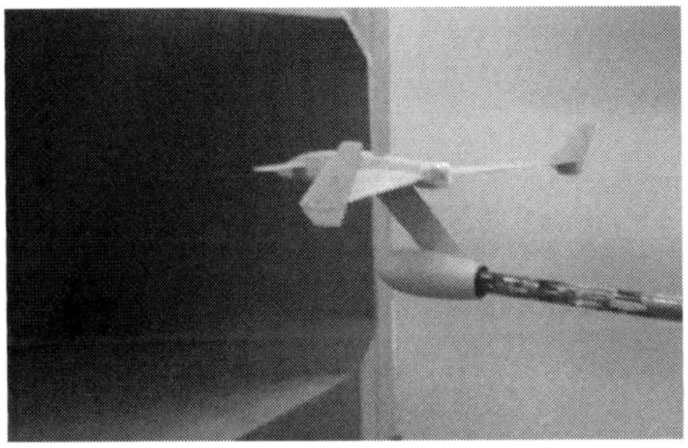

Fig. 5. Finished model in the test area.

3 Methodology

3.1. Wind tunnel

The ULAK-1 wind tunnel located at the Laboratory of Aerodynamics in the Plovdiv Branch of TU-Sofia has a low Reynolds number, an open test area, a closed-loop research facility (Fig. 6).

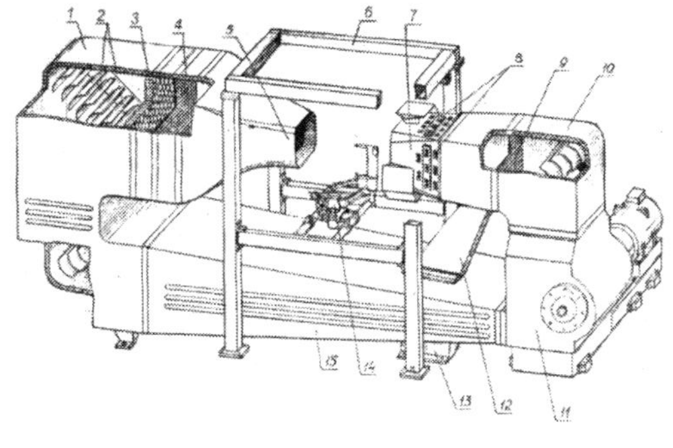

Fig. 6. ULAK-1 wind tunnel.

The maximum airflow speed in the test area is 50 m/s, the dimensions of the cross section of test area are 600×400 mm. Aerodynamic forces and moments are measured by internal six-component tensiometric aerodynamic balances. The angle of attack and sideslip angle are controlled by α-β mechanism. The experiment control and measurements are fully automated and computerised.

 DOI: 10.1051/matecconf/201713301002

3.2. Measured properties

In the present study the aerodynamic characteristics in the plane of symmetry are analysed in terms of the angle of attack – α at a given Reynolds number – *Re*:

$$\{L, D, M, c_L, c_D, c_M\}^T = f(\alpha, Re), \quad (1)$$

where: L is lift force; D is drag force; M is pitch moment and c_L, c_D, c_M respectfully are their aerodynamic coefficients.

Aerodynamic balances measure directly the forces and moments in the coordinate system of the balances: axial force X, normal force Z and pitch moment M_y. The aerodynamic coefficients in the coordinate system of the balance are calculated using the following corelations [1]:

$$c_x = \frac{X}{qS}; c_z = \frac{Z}{qS}; c_{m_y} = \frac{M_y}{qSc}, \quad (2)$$

where S is the reference area; c is reference length (mean aerodynamic chord) and q is dynamic pressure ($q = \frac{\rho V_\infty^2}{2}$), where V_∞ is free stream velocity and ρ is density of the air.

Once the coefficients (2) are measured the aerodynamic coefficients in the wind coordinate system can be calculated using the transformations:

$$c_D = c_x \cos\alpha + c_z \sin\alpha;$$

$$c_L = c_z \cos\alpha - c_x \sin\alpha; \quad (3)$$

$$c_m = c_{m_y} + \frac{(z_0 - z_G)c_x}{c} - \frac{(x_0 - x_G)c_z}{c},$$

where index "0" and "*G*" indicate the location of the coordinates of the balances moment centre and the model respectfully.

3.3. Aerodynamic corrections

The "raw" results from the aerodynamic experiment have to be corrected to eliminate various sources of errors [2]. The most significant of them for the ULAK-1 wind tunnel are corrections of the turbulence factor (TF) and correction from the downwash of the flow in the test area.

The first correction defines the ratio between the intensity of the turbulence in the test area and the free atmosphere turbulence [2]. Previously conducted experiments in ULAK-1 show that the turbulence factor is $TF = 2{,}14$ [3]. Hence the measured Reynolds number has to be corrected so that it yields the effective Reynolds number

$$Re_{eff} = TF.Re\,. \quad (4)$$

The second correction eliminates the influence of the downwash in the test area flow. Different methods are used to define the downwash angle in the test area. In the present study the model is tested in two different configurations at various angles of attack – normal and inverted (the model is positioned upside down) flight. Provided the downwash angle in the test area is zero the results will be the same. Otherwise, the asymmetry will cause different results for each angle of attack and thus the zero-lift angle of attack will be different in normal and inverted flight. Finally, the downwash angle in the test area will be

$$\varepsilon_{TA} = \frac{\alpha_{0_{nor}} - \alpha_{0_{inv}}}{2} \quad (5)$$

and the true angle of attack in the test area is

$$\alpha_{true} = \alpha - \varepsilon_{TA}. \quad (6)$$

The measured aerodynamic coefficients then have to be corrected:

$$c_{D_{true}} = c_D \cos\varepsilon_{TA} + c_L \sin\varepsilon_{TA};$$

$$c_{L_{true}} = c_L \cos\varepsilon_{TA} - c_D \sin\varepsilon_{TA}. \quad (7)$$

4 Results and discussion

Three separate experiments were conducted following the above-mentioned methodology: normal flight and inverted flight to determine the downwash angle in the test area and drag polar measurement of the support. The drag of the support is taken in account and is subtracted from the total drag of the model cum support. Lift and drag coefficients of the support are given in Table 2. The interference drag between the aircraft and the support is neglected.

For all experiments the reference area of the model and the reference length are: $S = 0{,}018\ \text{m}^2$; $c = 0{,}053$ m. The air density is $\rho = 1{,}1652\ \text{kg/m}^3$. The effective Reynolds number based on the mean aerodynamic chord c calculated using (4) is $Re_{eff} = 190\,000$.

Table 2. Aerodynamic characteristics of the support

α, deg	c_D	c_L	α, deg	c_D	c_L
-7,91	0,0072	-0,0049	6,07	0,0067	0,0052
-5,91	0,0070	-0,0051	8,07	0,0082	0,0105
-3,92	0,0068	-0,0054	10,06	0,0110	0,0158
-1,91	0,0065	-0,0055	12,06	0,0102	0,0210
0,08	0,0061	0,0059	14,05	0,0126	0,0207
2,08	0,0079	0,0056	16,04	0,0147	0,0258
4,08	0,0065	0,0054	18,04	0,0156	0,0254

Calculations based on formula (5) yield downwash angle in the test area $\varepsilon_{TA} = 1{,}2$ degrees, which is consistent with similar previous measurements [4]. After corrections (5) and (6) the final results for aerodynamic characteristics are shown graphically in Figures 7 to 10.

Fig. 7 depicts the variation of the lift and drag coefficients of the aircraft in terms of the angle of attack. It can be seen that the lift coefficient curve does not reach its maximum value and the stall angle of attack is not achieved. This is due to the limited maximal angle of attack for ULAK-1 (18 deg) and the downwash angle additionally reduces it. Also the swept wing needs higher angle of attack to stall, so a new support has to be designed to allow higher angles of attack. The minimum drag coefficient is $c_{Dmin} = 0{,}0407$, which is quite high for a "clean" configuration of aircraft but this can be explained with the low Reynolds number.

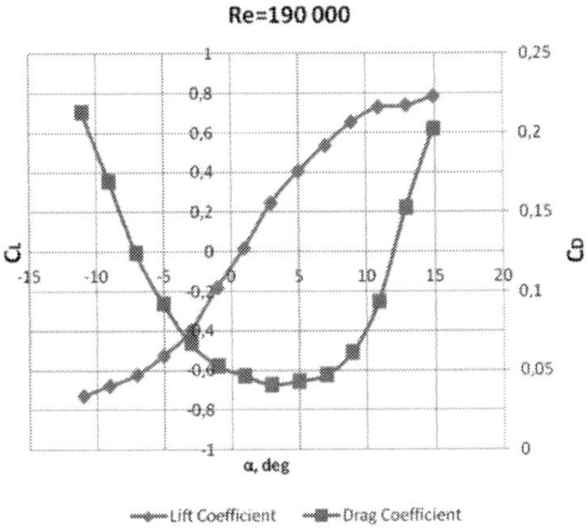

Fig. 7. Lift and Drag Coefficients vs. Angle of Attack.

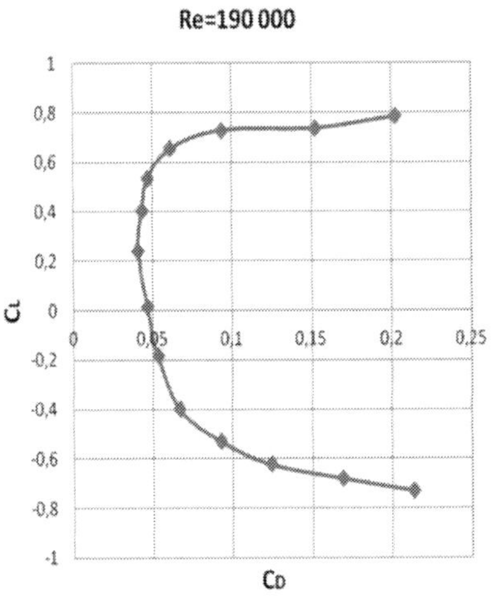

Fig. 8. Drag Polar.

The relatively high drag coefficient influences the values of the Lift-to-Drag ratio (Fig. 9). The maximum value is close to 12 at 7 deg. angle of attack. The pitch moment coefficient (Fig. 10) is calculated about the aircraft centre of gravity and shows typical pattern for a statically stable aircraft.

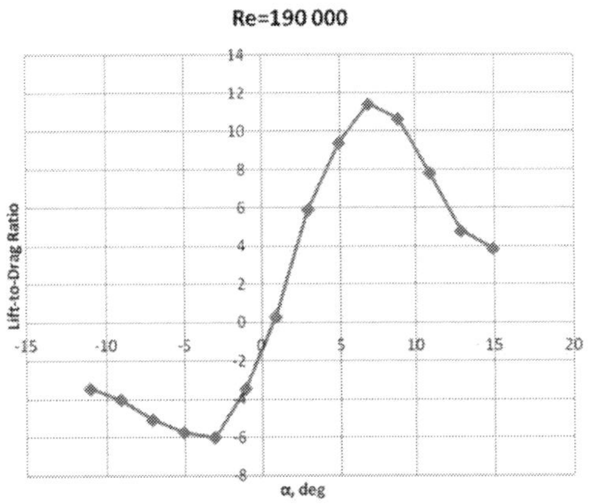

Fig. 9. Lift-to-Drag Ratio vs. Angle of Attack.

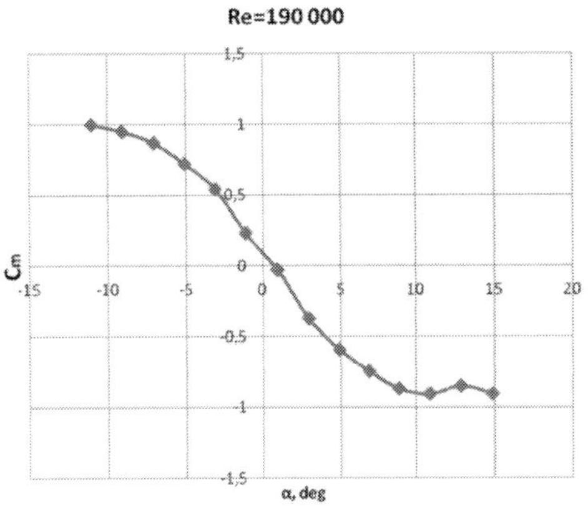

Fig. 10. Pitch Moment Coefficient vs. Angle of Attack.

5 Conclusion

Aerodynamic experiments were held with a specially designed model of TERES-02 UAV. The results will be used for calculation of aircraft performance and validation of numerical simulations.

References

1. J. Anderson Jr., *Fundamentals of aerodynamics*, (McGraw-Hill, 2001)
2. J. Barlow, W. Rae, A. Pope, *Low-Speed Wind Tunne Testingl*, (John Wiley & Sons, 1999)
3. S. Penchev, D. Seyzinski, D. Kolibarov, Flow parameters investigation in the test area of wind tunnel ULAK-1 (UT-1), Mechanics of Machines **62** 9-12 (2006)
4. D. Zafirov, H. Panayotov, Joined-wing test bed UAV, CEAS Aero J. **6** 137-147 (2015)

A study of sector configurations capacity for air traffic service

Teodosiy Todorov[1,*], and *Plamen* Petrov

[1]Technical University – Sofia, Department of Air Transport, 8 Kliment Ohridski Blvd., 1000 Sofia, Bulgaria

Abstract: A study of sector configurations capacity in the upper air space of the Republic of Bulgaria is presented. The method used to determine the capacity of sectoral configurations is based on the air traffic control workload. The results obtained can be used to assess and improve airspace capacity and for efficient planning and use of air traffic service resources.

1 Introduction

In the literary sources there are definitions for Airspace Capacity, which differ in their essence. Capacity is the maximum number of aircraft which can be serviced for a particular period of time by the system or one of its components [1]. The capacity of the Air Traffic Management (ATM) system is the maximum number of aircraft which can be simultaneously serviced safely and efficiently, while the Air Traffic Controller (ATC) is experiencing acceptable workload [2]. According to others, capacity is defined by the maximum number of aircraft which can be serviced in a given sector for ATS for a particular period of time while the ATC is experiencing his/her upper limit of workload [3]. Each ATM system evaluates the degree of workload its airspace can handle. This is the reason why there are so many models developed for determining the capacity which must meet the needs of every Air Navigation Service Provider (ANSP). We can summarize that all the methods are based on the workload of ATCs.

The capacity of airspace acquires new aspects and timeliness in free route airspace (FRA). As a result, the character of air traffic flows will change in airspace in each country. For this, it is necessary to assess air traffic, which can be serviced by each ATS sector.

2 Task

The present task is to study the capacity of sector configurations in the Bulgarian Flight Information Region (FIR). Examination of the capacity of sector configurations in the Upper Airspace.

The range of the study is limited to reviewing the workload of an ATC for allocating the capacity of sector configurations of family sectors Varna. The task is to select the categories of work tasks decided by the ATCOs and to define the capacity of an ATS sector. Statistical data from the Eurocontrol database have been used for the purpose of the study.

The study is conducted for airspace, consisting of four sectors -VA, VB, VC, and VD [4], with limits to their upper boundaries up to flight level (FL) 450. The airspace is configured in different sector configurations, vertically and horizontally. In height, the airspace can be configured without division from FL 245 to FL 660, with one division at FL 245, or with two divisions at FL 345 and FL 375. Sector VC is not used by itself, but in a combination with different sectors. An average speed of Vm = 460 kt is defined for commonly used types of airspace for the time of the study.

3 Literary Analysis

A study of the literature shows that there are no documents pertaining to the dynamic management of the airspace. Some of the software provides functions related to the dynamic sectors division and should be applied. When calculating the ATM sector capacity different models and methods are used.

The DORATAK is an analytic method used for estimating working tasks and analysis of workload [5]. It is based on simulation (Fast-Time Simulation - FTS). An ATC's workload is defined by the time needed for executing "observed" and "non-observed" tasks, while the capacity is defined as the workload level by which the ATC has enough time for "recovery".

Another method is presented in [6]. The ATM sector capacity evaluation is the maximum number of aircraft which a single ATC can handle simultaneously in one sector. With this method different working tasks done by the ATC are considered all together.

The ATC Capacity Analyzer (CAPAN) method, is a Fast Time Simulation (FTS), developed by EUROCONTROL. This method uses an airspace model in which workload is generated for the simulated working position with given air traffic. After the exercise CAPAN evaluates the workload. The result is compared to a preset upper limit to traffic-handling capacity. In order to make the simulated model of airspace as similar to real airspace as possible, the correlation between qualitative (high and low workload) and quantitative evaluation is determined experimentally.

* Corresponding author: teodosiy.todorov@bulatsa.com

NEST is a software product for modeling through a set of scenarios, used by a EUROCONTROL Network Manager and an ANSP for design and development of airspace structure, capacity planning and analysis. NEST uses database for the European airspace and route network as well as traffic forecasts by database STATFOR, which are provided by EUROCONTROL.

The evaluation of the methods examined shows that they are not suitable for FRA.

4 Macroscopic model of workload for capacity evaluation

For the purposes of this study, a macroscopic model of workload for capacity assessment has been selected. The macroscopic model performs quantitative evaluation of workload depending on the geometry of the ATS sector, on the direction of air traffic flow and on the frequency of conflicts between airborne aircraft. This analytical approach is easily extrapolated in new conditions and its parameters can be selected and defined by comparison to real traffic.

Using this model, we can investigate the correlation between the capacity of the ATM sectors and the effectiveness of the air traffic controllers' work.

When a "J" number of separate tasks are present, the "intensity of the workload" takes the following form:

$$G = \sum_{j=1}^{J} \tau_j \lambda_j , \qquad (1)$$

where τ_j e is the time necessary to fulfill a task j, while λ_j is the rate at with which it is fulfilled.

The determined value Gm, at which an air traffic controller will no longer be able to perform his work safely if he has to handle additional traffic, is the maximum level of workload and determines the capacity of the ATM sector.

Using this model, the work tasks are summarized in four categories – background tasks, transition tasks, recurring tasks and conflict tasks.

The background tasks include routine actions and occupy a constant, small part of the air traffic controller's time. The workload from the background tasks can be presented in the following form:

$$G_b = \tau_b \lambda_b . \qquad (2)$$

Transition tasks are performed every time an aircraft passes through the sector. These tasks require a moderate amount of time to be performed τ_t and a moderate rate at which they are performed λ_t that can be presented in the following way:

$$\lambda_t = E[N]/T , \qquad (3)$$

where $E[N]$ is the average number of aircraft in a particular sector and T is the average time it takes an aircraft to fly through the aforementioned sector.

Recurring tasks are performed periodically, while the observed aircraft is present within the particular ATM sector.

These tasks require an average amount of time to do τ_r and are performed at a rate of λ_r, where:

$$\lambda_r = E[N]/P. \qquad (4)$$

In this formula P is the average time period, during which the tasks are performed.

For conflict tasks it is accepted, that an average amount of time τ_c is necessary for their completion and an average rate of completion λ_c. For a current sector with a defined number of aircraft – N_s, the rate of completion of conflict tasks can be presented as:

$$\lambda_c = B N_s^2 / Q , \qquad (5)$$

where B is the physical constant, dependent on the velocity of the converging speed between 2 aircraft and the separation norms; N_s – number of aircraft; Q – sector volume.

The number of aircraft N in the considered sector is a random quantity and λ_c is proportional to the expected value of $E[N]$. Representing the bulk density of aircraft as k:

$$k Q = E[N], \qquad (6)$$

and assuming the distribution of Poisson [2], the following frequency for conflict tasks is received:

$$\lambda_c = B k (k Q + 1). \qquad (7)$$

Substituting it in (1), it can be said that the intensity of the overall workload G is the following:

$$G = G_b + \tau_t kQ/T + \tau_r kQ/P + \tau_c Bk(kQ+1). \qquad (8)$$

For this method there is no need for a large computational resource or access to statistical information for the air traffic flow. An expert opinion is needed for the correct selection and determination of the empiric parameters.

5 Determining the Capacity of an ATS Sector

For this purpose, the parameters determining the frequency of the work tasks (k, T, B and Q) are selected from geometric characteristics for the ATS sector under consideration and Air Traffic. G_b, τ_t , τ_r, P and τ_c have empirical values.

The workload is calculated by the macroscopic model for an ATS sector with a volume $Q = 10000$ NM^3, height $H = 10000$ ft and average aircraft sector flight time – $T = 1200$ s. An average rate of convergence 380 kt is adopted, with randomly located aircraft, all flying at a speed $V_m = 460$ kt.

Model Parameters: $G_b = 0{,}1$; $\tau_c = 50$ s; $\tau_r = 2$ s; $\tau_t = 15$ s; $P = 300$ s; $T = 1200$ s, $M_h = 7$ nm; $M_v = 1000$ ft; $E[V_{21}] = 380$ kt; $Q = 10000$ NM^3.

The graph in Fig. 1 shows the workload intensity of the four categories of tasks as a function of the number of aircraft.

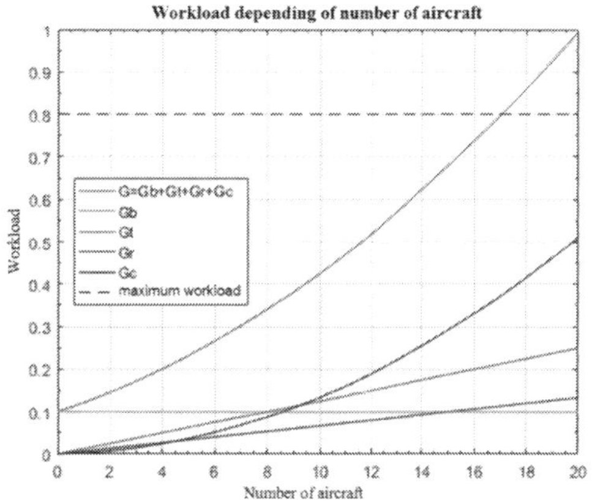

Fig. 1. Workload Intensity depending on the number of aircraft.

For the maximum workload Gm, a value of 80% of an hour of ATCO operational work is assumed, based on the basic tasks performed and the presence of an automated air traffic management system. From Fig. 1 it follows that the sum total of the work tasks reaches a value of Gm at 80% when traffic reaches 17 aircraft (density of 17 aircraft for a volume of 10 000 NM³).

5.1 Analytical relation between the calculated capacity and the volume of the ATS sector

The workload from recurring tasks and such related to conflicts, depends directly on the volume of the sector. The boundaries of the sector are extended along the main direction of the air traffic flow, in order to reduce the workload from transition tasks. For this purpose, three sectors are defined with a ratio of 1:1, 1:2 and 1:4. A relation can be found between the volume of the sector and its longest side *a*. For a rectangular sector with a height of *H*, side *a* and *a* ratio between the sides 1:*u* the formula is as follows:

$$Q = u a^2 H. \tag{9}$$

If the air traffic flow is along the length of the sector and Lm is the average distance, which an aircraft flies through the sector, the capacity of the sector can be calculated from its volume and the ratio between the length and width:

$$L_m = u a; \tag{10}$$

$$T = \frac{L_m}{V_m}. \tag{11}$$

The results from equations (9), (10) и (11) are presented in Fig. 2 with the correlation between the capacity and volume of the ATS sector by using the parameters from Fig. 1.

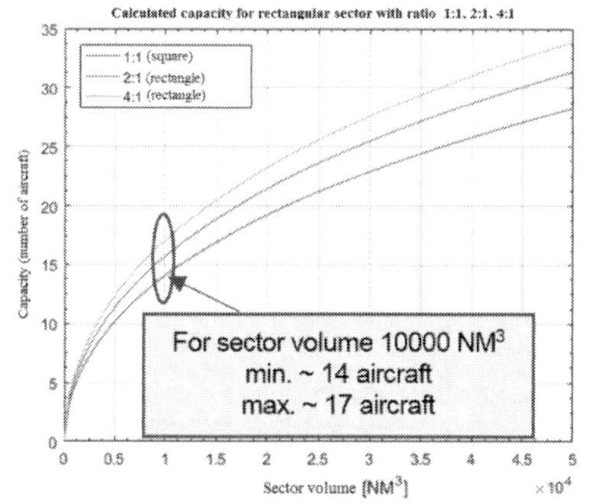

Fig. 2. Calculated capacity of rectangular sectors with a ratio of 1:1 (square), 2:1 and 4:1 depending on the volume.

For sectors with a volume of 10000 NM³ the graph shows that the calculated capacity increases from 14 aircraft for a square to 17 aircraft for a rectangle with a ratio of 4:1. In conclusion, for a sector with a maximum allowed workload, the maximum density of the air traffic decreases with the increase of the volume.

6 Results from the calculation of the capacity of sector configurations

When observing configurations of three and four sectors, united into one, the ATC team has to provide services in a relatively big airspace. For that reason configurations [*VA*, *VC*, *VD*], [*VB*, *VC*, *VD*] and [*VA*, *VB*, *VC*, *VD*] have the following values for parameters $\tau_r = 3$ s and $P = 200$ s. In all other configurations: *VA*, *VB*, *VD*, [*VA*, *VC*], [*VB*, *VD*] and [*VC*, *VD*], the following values are applicable $\tau_r = 2$ s and $P = 300$ s.

Table 1 presents the results from the calculation of the different configurations, on the basis of the defined parameters of every configuration and gives the number of aircraft every ATC has to provide service to simultaneously.

Table 1. Calculated capacity for sector configuration.

Sector Configuration	Area [nm²]	FL245 to FL455	FL245 to FL365	FL365 to FL455	FL245 to FL345	FL345 to FL375	FL375 to FL455
VA	4648	14	12	13	12	8	11
VB	5106	14	12	13	12	8	11
VD	5656	17	14	15	14	9	13
VA&VC	7170	19	15	15	15	10	14
VB&VD	10762	21	18	19	18	12	17
VC&VD	8178	20	17	18	17	11	16
VA&VC&VD	12826	20	18	18	18	12	17
VB&VC&VD	13284	17	16	17	16	12	16
VA&VB&VC &VD	17932	21	19	20	19	14	18

 DOI: 10.1051/matecconf/201713301003

In Table 2 the calculated air traffic density is presented for every configuration.

Table 2. Results for air traffic density for an airspace with a volume of 10000 NM^3.

Sector Configuration	Area [nm^2]	FL245 to FL455	FL245 to FL365	FL365 to FL455	FL245 to FL345	FL345 to FL375	FL375 to FL455
VA	4648	8	13	18	15	34	17
VB	5106	7	11	17	14	31	16
VD	5656	8	12	17	15	32	17
VA&VC	7170	7	10	14	12	28	14
VB&VD	10762	5	8	11	10	22	11
VC&VD	8178	7	10	14	12	27	14
VA&VC&VD	12826	4	7	9	8	18	10
VB&VC&VD	13284	3	6	8	7	18	9
VA&VB&VC &VD	17932	3	5	7	6	15	7

Fig. 3 shows the results graphically for sector configurations *VA* and *VC* and *VD*.

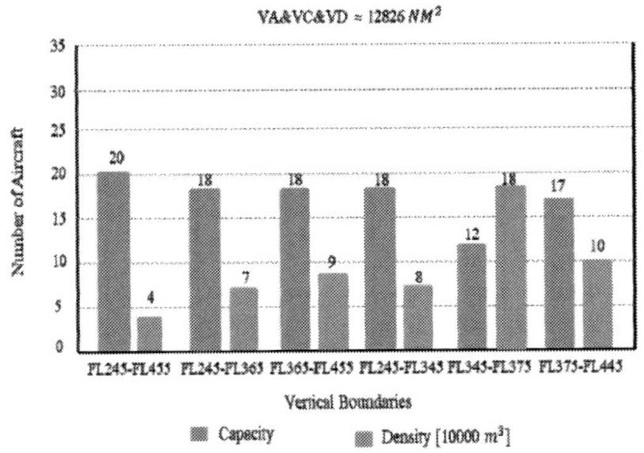

Fig. 3. Results for the capacity and density for 10000 NM^3.

A conclusion can be made that in order to increase the maximum number of aircraft for a certain sector configuration, the maximum allowed air traffic density has to be increased. This can be achieved when the airspace is divided into smaller sectors, in order to provide service to air traffic with a greater density. As a result, within the whole airspace the maximum number of aircraft increases.

7 Conclusion

The present work on capacity determination of sectoral airspace configurations is a study of the workload of the ATC. Determining the capacity of sector configurations is a subtask of the unified task of increasing the efficiency of airspace. The results obtained can be used to estimate the workload of the ATC. The formulas shown may be used for automated implementation of basic ATM features that provide benefits for flight safety and efficiency of air transport. For a maximum workload value, Gm can be assumed 80% with an automated air traffic management system. Determining the workload of the ATC has practical significance for Strategic capacity planning of the sectoral configurations.

References

1. ICAO, doc. 9882, *Manual on air traffic management system requirements*, (2008)
2. P. Petrov, *Automation of management of the air traffic - part one*, Publishing House of Technical University - Sofia, (2013)
3. A. Majumdar, W. Ochieng, G. McAuley, J. Lenzi, C. Lepadatu, The factors affecting airspace capacity in Europe: *A framework methodology based on cross sectional time-series nalysis using simulated controller workload data*, (2004)
4. BULATSA, AIP Republic of Bulgaria - ENR 6.2-3, (2016)
5. R. Jaurena, *Guide for the application of a common methodology to estimate airport and ATC sector capacity for the SAM region*, Lima, Peru, July (2009)
6. F. Dowling, EEC *Note No. 4/98 – Sector capacity assessment for Dublin ACC*, Eurocontrol, February (1998)

MATEC Web of Conferences **133**, 01004 (2017) DOI: 10.1051/matecconf/201713301004
BulTrans-2017

Measuring the efficiency of air traffic management systems

Teodosiy Todorov[1,*], and *Plamen* Petrov

[1]Technical University – Sofia, Department of Air Transport, 8 Kliment Ohridski Blvd., 1000 Sofia, Bulgaria

Abstract: Global demand for more capacity in air traffic management systems is increasing. At the same time there is growing demand for quality of the service provided by the Air Navigation Service Providers as well as for lower charges. This has led to efforts by the European Union to increase the efficiency of Air Traffic Management in Europe. An analysis and evaluation of the metrics that are currently used to measure the performance of Air Navigation Service Providers and to develop a system of performance metrics for the Air Traffic Management System is presented.

1 Introduction

The performance scheme and network functions introduced by the European Council contribute to the sustainable development of the air transport system by improving the overall performance of air navigation services (ANS). The European Air Traffic Management System (ETS) is so complex and sophisticated that the smallest improvement in efficiency saves aviation operators millions of euros.

For every air navigation service provider (ANSP) it is necessary to measure the performance of its processes and business activities in order to be able to assess whether the ANSP is achieving its goals and heading in the right direction towards its future vision for the ANSP. Performance measurements are also used to evaluate how an ANSP is performing with respect to similar ANSPs. It is necessary for ANSPs to be able to know their strengths and weaknesses in comparison to other ANSPs.

The objectives of this research are:

- to analyze and measure the indicators used to assess the performance of an ANSP;
- to make a selection of the main indicators to measure the performance of the ATM system;
- to process data on existing performance measurements that are being used or offered for use by ANSP in Europe.

The scope of the study is limited to examining the key areas of safety, capacity and environmental protection as well as their performance indicators. Statistical data from the Eurocontrol database is used for the purpose of the survey.

2 Literary Analysis

Air Navigation Services (ANS) includes five broad categories of services provided to air traffic during all phases of operation (area control, approach control and aerodrome control). These services are: Air Traffic Management (ATM), Communication services, Navigation services and Surveillance services (CNS), Meteorological services for air navigation (MET) and Aeronautical Information Services (AIS) and Search and Rescue (SAR) (ICAO, 2001) [1]. ANSPs provide a package of aeronautical services, excluding SAR, which can be performed by the government or private organizations. The main focus of this study is on ATM with a further focus on service activities for Air Traffic Control (ATC) including CNS services. One of the tasks of the ATM system is to develop a quality indicator, which should quantify the quality of the system functioning.

The study of the literary sources has shown that there is literature for assessing the efficiency of ATM processes and optimal management, according to a previously adopted optimization concept. It comes down to the compilation of a quantitative quality indicator, most often in a mathematical form, which is called the criterion of optimality of ATM system. The various allowances can be quantified in the "better-worse" scale at the value of this quality indicator determined for them.

The study shows that the definition of the concepts of "efficiency" and "effectiveness" of processes in the ATM system in Bulgaria is ambiguous.

3 Efficiency, effectiveness and performance of the systems

The ambiguous definition of the concepts of "efficiency" and "effectiveness" stems from the difference in the English translation of the words and what is popular in Bulgarian. Efficiency reveals the relationship between the resources used and the result achieved. As an indicator of performance measurement, the productivity of the ATM system is used.

Effectiveness is a concept that reveals the degree of achievement of the objectives set in the ANSP and,

* Corresponding author: teodosiy.todorov@bulatsa.com

DOI: 10.1051/matecconf/201713301004

above all, whether the services offered are adequate to the needs of the aviation operators.

The usefulness of a system is determined by the concepts of efficiency and performance. Both terms have common features but are not identical. The performance of the system is the full set of its features related to its ability to meet established and presumed needs. Each of these features is called a measure of performance (MOP). They represent quantitative performance measures that can be measured and dimensioned. Only the most representative features reflecting the main properties of the system are selected. They combine into a comprehensive performance index. While performance characterizes the system itself, effectiveness is a feature of usage (action, operation). Effectiveness characterizes the system from the point of view of its users. Effectiveness can be considered to measure the external characteristics of the system, while the performance - internal. Efficiency as well as qualities can be presented in a mathematical form and have Means of Efficiency (MOE).

In Commission Implementing Regulation (EU) No 390/2013 of 3 May 2013, the English term "Performance Scheme" is called "efficiency". Therefore, for the purposes of this study, it is assumed that the performance scheme introduced by the European Commission is used in the ANSP on both criteria - efficiency and effectiveness. This allows both concepts to be used at the same time to characterize ANS.

4 Process efficiency in the ATM system

The Single European Sky (SES) performance scheme was introduced with the second Single European Sky (SES II) package. The aim of the scheme is to improve the effectiveness of ANS and network functions in the Single European Sky (European Parliament and Council of the European Union, 2009) [2].

The following key performance areas are identified: safety, environment, capacity and cost-efficiency [2]. For each key area there are goals and one or a limited number of Key Performance Indicators (KPE) as shown in Fig. 1.

Evaluation of the overall ANS performance can be done using a quantitative performance indicator. One way of forming an indicator is to benchmark the system or process with requirements that require a unified approach to assessing efficiency [3]. For each key performance area, indicators are set. Recorded in analytical form are called KPI and can be represented as I_s ($s = 1, 2 \ldots \mu$), where s is the performance area and the μ is the KPI.

5 Measurement of efficiency in air traffic management systems

The efficiency framework has been expanded into seven areas: capacity, cost-efficiency, efficiency, flexibility, predictability, safety and environmental protection.

5.1 Safety

Review of Safety ANS performance in terms of accidents, ATM related incidents and the level of safety occurrence reporting in the EUROCONTROL area[4].

Safety is of upmost importance in aviation. For every change that is made in ATM systems, safety must always be considered as it cannot be jeopardized in pursuit of greater efficiency or effectiveness. However, it is not possible to optimize safety as such. If one seeks to ensure 100% safety in ATM that could only be achieved by grounding all aircraft. The Performance Scheme introduces the following Safety Measurement Indicators shown in Fig. 1.

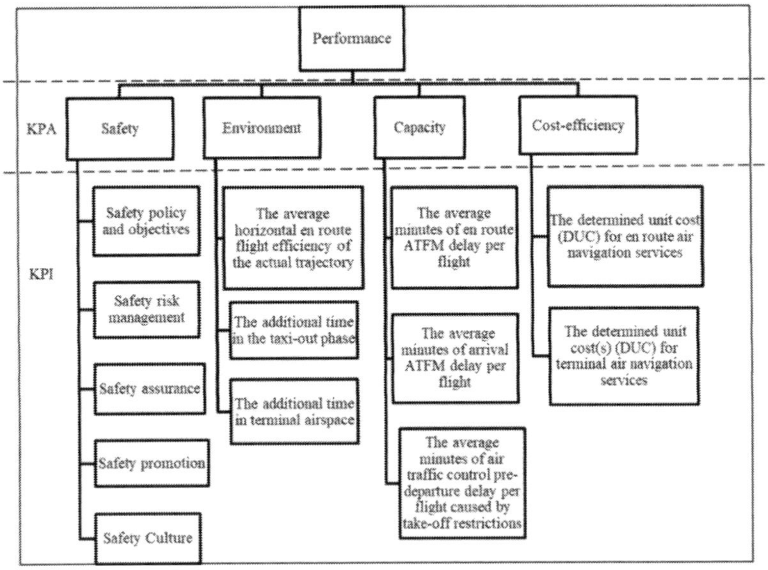

Fig. 1. Key performance areas and indicators for their measurement.

The application of the severity classification is done using the Risk Analysis Tool (RAT) based on Eurocontrol methodology for all operational events classified as SMI (Separation Minima Infringements) and RI (Runway Incursions) and 80% for specific events [4].

The safety levels set by the Competent Authority for 100,000 served aircraft are determined on the basis of statistical analysis of empirical data using Poisson probability distribution and are shown in Table 1.

Table 1. Target safety levels for 100,000 served aircraft.

Category Aviation Incident (AI)	2017
AA (crash)	0
A (serious incident) + **B** (big accident)	≤ 0,43
C (significant incident)	≤ 2

5.2 Productivity

Productivity can be defined as the ratio of the useful result (output of the system) to the resource input (input into the system). When looking at the ATM system, productivity is the output of the system divided by all inputs into the system and can be adopted:

- the number of controlled aircraft;
- flight time for controlled flights in hours;
- kilometers passed by controlled aircraft;
- number of flight paths required by ATCOs providing area control service.

Input data can be personnel, facilities, equipment and systems. When examining the input to the ATM system, it is logical to consider the contribution of the work of the ATCO. This contribution may be measured in number of ATCOs (in full-time equivalent) or in working hours (hours worked) [5]. Hourly ATCO productivity measures how many flight hours (output) per working hour in the ATCO (input) will the ATCO produce, so the goal is to increase this indicator.

Productivity can be defined as follows:

$$\text{IFR Flight Hours per ATCO} = \frac{\text{IFR Flight Hours}}{\text{Number of ATCOs}}. \quad (1)$$

The flight time or distance in kilometers is used for the exit, the number of ATCOs is used for entry. The average value of the working hours on an annual basis is:

$$\text{Average Annual Working Hours for ATCOs in OPS} = \frac{\text{ATCOs in OPS Hours}}{\text{ATCOs in OPS}}, \quad (2)$$

such as:

$$\text{IFR Flight Hours per ATCO in OPS Hour} = \frac{\text{IFR Flight Hours}}{\text{ATCOs in OPS Hours}}. \quad (3)$$

In this case, the productivity of the ATCO per hour will be defined as:

$$\text{Proroductity of ATCO per Hour} = \frac{\text{Total Fligh Hours}}{\text{ATCOs in OPS Hours}}. \quad (4)$$

5.3 Flight efficiency and environmental factors

For each flight, the goal is to define a trajectory with a minimum deviation from the optimal one [5]. The aviation operator (AO) has determined the optimal trajectory and it is included in the flight plan. There are two main types:

- horizontal inefficiency where the trajectory is longer in kilometers than in the optimal trajectory;
- vertical inefficiency where the height at which the aircraft flies is not optimal.

The horizontal flight efficiency H can be represented as:

$$H = \frac{\text{Actual Route} - \text{Great Circle Distance}}{\text{Great Circle Distance}}. \quad (5)$$

The track extension is therefore a number without unit with the optimal value is zero.

Vertical flight efficiency V can be represented in the form of:

$$V = \sum_s V_s = \sum_s \frac{T_S}{T}\left(\frac{L - L_S}{L}\right), \quad (6)$$

where T is the total flight time; T_S – the flight time of one flight level; L – the cruising level; L_S – the actual flight level.

The optimal trajectories are modeled using standard Eurocontrol aircraft databases [5]. The actual trajectory was simulated using the same aircraft performance data and atmospheric assumptions as the reference trajectory. The fuel burn calculations were performed using a model called KERMIT (Kerosene Emissions Research Model):

$$\varphi = \beta_1 H + \beta_2 V_{CL} + \beta_3 V_{CR} + \beta_4 V_D + \beta_5 H V_{CL} + \\ + \beta_6 H V_{CR} + \beta_7 H V_D, \quad (7)$$

where H is horizontal inefficiency, V_{CL} – vertical inefficiency of the climb, V_{CR} – vertical inefficiency of the cruise, V_D – vertical inefficiency of the descent and β are constants.

The constants, which were found by regression analysis, are given in Table 2.

Table 2. β coefficients.

Parameter	Value	Parameter	Value
β_1	0,8973	β_5	0,4265
β_2	0,8233	β_6	0,469
β_3	1,1095	β_7	2,9618
β_4	1,6788	-	-

The components of φ are additive, φ is therefore not limited from zero to one. As the 3Di Score calculates fuel inefficiency its optimal value would be zero.

The following annul target levels are accepted on the Functional Airspace Block (FAB) level for the Key Performance Indicator "Average horizontal efficiency of the actual trajectory in a flight along a route" (KEA) are shown in Table 3.

Table 3. Performance Indicator Values – Environment.

	2015	2016	2017	2018	2019
Reference level for FAB	1,55%	1,50%	1,46%	1,41%	1,37%
Target level for FAB	1,55%	1,50%	1,46%	1,41%	1,37%

5.4 Capacity

When the capacity for ATC service does not meet the required number of aircraft, they are delayed at the airport. For the key performance area capacity, the indicators are defined as the average delay in minutes, as shown in Fig. 1.

The delay in the ATFM route is the delay expressed as the difference between the take-off time requested by the aircraft operator in the last flight plan submitted and the estimated take-off time determined by the ATFM central authority. The maximum values for route delays for the "Capacity" are shown in Table 4.

Table 4. Performance Indicator Values "Capacity".

	2015	2016	2017	2018	2019
Reference level for FAB	0,04	0,04	0,04	0,05	0,06
Target level for FAB	0,08	0,08	0,08	0,09	0,09
Target level for ANSP	0,12	0,12	0,12	0,14	0,14

In order to ensure the necessary capacity of the airspace, it is allowed to revise the target levels in the increase of the air traffic flows and forecast for their lasting preservation.

6 Conclusion

The present work is an analysis and selection of the main indicators for measuring the performance indicators of ATM system. The set target levels are determined on the basis of statistical data analysis. The research is a sub-task of the single task of identifying and measuring performance indicators. It is of practical importance to determine the productivity of the strategic resource planning system. The results obtained can be used to assess the performance of the ATM system.

References

1. ICAO, doc. 9882, *Manual on air traffic management system requirements*, (2008)
2. Commission implementing regulation (EU) No 390/2013 of 3 May *2013 laying down a performance scheme for air navigation services and network functions*, (2013)
3. P. Petrov, *Automation of management of the air traffic - part one*, Publishing house of Technical University - Sofia, (2013)
4. Performance review report an assessment of air traffic management in europe during the calendar year 2016, Performance Review Commission, (2016)
5. Hulda Ástþórsdóttir, *Performance Metrics in Air Traffic Management Systems*, Thesis for MsC, School of Science and Engineering at Reykjavik University, (2013)

Stand testing of energy saving motor oil in the internal combustion engine

Anatoliy Ponomarenko[1,*], *Vitaliy* Dudnik[2], *Nadiia* Afanasieva [2], *Mikhail* Boiko[3]

[1]SFEDU, 344006, Rostov-on-Don, st. B. Sadovaja, 105/42, Russia
[2]DonSTU, 344000, Rostov-on-Don, sq. Gagarina, 1, Russia
[3]Don-Invec LTD, 344090, Rostov-on-Don, av. Stachki, 194/2, Russia

Abstract. The problem of increasing the power efficiency of internal combustion engines by the method of its tribochemical treatment was considered. Concentrate additives Komol on the base of molybdenum complexes was developed. This concentrate reduces fuel consumption and increases engine power. In this paper the Subaru EJ-25 engine used in aviation and automotive industry was investigated before and after the tribochemical treatment by the bench tests. It's shown that adding tribochemically active additives into motor oil increases maximal engine power to 10,7% which is 8,7% on comparable engine rotation speeds. The effect was caused by reduced frictional losses as a result of the tribopolymer films formed on friction surfaces.

1 Introduction

Nowadays four-stroke piston internal combustion engines (ICE) are often used for general aviation. The ICE operation is followed by a great heat generation due to high temperatures of the burnt gases and contact stresses in the engine friction pairs. Therefore, the parts that form friction pairs are lubricated and cooled down. The quality of the motor oil impacts the engine power efficiency, especially its lifetime, wear resistance and friction energy losses.

One of the most successful ways to prevent the wear of engine moving parts, such as pistons and cylinders, is the use of tribochemical additives. Tribochemical reactions during the engine operation lead to the formation of films with a thickness of some tens and hundreds nanometers. The tribochemical films protect the covered surface from wear and degradation. Modern lubrication systems use oils that contain chemical additions of different composition.

In recent years the active compounds of lubricating oil additives include such elements as sulfur [1] and phosphorus [2] which improve antifriction, antiwear and other characteristics. In the case of additives that contain several active compounds, the zinc dialkyldithiophosphate is most widely used [3]. Our previous studies of tribochemical processes on friction surfaces have demonstrated that under certain conditions these surfaces get covered with polymer film capable of reducing the friction coefficient up to 10 times [4]. The bench tests show that the significant decrease in friction losses lowers the fuel consumption of ICE by up to 10% and increases its capacity [5].

Traditionally, oils modified by tribochemical additives (energy-saving oils) include coordination compounds of transition metals and complex esters of organic acids, which participate in tribochemical processes of forming a tribopolymer film.

The results of comparative tribometric tests of additive concentrates with coordination compounds of different transition metals in SAE10W40 motor oil are shown in Table 1.

Table 1. The results of tribometric tests of SAE10W40 oil with concentrates of additives of transition metals in an amount of 2% using end friction machine.

Metal in coordination compound of additive	Friction coefficient	Load capacity, MPa
-	0,083	68
copper	0,011	153
nickel	0,014	138
cobalt	0,018	136
zinc	0,017	134
molybdenum	0,010	158

The results demonstrate that injected additives reduce the value of friction coefficient by the rate of 4,9-8,3. Additionally, load capacity increases by the rate of 2,0-2,3 in comparison to initial motor oil. This effect can be explained by the tribopolymer film formed on friction surfaces (Fig. 1 b, c). The film eliminates direct contact of metal surfaces and replaces it with friction between two polymers, which in the presence of oil reduces the value of friction coefficient significantly. The presence of tribopolymer film on conjugated surfaces considerably increases the real area of contact and thus the load capacity rises.

* Corresponding author: invec@mail.ru

 DOI: 10.1051/matecconf/201713301005

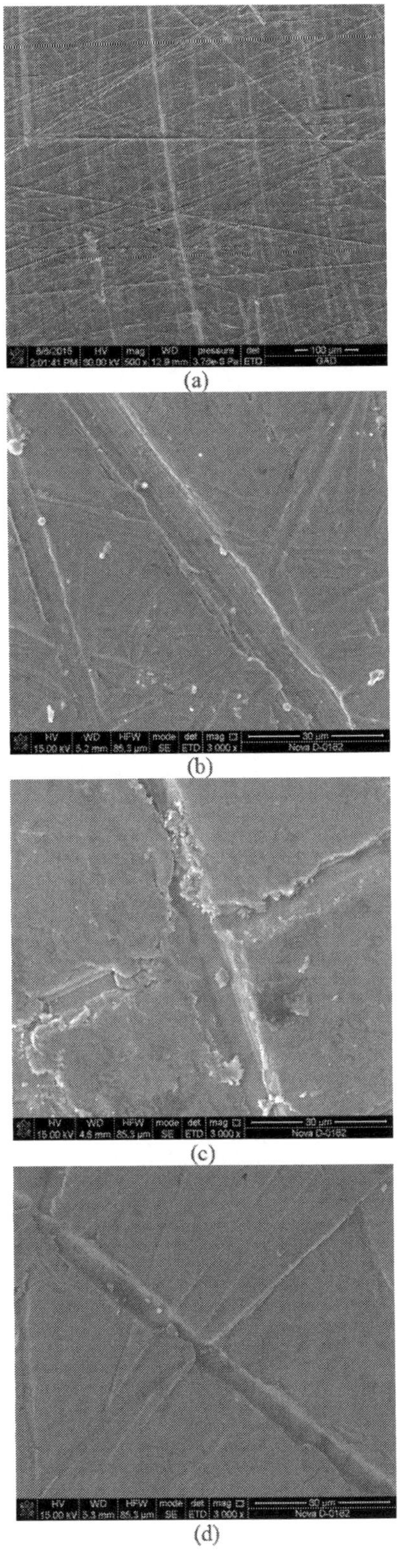

Fig. 1. SEM picture of surface (a) after the friction test in initial oil; in modified oil with (b) copper complex, (c) nickel complex, (d) molybdenum complex.

Field tests of SAE5W30 motor oil intended for ICE used in passenger cars were conducted. The oil composition was enriched with oil-soluble coordination compound of molybdenum in the form of additives concentrate Komol. The surface pictures taken after the friction test in initial oil (Fig. 1a) and in modified oil (Fig. 1 (b-d)) are presented in Fig. 1.

The bench tests of effects caused by motor oil modification were performed on an engine unit based on the Subaru EJ-25 conversion engine. The tested engine was taken from the aircraft "Shmel" produced in Russia (fig. 2) after 410 flight hours on board. The Subaru EJ-25 conversion engines are used in the automotive industry and can be transformed for aviation by replacing the control and transmission units. These engines are applied on a number of aircrafts and helicopters, such as Dinali H2S, Cikwere CH-12, AeroCopter AK1-3 and others.

Fig. 2. The aircraft "Shmel" whose engine was used for motor oil modification bench tests (Source: Author's own figure).

2 Measuring equipment

The measurements of engine unit parameters after the motor oil modification were performed by using the specially developed test bench capable of imitating the activity of aircraft engine at the starting. The test bench (Fig. 3a) includes: the metal frame, fuel tank, engine support, the engine itself, cooling system and the measurements panel that includes a set of devices for measuring the standard engine parameters.

The propeller with variable blade step was used as loading device. Engine load was controlled by the way of blades pitch angle variation. The intermediate shaft with two flanges was installed between propeller and the engine reduction gear.

Propeller power consumption measurements were based on intermediate shaft deformation. The measurements were performed by using the four tensometers glued to the shaft at an angle of 45° to each other and with the bridge switching circuit. The usage of bridge circuit reduces the influence of temperature on the measurement results. The data from the tensometers were received by a small-size radio transmitter installed directly on the rotating shaft. The transmitter was controlled by using a serial peripheral interface (SPI) that synchronizes any transmission with common clock signal generated by processor.

The torque meter installed on the intermediate shaft of tested engine is shown in fig. 3b. The battery was installed opposite to the transmitter in order to provide mass balance. The tensometers were installed in the middle of intermediate shaft and thus the shaft did not change in thickness and shape at the distance of 40 mm before and after that.

(a) (b)

Fig. 3. Test bench for measuring the characteristics of aviation-used engines (a) and the torque meter installed on intermediate shaft (b) (Source: Author's own figure).

3 Research methodology

The bench tests aim to detect the changes of engine unit performance caused by the motor oil modification. The tests assumed two step measurements:

1) engine with initial motor oil;

2) engine with modified motor oil after 2 hours of operation.

The tests were performed by increasing the rotation frequency to its maximum possible value at the current blade pitch angle of the propeller. The step of blade pitch angle change was 3°. The zero blade pitch angle corresponds to the minimum angle of operating range and the maximum value was 15°.

The Komol additive was injected after the tests with the initial oil within the operating range of the blade pitch angles. After two hours of idle engine operation the tests were repeated for the same range of blade pitch angles.

Fig. 4 demonstrates an example record of the engine torque value. It is clear that in the case of four-stroke engine the values of torque are strongly nonlinear during the recording time and should be time-averaged.

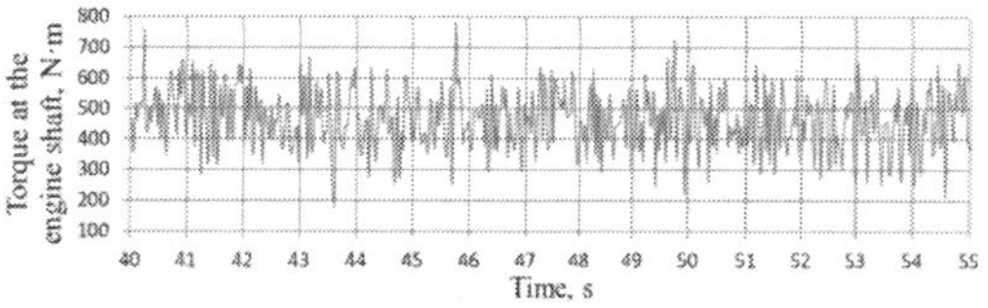

Fig. 4. An example record of the torque value at the engine shaft after 2 hours of operation with modified motor oil at blades pitch angle of 0°.

4 Measurement results

The bench test results demonstrate a general tendency for increasing the values of torque and maximum rotation frequency after the motor oil modification. Thus the engine power increased when the modified motor oil was used.

After 2-hour operation with modified motor oil, the observed average values of rotation frequency increased by 4,1%, torque – 6,4% and the power – 10,7%. The effect was observed for the full operating range of the blades pitch angles (Table 2).

The engine power and torque depending on rotation frequency and the propeller blades pitch angle are shown in Fig. 5-7.

Table 2. The results of bench tests of the engine motor oil in the initial state and after 2-hour operation with Komol additive.

The propeller blades pitch angle, °	Average engine torque, N m	Rotation frequency, min-1	Power, h. p.	Average engine torque, N m	Rotation frequency, min-1	Power, h. p.	Relative torque increase, %	Relative power increase, %
15	195,3	4800	133,3	181,3	4617	119,0	7,7	12,0
12	187,4	5260	140,2	176,6	5020	126,1	6,1	11,2
9	178,6	5660	143,8	168,3	5390	129,0	6,1	11,4
6	167,7	6095	145,4	156,8	5900	131,6	7,0	10,5
3	152,3	6660	144,2	144,2	6450	132,3	5,6	9,1
0	140,0	7295	145,2	132,3	7000	131,7	5,8	10,3
	Modified oil (2-hours operating time)			Initial oil			**Total**	
							6,4	**10,7**

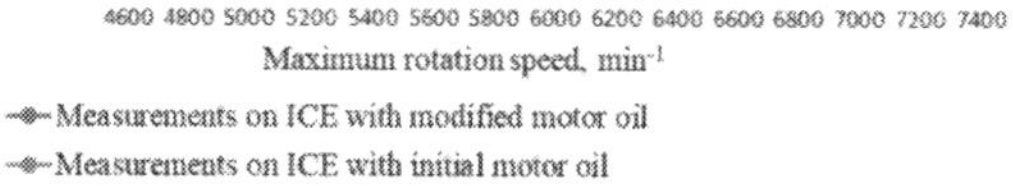

Fig. 5. The ICE power depending on the maximum rotation frequency at the various propeller blades pitch angle for the conditions of initial motor oil and after 2-hour operation with modified one.

DOI: 10.1051/matecconf/201713301005

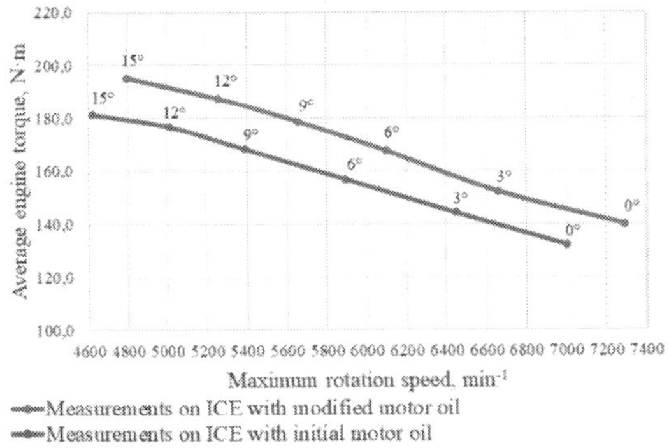

Fig. 6. The torque on ICE shaft dependence on the maximum rotation frequency at various propeller blades pitch angle for the conditions of initial motor oil and after 2-hour operation with modified oil.

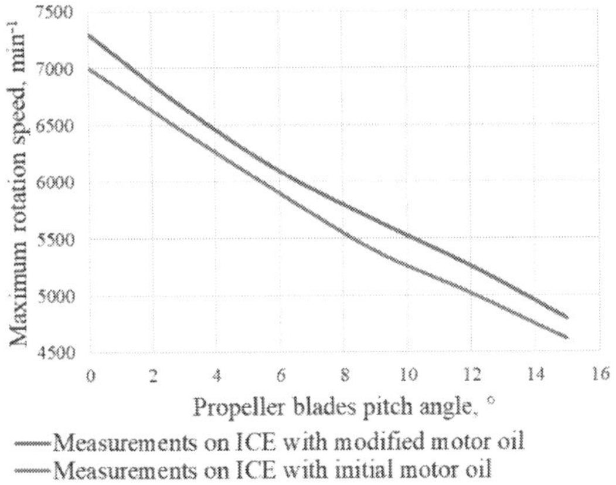

Fig. 7. Maximum rotation frequency dependence on the propeller blades pitch angle for the conditions of initial motor oil and after 2 hours of operation with modified oil.

The results of the measurements show that despite the increase in the engine rotation frequency the power almost stops to grow. The real power on the engine shaft was higher due to losses occuring in one-step reduction gear. However, these losses were negligible and do not influence the relative value of power increase.

Finally, the approximate functions that describe the relation between the maximal power of the engine and the rotation frequency have been obtained. Thus, in the case of initial oil the following equation is obtained:

$$P = 1{,}573 \times 10^{-9}\,\omega^3 - 31604{,}555 \times 10^{-9}\,\omega^2 + \\ + 211000989{,}904 \times 10^{-9}\,\omega - 336{,}103 \qquad (1)$$

where:
P – engine power, h. p.;
ω – maximum engine rotation frequency, min^{-1}.
For the case of modified oil:

$$P = 2{,}404 \times 10^{-9}\,\omega^3 - 47252{,}830 \times 10^{-9}\,\omega^2 + \\ + 308814304{,}368 \times 10^{-9}\,\omega - 526{,}303 \qquad (2)$$

The obtained approximation functions allow us to find out the engine power at comparable rotation speeds. The increase in power at the same rotational speeds is 8,7%.

Table 3. Engine power variation at comparable rotation speeds.

Propeller blades pitch angle, °	Maximum rotation speed, min^{-1}	Initial oil	Modified oil (2-hours operating)	Relative power increase, %
		Engine power, h. p.	Engine power, h. p.	
15	4800	122,494	133,164	8,0
12	5200	127,691	139,8364	8,7
9	5600	130,628	143,3894	8,9
6	6000	131,907	144,745	8,9
3	6400	132,133	144,827	8,8
0	6800	131,911	144,558	8,7
Total:				**8,7**

5 Conclusion

The results of comparative tribometric tests of additives concentrates with coordination compounds of different transition metals in SAE10W40 motor oil demonstrate that injected additives reduce the value of friction coefficient by the rate of 4,9-8,3. This effect can be explained by the antifriction tribopolymer film formed on friction surfaces.
The most effective one is the concentrate additive Komol containing a complex of molybdenum.
The results of propeller engine unit bench tests demonstrate that motor oil modification by the additive compound Komol increases produced power by 8,7% at comparable rotation speeds.
The proposed technology can be used in gasoline and diesel internal combustion engines in road, rail and other modes of transport.

This work was supported by the Ministry of Education and Sciences of the Russian Federation (project no. 14.575.21.0095). Unique identifier of applied scientific research (project) RFMEFI57514X0095.

References

1. G. Colas, A. Saulot, C. Godeau, Y. Michel, Y. Berthier, Wear **305**, 192-204 (2013)
2. N. Gosvami, J. Bares, F. Mangolini, A. Konicek, D. Yablon, R. Carpick, Science **348**, 102-106 (2015)
3. M. Nicholls, T. Dob, P. Norton, M. Kasrai, G. Bancroft, Tribology International **38**, 15-39 (2005)
4. A. Ponomarenko, A. Burlov, M. Boiko, T. Shiryaeva, A. Kalmykova, S. Zaichenko, M. Milutka, Journal of Friction and Wear **36-1**, 15-22 (2015)
5. A. Ponomarenko, M. Boiko, A. Burlov, A. Kalmykova, T. Bozhenko, *Tribochemical nanotechnology for energy-saving lubricants*, SGEM 2015 Proceedings, 235-242 (2015).

 DOI: 10.1051/matecconf/201713301006

Mobile test stand for evaluation of electric power plants for unmanned aircraft

Vladimir Serbezov[1,*], *Stamen* Dimitrov[2], and *Konstantin* Rangelov[1]

[1]Dronamics Ltd., 4 Iskar St., 1000 Sofia, Bulgaria
[2]Department of Aeronautics, Technical University - Sofia, 8 Kl. Ohridski Blvd., 1000 Sofia, Bulgaria

Abstract. The absence of accurate performance data is a common problem with most civilian unmanned aerial vehicle (UAV) power plant producers. The reasons for this are the small size of most of the manufacturers and the high price of precise wind tunnel testing and computer simulations. To overcome this problem at Dronamics Ltd., with support from the Department of Aeronautics of TU-Sofia, a mobile test stand for evaluation of electric power plants for unmanned aircraft was developed. The stand may be used statically, or may be installed on the roof of an automobile. The measurement system of the stand is based on popular hardware that is used in radio controlled models and in general automation. The verification of the measurement system is performed by comparing static test results with data published by the manufacturer of the tested electric motor. Tests were carried out with 2 different types of propellers and the results were compared with published results for common propellers as well as with results of theoretical studies. The results are satisfactory for practical applications. The use of this type of test stands can be a cheap and effective alternative for research and development start-up companies like Dronamics.

1 Introduction

The absence of accurate performance data is a common problem with most civilian UAV propeller and engine producers. The reasons for this are the small size of most manufacturers and the high price of precise wind tunnel testing and computer simulations. In recent years there has been some academic effort to cover this gap. A good example is the survey performed by Brandt and Selig [1]. They tested different small propellers (up to 11" diameter) in a wind tunnel. In addition, there were some attempts by UAV enthusiasts to overcome this issue by deriving empiric thrust equation for a propeller with a given diameter and pitch [2]. The latter may serve as a rough approximation for preliminary design, but they do not account for important factors such as widely varying Reynolds number and propeller geometry.

Research in the field of UAV propeller performance is carried out also in Bulgaria [3, 4]. It yielded promising results but it is focused mostly on multi copter rotors and it is not applicable for fixed wing UAVs with relatively large propellers with diameters of up to 30" (762 mm).

To overcome this problem in the interest of Dronamics experimental prototypes, with the support of the Department of Aeronautics of Technical University – Sofia a test stand for testing propeller/electric motor powerplants for UAVs was constructed. In order to make testing at high propeller advance ratios possible it was decided to mount the test stand on the roof of a passenger automobile. Partially, this decision was motivated by tests carried at NASA Armstrong Flight Research Center [5]. Another feature of the project was that its target was to solve a practical problem at affordable cost. For this reason, the measurement and data acquisition systems of the test bench were based on popular hardware that is used in radio controlled models and in general automation.

2 Test stand description

The general arrangement of the test stand (set for static test on the ground) is shown in Fig. 1.

Fig. 1. Static installation of the test stand.

It was assumed that thrust measurement is most important and the shaft power can be estimated from the electrical power of the engine, given its efficiency is

* Corresponding author: vserbezov@dronamics.com

 DOI: 10.1051/matecconf/201713301006

relatively constant. This allowed significant simplification of the stand structure and use of strain gauges placed on a bending loaded beam.

An Arduino Uno board, connected by a serial port with a laptop, was used as the data acquisition system. The Arduino board allowed for stable operation with a sample rate of approximately 2 samples per second, which was judged to be satisfactory for the stand needs. The Arduino board was placed in a box right behind the thrust beam of the stand. The sensors used are listed in Table 1.

Table 1. List of sensors.

Sensed quantity	Sensor	Data acquisition
Thrust	Sparkfun HX711 load cell amplifier	Arduino, digital input
Rotational speed	US1881 latch hall sensor	Arduino, digital input
Ambient pressure	BMP-183 Barometric Pressure and Altitude Sensor	Arduino, digital input
Ambient Temperature	BMP-183	Arduino, digital input
Engine input voltage	60V to 5V voltage splitter	Arduino, analogue input
Engine input current	ACS709 hall sensor up to 75A	Arduino, analogue input
Airspeed	HD350 pitot tube anemometer	Serial connection with the computer
Ground speed	Garmin GPSMAP 60CSx	Internal memory storage

2.1. Propeller thrust measurement

Some designs of the load cell were evaluated in ANSYS. Finally, a standard I beam was chosen. The boundary conditions and results of the strain analysis are shown in Figures 2 and 3. The 350 Ohm strain gauges, connected in half-bridge, have a resistance change of 0,3566 Ohm at 100 N load, which results in 1,885 mV output at 3,7 V bridge input voltage. The output is sensed by a Sparkfun HX711 load cell amplifier.

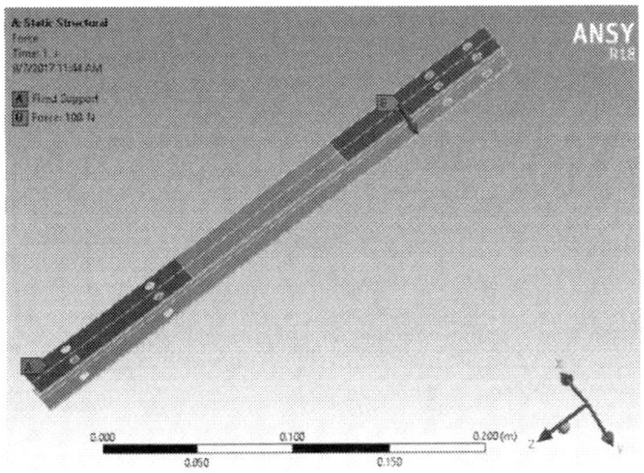

Fig. 2. Boundary conditions and loading applied on Thrust beam in ANSYS.

The calibration of the load cell was performed on the test stand, by loading it and measuring the load with a general purposes electronic scale with 10 g accuracy. Two cycles of step loading/unloading from 0 to 15 kg were performed. The results are shown in Fig. 4. A hysteresis of 0,3 kg was observed at maximum load which was assumed to be satisfactory. Also, there was a slight nonlinearity due to the imprecise placement of the gauges on the beam, but it was corrected through the use of a second order polynomial correction performed in the Arduino.

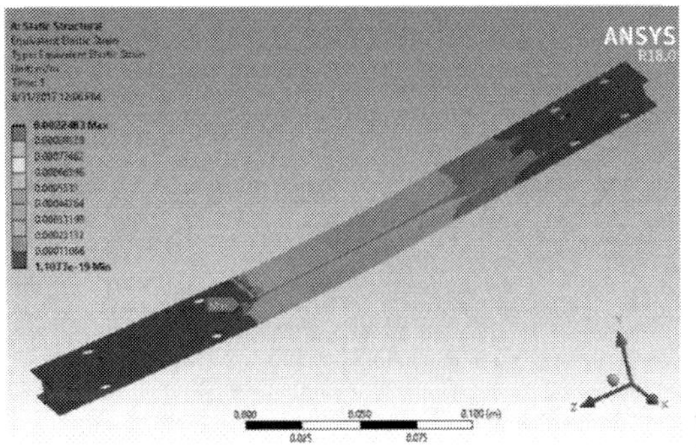

Fig. 3. Strain analysis of the thrust beam in ANSYS.

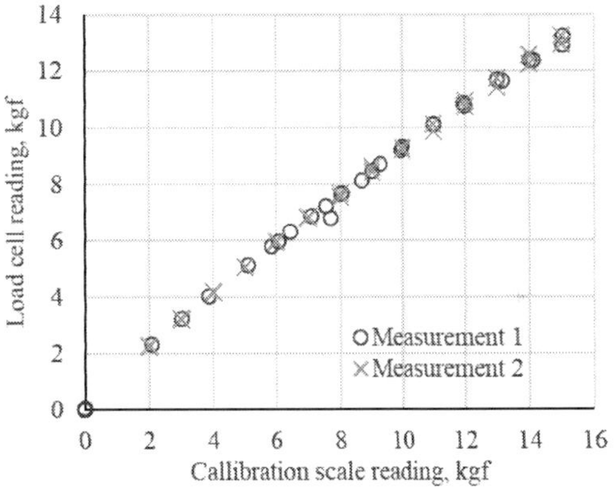

Fig. 4. Thrust measurement system calibration results.

2.2 Engine speed measurement

The T-motor U11KV90 DC engine is an 'outrunner' type, e.g. it has a rotating case with permanent magnets, placed on its inner surface. This configuration results in strong magnetic field around the engine, allowing for the use of a latch hall sensor placed over the engine rotor, connected to an interrupt digital input of the Arduino board. The rotational speed was calculated by counting the number of interrupts per data sample interval. The System was tested and approbated with the help of a manual laser RPM reader.

2.3 Electric signals measurement

The voltage and current were measured at the input of the engine speed controller. Voltage up to 60 V DC was sensed through a voltage splitter at a 5 V analogue input of the Arduino board with 10-bit resolution. The ACS 709 current sensor was connected to Arduino in a similar way. It was connected at the minus input of the engine speed controller.

2.4. Ambient conditions and airspeed measurement

The ambient pressure and temperature were measured by a .BMP-183 barometric pressure sensor, placed inside the Arduino box. A disadvantage of this position was that the temperature reading was influenced by the heat from the sun and propeller airstream cooling. For this reason, an additional temperature reference was needed for the after calculations.

The airspeed, when the stand was placed on the roof of an automobile, was measured by a HD350 Pitot tube anemometer, connected through a USB connection to the laptop. The data logger programs of the Arduino and the anemometer used the computer clock, allowing for an easy merger of the data. The Pitot tube was mounted approximately 0,5 m off the right side of the car (visible on Fig. 5). It was assumed that this position will be unaffected by the car boundary layer and the propeller stream.

Fig. 5. The test stand mounted for runway tests.

For additional ground speed reference an Garmin GPSMAP 60CSx handheld system was mounted inside the automobile. The GPS system stored the data in its internal memory, requiring preliminary synchronization of the GPS and the data logging computer.

3 Static testing

The static tests were performed with T-motor U11KV90 DC motor with T-motor G29×9,5 propeller, which is optimised for multirotor UAVs. The aim was to test the data acquisition system and to approbate it using existing performance data from the motor and propeller manufacturer [6]. An additional task was to test the endurance of the batteries that supply the engine. The data from a typical test run is shown in Fig. 6.

Several static test runs were performed. The last three runs were used for stand qualification. The data acquired was processed in Matlab and the thrust data was corrected for the ambient pressure using equation (1):

$$Thr\_c = Thr \times 10325 \text{ Pa} / Pamb, \qquad (1)$$

where the *Thr* is the measured thrust, and the *Pamb* is the measured ambient pressure. Then a second-degree polynomial was fitted to the data and was compared to the published propeller data (Fig. 7). The similarity achieved is fine. The goodness of fit figures are as follows in Table 2.

Table 2. Goodness of fit of thrust measurement.

Figure of merit	value
Coefficient of determination (R-squared)	0,98056
Root mean square error (RMSE), N	3,97

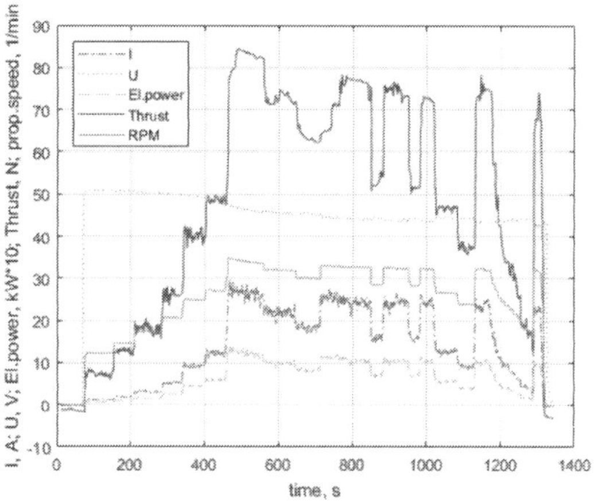

Fig. 6. Typical static test run data plot for G29×9,5 propeller.

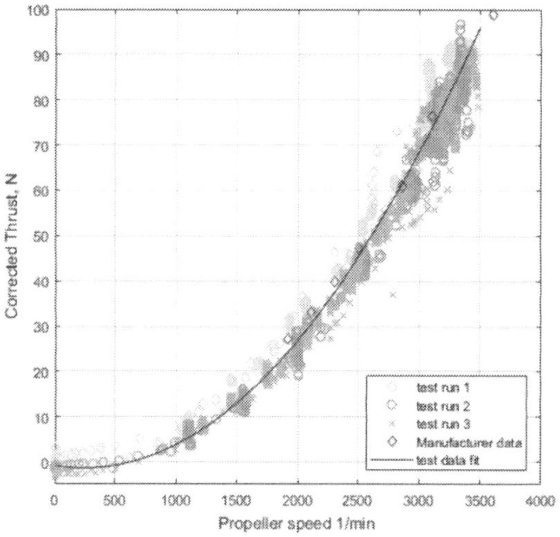

Fig. 7. Comparison of manufacturer and experimental data for T-Motor U11KV90 engine with T-Motor G29×9,5 propeller.

DOI: 10.1051/matecconf/201713301006

4 Runway testing

The runway tests were performed with Menz-E 26x20 2-blade wooden propeller at Sapareva Banya runway (Fig. 5). The duration of the tests was approximately 50 min and was limited by the endurance of the battery packs used.

The propeller was tested at speeds ranging from 10 to 25 m/s. As accurate keeping of the airspeed was impossible due to the uneven motion of the automobile and the open-air turbulence, the focus was on postprocessing the acquired data in Matlab. The test data was filtered and only data points with small rate of change of the speed, rpm and thrust readings were picked. The data was then used to estimate the propeller advance ratio and coefficients of thrust and power in accordance with the common formulae for propeller performance, that can be found in the literature, for instance in [7].

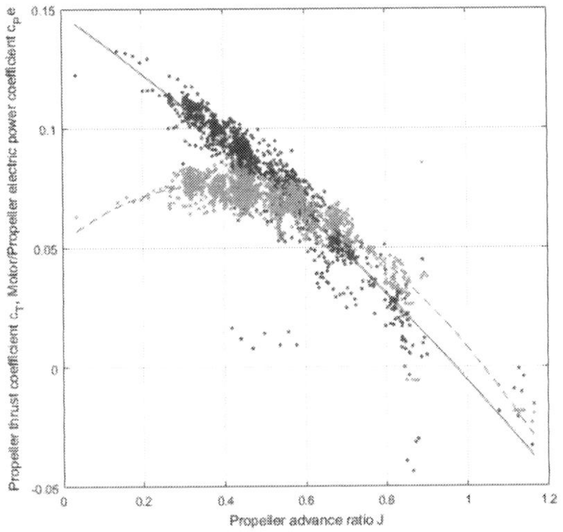

Fig. 8. Estimation of thrust (solid line) and power at electric motor input (dashed line) characteristics for Menz-E 26×20 propeller.

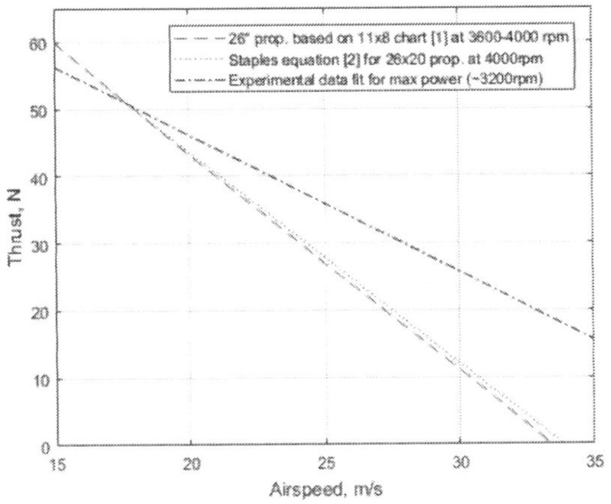

Fig. 9. Thrust comparison between U11KV90 engine with Menz-E26×20 propeller and estimates based on [1] and [2].

The results and the fitted curves for the coefficients of thrust C_T and power C_P are presented in Fig. 8.

Because experimental data for engine shaft power did not exist, electrical power was used instead. The results are still adequate, since the efficiency of the engine is relatively constant and the shaft power can be estimated through further calculations.

A comparison between the experimental results and the estimates for the 26×20 propeller thrust, based on [1] and [2] is shown on Fig. 9. It is obvious that the differences are quite significant, which fully justifies the efforts for having the test bench.

5 Conclusions

Although aimed at solving practical problems at relatively low cost, the test stand showed a great potential for solving research problems in the field of powerplants for middle sized fixed wing UAVs.

The data collected during the experiments was successfully used for estimation of the powerplant performance characteristics with acceptable accuracy.

The idea of mounting the test bench on the roof of a passenger car was practical and consideration was given to performing the experiments in calm weather conditions.

The experience gained allows for significant improvements in the data acquisition system of the test bench and in the accuracy of the collected data without significant rise in equipment costs.

The authors would like to thank Professor Michael Todorov – head of the Department of Aeronautics at Technical University – Sofia for supporting the work on this project.

References

1. J. Brandt, M. Selig, Propeller Performance Data at Low Reynolds Numbers, *AIAA 2011-1255*, (2011)
2. G. Staples, Propeller Static & Dynamic Thrust Calculation, http://www.electricrcaircraftguy.com/2014/04/propeller-static-dynamic-thrust-equation-background.html
3. F. Panayotov, I. Dobrev, F. Massouh, M. Todorov, Numerical model for rapid computatuon of the flow-field of a rotor in hover, *ICMT 2017*, Brno, Czech Republic (2017)
4. H.Panayotov, S. Penchev, Determination of propeller characteristics for multirotor drone design, *TECHSYS 2017*, Plovdiv, Bulgaria (2017)
5. LEAPTech to Demonstrate Electric Propulsion Technologies https://www.nasa.gov/centers/armstrong/Features/leaptech.html (2015)
6. T-Motor U11 KV90 electric motor specifications: http://store-en.tmotor.com/goods.php?id=328
7. E.Torenbeek, *Synthesis of Subsonic Aircraft Design*, Delft Univerity Press, Delft, The Netherlands, 191, (1982)

MATEC Web of Conferences **133**, 01007 (2017)
BulTrans-2017
DOI: 10.1051/matecconf/201713301007

Modeling of shimmy oscillations in aircraft landing gear

Hristina Georgieva[1,*]

[1]Technical University – Sofia, Aeronautical Department, 8 Kliment Ohridski Blvd., 1000 Sofia, Bulgaria

Abstract. The paper proposes an analysis of shimmy oscillations in one of the most used aircraft in general aviation. The main goal of the research is to develop a mathematical model with three degrees of freedom created in a Matlab environment on a Cessna Citation M2. The parameters of the model are obtained from Cessna Textron Aviation, Goodyear Aviation tires and scientific studies. Present research focuses on three main areas: 1) mathematical model, 2) numerical application and results, and 3) discussion and conclusion. The results obtained are presented graphically to facilitate the study of landing gear shimmy oscillations. The effects observed present the subject of discussions in the paper. The model makes the optimisation of the amplitude and time of shimmy oscillation possible.

1 Introduction

Aircraft landing gears may experience self-excited oscillation in lateral and torsional direction, known as shimmy oscillations, caused by the interaction between the tyres and the ground. Shimmy oscillations appear in aircraft ground dynamics during taxing, take-off and landing. They can be found on both the nose and main landing gear, but the first occurrence is more common. Shimmy oscillations are usually not catastrophic but they can reduce landing gear stability and result in serious problems (excessive wear, shortened life cycle of gear parts, safety concerns, discomfort for pilots and passengers or eventually lead to an accident) [1-3]. Shimmy oscillations may occur due to wheel imbalance, torsional freeplay, inadequate torsional stiffness, etc.

In order to increase the understanding of these oscillations on aircraft landing gear researchers have elaborated different shimmy models. One of earliest contribution to shimmy dynamics was reported by Broulhiet in 1925 [4]. In the study the role of tyre flexibility and side slip in the shimmy mechanism was recognized. Smiley [5] investigated shimmy oscillations for three different landing gear structures. The nonlinear geometric effects were not included in the model while one of the cases had a non-zero rake angle. The paper included linear stability of a landing gear model and discussed a systematic way of modelling the strut geometrical aspects. Somieski described shimmy oscillations as a nonlinear phenomenon for a nose landing gear with zero rake angle [6]. Woerner and Noel [7] explain the main cause of shimmy oscillations is the energy transfer from the contact force between the tyres and the ground to the vibrational modes of the landing gear system. This stability depends on the damping and stiffness characteristics of tyres and the supporting structure.

The developed models of shimmy oscillations are mainly based on different number of degrees of freedom, different analysis methods and different ways of modelling the elastic tyres [8-14]. These models allow for the analysis of the stability and the response of shimmy oscillations.

Shimmy oscillations are a problem in the design and operation of aircraft landing gears, especially in general aviation aircraft. This aircraft can be used as a business jet or aircraft for special missions such as aerial survey, air ambulance, flight inspection and surveillance. This required an accurate and appropriate analysis in order to increase their safety during ground motion. Cessna is a representative of this type of aircraft in general aviation.

The aim of this research is to study the impact of the vertical force on the gear, taxiing velocity of the aircraft and torsional angle on the amplitude and the time of the shimmy oscillations in the aircraft nose landing gear and their optimisation. To do this, the paper proposes a mathematical model on a Cessna Citation M2 with three degrees of freedom created in a Matlab environment.

2 Mathematical model of shimmy oscillation in aircraft landing gear

2.1 Aircraft nose landing gear

The geometry of aircraft nose landing gear is shown in Fig. 1, where positive X-axis is the rear direction of the aircraft, the Z-axis is the upward to the ground and the Y-axis completes the right-handed coordinate system. The nose landing gear consists of a wheel with a pneumatic tyre with radius R mounted on an axle that is attached to a strut via a mechanical caster with length e. The strut axis is inclined to the vertical at the rake angle φ. The applied vertical force F_z on the gear represents the combined

* Corresponding author: hgeorgieva@tu-sofia.bg

 DOI: 10.1051/matecconf/201713301007

effect of the fuselage and wings. The aircraft moves with a forward velocity V_x.

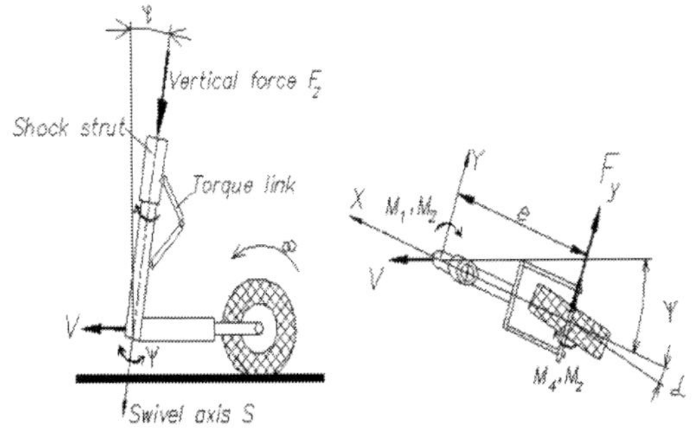

Fig. 1. Side and top view of the nose landing gear.

The model has three degrees of freedom:

- The rotation of the gear around the strut axis S (torsional mode) described by the torsional angle ψ;
- The bending of the gear around the X-axis, (lateral mode) described by angle δ;
- The two modes are coupled through the tyre-ground interaction. The present work uses the stretched string tyre model developed by von Schippe and Dietrich [15]. The model defines the nonlinear kinematic relationship between the torsional angle ψ, lateral bending angle δ and the lateral deformation λ of the leading edge of the tyre contact patch.

The presence of a non-zero rake angle φ has several important geometrical effects in an aircraft nose landing gear [8]. It involves an effective caster length e_{eff} that is given by:

$$e_{eff} = e\cos\varphi + R\tan\varphi + e\sin\varphi\tan\varphi. \tag{1}$$

The wheel swivel angle θ with the ground is different from the steering angle and it is expressed by:

$$\theta = \psi\cos\varphi. \tag{2}$$

When the steering angle ψ is non-zero there is also a tilt of the wheel:

$$\gamma = \psi\sin\varphi. \tag{3}$$

In this paper the effect of this angle is not considered.

The vertical force F_z is offset from the X-Z plane due to the effective caster length e_{eff}, which creates a moment that destabilizes the orientation of the gear.

The equations of motion for the model of the nose landing gear, taking into account torsional and lateral motion and their coupling via the elastic tyre can be written as:

$$\begin{cases} I_z\ddot{\psi} = M_{K\psi} + M_{D\psi} + M_{F1} + M_{D\alpha} + F_z\sin\varphi e_{eff}\sin\theta; \\ I_x\ddot{\delta} = M_{K\delta} + M_{D\delta} + M_{\lambda\delta} + M_{D\alpha} + F_z e_{eff}\sin\theta; \\ \dot{\lambda} = V\sin\theta + l_g\dot{\delta}\cos\delta - (V/L)\lambda + \left(e_{eff} - h\right)\cos\theta\dot{\psi}\cos\varphi, \end{cases} \tag{4}$$

where the dimensions are given in Table 1.

The moment $M_{K\psi}$ corresponding to the torsional stiffness of the strut is a function of the steering angle ψ and is defined by:

$$M_{K\psi} = k_\psi\psi. \tag{5}$$

The moment $M_{D\psi}$ due to the torsional damping of the strut is a function of the steering angular velocity $\dot{\psi}$ and is expressed by:

$$M_{D\psi} = c_\psi\dot{\psi}, \tag{6}$$

where k_ψ and c_ψ are the torsional stiffness and damping coefficients of the strut.

The self-aligning torsional moment $M_{K\alpha}$ about the strut is given by:

$$M_{K\alpha} = \begin{cases} F_{zn}k_\alpha\dfrac{\alpha_m}{\pi}\sin\left(\alpha\dfrac{\alpha_m}{\pi}\right) & if\,|\alpha| \le \alpha_m; \\ 0 & if\,|\alpha| > \alpha_m, \end{cases} \tag{7}$$

where k_α is the self-aligning coefficient of the tyre and α_m is the limit of the self-aligning moment.

The lateral restoring force $F_{K\lambda}$ of the tyre is defined as:

$$F_{K\lambda} = k_\lambda \tan^{-1}\left\{7\tan(\alpha)\cos\left[0{,}95\tan^{-1}7\tan(\alpha)\right]\right\}F_{zn}, \tag{8}$$

where k_λ is the restoring coefficient of the tyre.

The tyre slip angle α is related to the lateral deformation λ by:

$$\lambda = \mathrm{atan}\frac{\lambda}{L}, \tag{9}$$

where L is the relaxation length of the tyre.

The combined moment M_{F1} due to the tyre restoring force $F_{K\lambda}$ and self-aligning moment $M_{K\alpha}$ that are functions of the tyre lateral deformation λ is given by

$$M_{F1} = M_{K\alpha} - e_{eff}F_{K\lambda}. \tag{10}$$

The moment $M_{D\lambda}$ due to the tyre lateral damping is given by:

$$M_{D\lambda} = c_\lambda\dot{\psi}\cos\varphi, \tag{11}$$

where c_λ is the lateral damping coefficient of the tyre.

The moment $M_{K\delta}$ due to the lateral bending stiffness of the strut is function of δ and is expressed by:

$$M_{K\delta} = k_\delta\delta, \tag{12}$$

where k_δ is the lateral bending stiffness of the strut.

The moment $M_{D\delta}$ due to the damping characteristics of the strut is function of $\dot{\delta}$ and is given as:

$$M_{D\delta} = c_\delta\dot{\delta}, \tag{13}$$

where c_δ is the damping coefficient of the lateral bending of the strut.

DOI: 10.1051/matecconf/201713301007

The moment $M_{\lambda\delta}$ is the result of the force created from the lateral deformation of the tyre and is given as:

$$M_{\lambda\delta} = l_g F_{K\lambda} \cos\theta \cos\varphi, \tag{14}$$

where l_g is the distance between the nose landing gear attachment point in the fuselage and the ground.

2.2 Aircraft data source

The parameters of the model are obtained from G. Somieski, 1997 [6], P. Thota et al., 2008 [8], Cessna Textron Aviation [16], Goodyear Aviation tires [17] and are presented in Table 1.

Table 1. System parameters for modelling.

Parameter	Value	Unit
Structure parameters		
Caster length, e	0,1	m
Gear height, l_g	1,0	m
Strut torsional stiffness, k_ψ	$-1{,}10^5$	Nm.rad^{-1}
Strut torsional damping, c_ψ	-45	Nm.s.rad^{-1}
Strut moment of inertia about z-axis, I_z	1	kg.m^2
Strut later bending stiffness, k_δ	$-3{,}24.10^6$	Nm.rad^{-1}
Strut later bending damping, c_δ	-150	Nm.s.rad^{-1}
Strut moment of inertia about x-axis, I_x	1	kg.m^2
Rake angle, φ	4	rad
Tyre parameters		
Radius of nose wheel, R	0,2286	m
Contact patch length, h	0,1	m
Elastic tyre damping coefficient, c_λ	-270	Nm^{2}rad^{-1}
Elastic tyre self-aligning coefficient, k_α	-2	m/rad
Elastic tyre restoring coefficient, k_λ	20	rad^{-1}
Relaxation length, L	0,2	m
Self-aligning moment limit, α_m	0,1745	rad
Continuation parameters		
Vertical force on the gear, F_{zn}	8-13	kN
Forward velocity, V_x	90-70	m/s

3 Numerical simulation and results

A numerical simulation with Matlab software is performed in order to analyse the behavior of the proposed model for studying the shimmy oscillations in the nose landing gear defined by equation (4).

Three different simulations are studied.

- Simulation 1: In this simulation, the system is supposed to have a constant vertical force of $F_z = 4491$ kg on the gear, representing values of the maximum landing weight (MLW);
- Simulation 2: The system in this simulation is subjected to a constant vertical force of $F_z = 3831$ kg on the gear corresponding to 80 % of MLW;
- Simulation 3: In this simulation, the system received a constant vertical force of $F_z = 3171$ kg on the gear, representing values of the basic operating weight (BOW);

The three simulation cases are run with taxiing velocity V_x changing from 70 to 90 m/s and disturbance yaw angle $\psi = 0{,}3$ rad.

To obtain the amplitude of the cycle, simulation over a time of 1 second is made. From the time histories we can record the amplitude as function of varied parameters: taxiing velocity V, vertical force of the gear F_z, damping constant of the strut c_ψ and disturbance yaw angle ψ.

Figures 2-10 show some of the results about the lateral tyre deflection λ, the lateral bending angle δ and the phase plot of shimmy angle ψ for the second case of simulation.

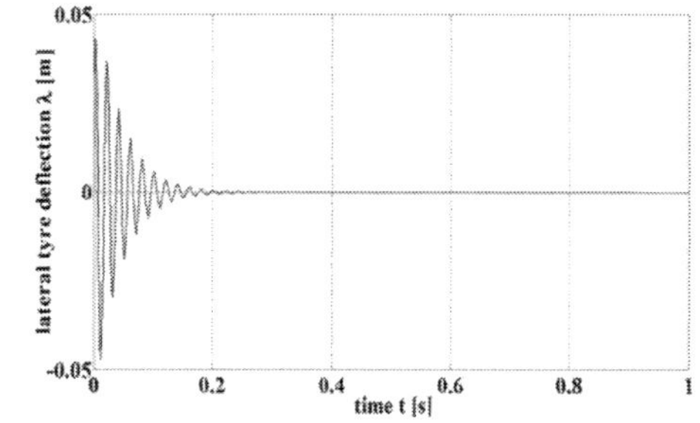

Fig. 2. Modelled lateral tyre deflection for: $F_{zn} = 10\ 690$ N, $V_x = 90$ m/s, $\psi = 0{,}3$ rad.

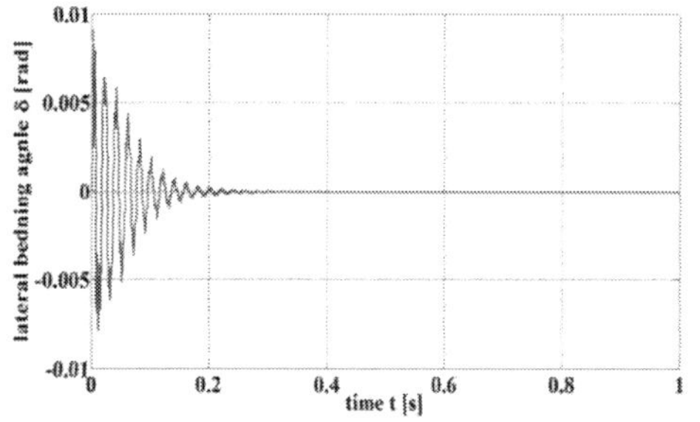

Fig. 3. Modelled bending angle of tyre: $F_{zn} = 10\ 690$ N, $V_x = 90$ m/s, $\psi = 0{,}3$ rad.

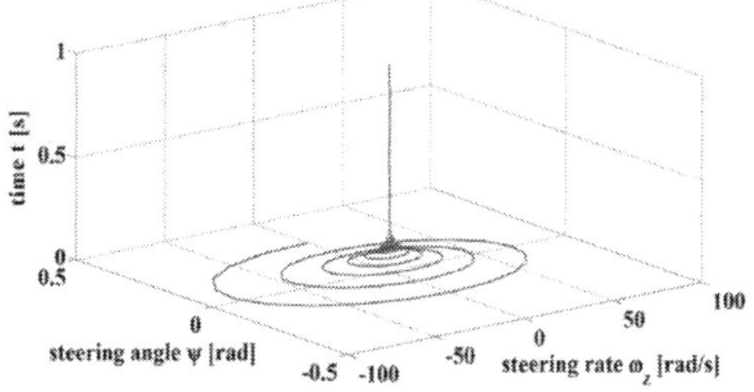

Fig. 4. Modelled phase plot of shimmy yaw angle for: $F_{zn} = 10\ 690$ N, $V_x = 90$ m/s, $\psi = 0{,}3$ rad.

 DOI: 10.1051/matecconf/201713301007

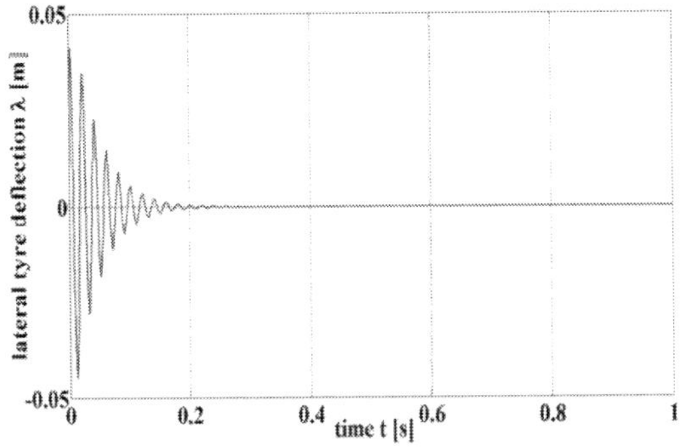

Fig. 5. Modelled lateral tyre deflection.
F_{zn} = 10 690 N, V_x = 80 m/s, ψ = 0,3 rad.

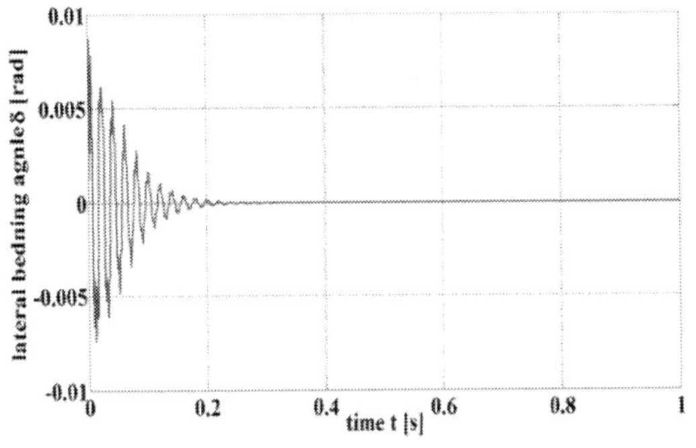

Fig. 6. Modelled bending angle of tyre:
F_{zn} = 10 690 N, V_x = 80 m/s, ψ = 0,3 rad.

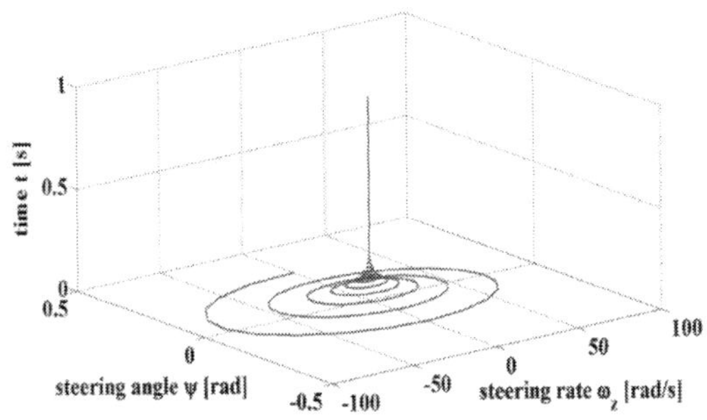

Fig. 7. Modelled phase plot of shimmy yaw angle.
F_{zn} = 10 690 N, V_x = 80 m/s, ψ = 0,3 rad.

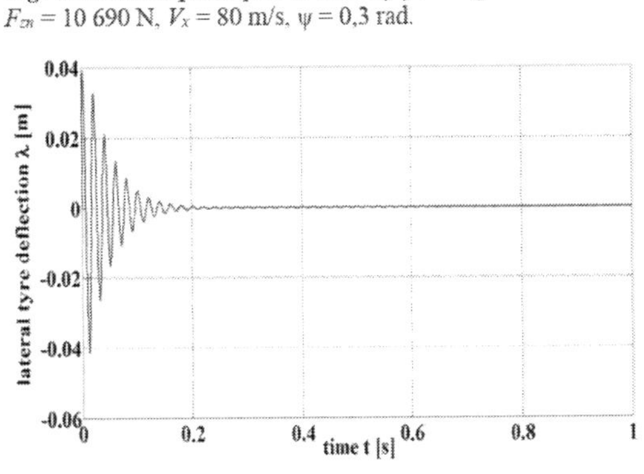

Fig. 8. Modelled lateral tyre deflection.
F_{zn} = 10 690 N, V_x = 70 m/s, ψ = 0,3 rad.

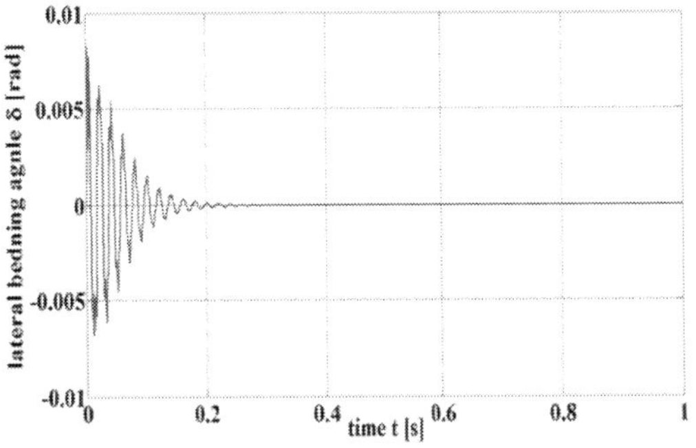

Fig. 9. Modelled bending angle of tyre:
F_{zn} = 10 690 N, V_x = 70 m/s, ψ = 0,3 rad.

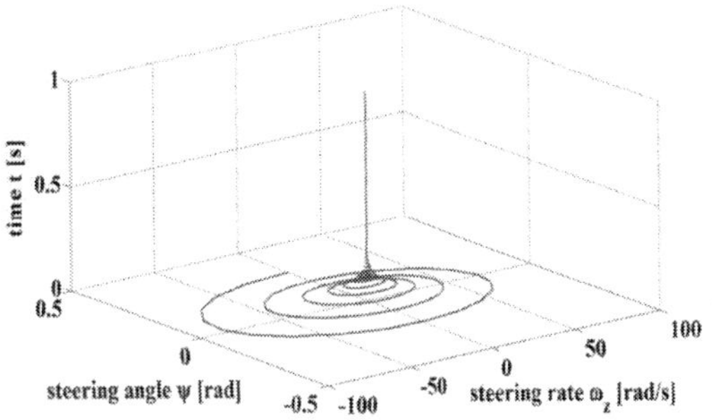

Fig. 10. Modelled phase plot of shimmy yaw angle.
F_{zn} = 10 690 N, V_x = 70 m/s, ψ = 0,3 rad.

Figure 11 shows unstable case (MLW and taxiing velocity V_x = 90 m/s) of the lateral tyre deflection λ.

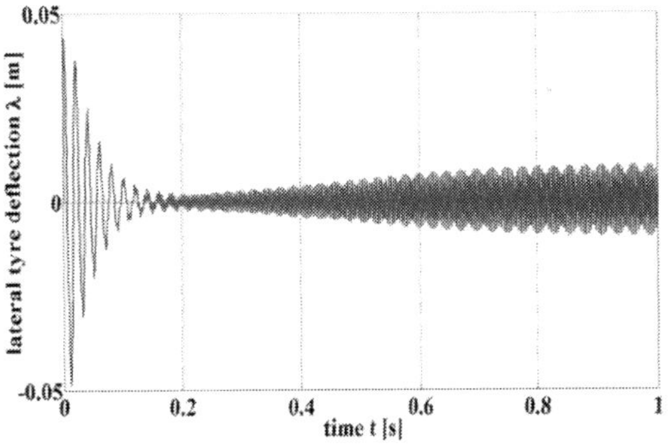

Fig. 11. Modelled lateral tyre deflection.
F_{zn} = 12 531 N, V_x = 90 m/s, ψ = 0,3 rad.

Table 2 presents the results about the amplitude λ_A and the time of shimmy oscillations t_d for all simulation cases.

Table 2. Obtained simulation results.

F_z, N \ V_x, m/s		90		80		70	
12 531	λ_A, m	0,043	unstable	0,042	stable	0,039	stable
	t_d, s	-		0,960		0,340	
10 690	λ_A, m	0,043	stable	0,042	stable	0,039	stable
	t_d, s	0,310		0,290		0,280	
8 847	λ_A, m	0,043	stable	0,042	stable	0,039	stable
	t_d, s	0,290		0,280		0,270	

4 Discussion and Conclusions

The paper proposes a mathematical model applied on a Cessna Citation M2 nose landing gear for shimmy oscillations. The parameters of the model are obtained from Cessna Textron Aviation, Goodyear Aviation tires and scientific studies (G. Somieski, 1997, P.Thota et al., 2008).

The model has three degrees of freedom and allows to study the influence of the gear vertical force, aircraft taxiing velocity and torsional angle on the amplitude and time of shimmy oscillations and their optimization.

The optimal value of shimmy oscillations are: the amplitude λ_A = (0,042-0,039) m and time t_d = (0,29-0,27) s can be found in the interval of taxiing velocity V_x = (70-80) m/s, aircraft weight interval F_z = (BOL-80 % from the MLW) N, torsional damping coefficient c_ψ = –45 Nms/rad and torsional angle ψ = 0,3 rad.

The model shows that in the case of *MLW* and taxiing velocity V_x higher than 80 m/s the system becomes instable.

The model gives also the possibility to study the impact of the rake angle, caster length and the strut torsional damping coefficient on the amplitude and time of shimmy oscillations and their optimization.

The proposed mathematical model is simple and this decreases computation times

The model that is presented in this work is generic and appropriated for the aircraft with single-wheel nose landing gear in the context of study of shimmy oscillations.

Future work will include study of more degrees of freedom like longitudinal bending, the effect of gyroscoping coupling of the tyre and duel-wheel configuration of the landing gear. Modelling of the change of the aircraft trajectory under the shimmy oscillations is also planned.

The final goal would be the formulation of an integrated mathematical model of a generic unmanned aerial vehicle.

References

1. Bureau d'acquêtes et d'Analysais pour la sécurité de l'aviation civile, Rapport accident (2011)
2. South African Civil Aviation Authority, Aircraft accident rapport and executive summary (2015)
3. La Comisión de Investigación de Accidentes e Incidentes de Aviación Civil, Technical Report A-068/1999 (2006)
4. M. Broulhiet. La suspension de la direction de la voiture automobile - Shimmy et dandinement, Bull. *Soc. Ing. Civ*., France, **78**, (1925)
5. R. Smiley, Correlation, evaluation, and extension of linearized theories for tire motion and wheel shimmy, *NACA tech. rep*, 1299, Washington, DC, (1957)
6. G. Somieski, Shimmy analysis of a simple aircraft nose landing gear model using different mathematical methods, *Aerospace Science and Technology*, **8**, 1270-9638, (1997)
7. P. Woerner, O. Noel, Influence of nonlinearity on the shimmy behaviour of landing gear, *81st Meeting of the AGARD Structures and Materials Panel, AGARD-R-800*, (1995)
8. P. Thota, B. Krauskopf, M. Lowenberg, Shimmy in a nonlinear model of an aircraft nose landing gear with non-zero rake angle, *ENOC-2008, Saint Petersurg, Russia, 30 June – 4 July*, (2008)
9. P. Thota, B. Krauskopf, M. Lowenberg, Interaction of torsion and lateral bending in aircraft nose landing gear shimmy, *Nonlinear Dynamics*, **57(3)**, 455-467, (2009)
10. P. Thota, B. Krauskopf, M. Lowenberg, Bifurcation analysis of nose landing gear shimmy with lateral and longitudinal bending, *J. Aircraft*, **47**, 87–95, (2010)
11. I. Besselink, Shimmy of aircraft main landing gears, Dissertation, University of Delft, The Netherlands, (2000)
12. E. Coetzee. Shimmy in Aircraft Landing Gear, *Technical report, Problem presented by Airbus* http://www.smithinst.ac.uk/Projects/ESGI56/ESGI5-AirbusShimmy/Report (2006)
13. C. Howcrof, B. Krauskopf, M. Lowenberg, S. Neild, Influence of variable side-stay geometry on the shimmy dynamics of an aircraft dual-wheel main landing gear. *SIAM Journal on applied dynamical systems*, **12(3)**, 1181-1209, (2013)
14. C. Howcroft, M. Lowenberg, S. Neild, B. Krauskopf, E. Coetzee, Shimmy of an aircraft main landing gear with geometric coupling and mechanical freeplay, *ASME. J. Comput. Nonlinear Dynam*, **10(5)**, (2014)
15. B. Schlippe, R. Dietrich, Shimmying of a pneumatic wheel", *NACA-TM 1365*, (1947)
16. Cessna Textron Aviation,www.cesssna.txtav.com/en, last accessed: 15 September, 2017
17. Goodyear Aviation Tires, www.goodyearaviation.com/index.html, last accessed: 15 September, 2017

MATEC Web of Conferences **133**, 01008 (2017)
BulTrans-2017 DOI: 10.1051/matecconf/201713301008

Flow field analysis of turbojet combustion chamber

Constantin Rotaru[1,*]

[1]"Henri Coanda" Air Force Academy, Aviation Department, 160 Mihai Viteazul Street, 500187, Brasov, Romania

Abstract. The paper presents some results regarding the identification of a turbojet engine's optimal combustion chamber geometry, characterized by small dimensions, with enhanced possibilities for air and fuel mixture circulation in the primary combustion zone in order to obtain good residence time in the flame tube, higher turbulence and optimal radial temperature distribution. Also, it deals with the identification of some solutions of the differential equation for reduced temperature under imposed initial conditions. Some models of combustion numerical computations have been applied, as presented by the specialized documentation, to establish the combustion rate and flame speed for laminar combustion. The geometry of a combustion chamber, with twenty-four injectors and size comparable to a real combustion chamber, has been created in order to obtain the speed and temperature flow field within the flame tube. The major conclusion of the current research is that the distribution of holes within the liner does not need to be symmetric, which means that they have to be positioned in such as a manner as to ensure a higher degree of turbulence in the combustion primary zone and a helical shape of stream lines in the proximity of the walls of the flame tube.

1 Introduction

The combustion chamber is one of the most important constructive elements of an aircraft engine and has an important role in the conversion of energy from chemical to kinetic. A reaction between oxygen in the air and fuel takes place upon the energetic collision between the molecules, which leads to a chemical reaction. The frequency of those collisions gives the rate of reaction, the time and length scale, and also the mixture properties. The curve of adiabatic temperature, regarded as a function of the equivalence ratio, φ, has a maximum value at $\varphi = 1$, but the actual temperature falls below the theoretical prediction, starting around an equivalence ratio of 0,4-0,5, due to the dissociation reactions [1]. This range of equivalence ratio corresponds to a fuel-to-air ratio of 2,5-3,5 and, taking into account the fact that gas temperature decreases after the point where $\varphi = 1$, a burner exit temperature can be reached from either the fuel-lean or the fuel-rich side of the stoichiometric point. The combustion process depends on the degree of turbulence and energy release leads to large alteration in turbulence intensity [2, 3]. The current flame tubes are based on experimental data and less so on theoretical, analytical and CFD calculation because the old models of the combustion chamber were improved slowly due to their complexity, and the advanced combustion algorithms being developed now are based on powerful computers. The chemical process of combustion normally takes place very quickly but the rate falls roughly as the square of the pressure. The burning of liquid droplets requires the liquid to evaporate and then for the fuel and air to diffuse together to form locally a near stoichiometric mixture [4]. This process is much slower than the main chemical process of combustion. A schemata of a combustion chamber airflow field is presented in Fig. 1. The air entering the dilution region creates a layer of cool air on the walls and changes the exit temperature so that it is suitable for the first turbine stage.

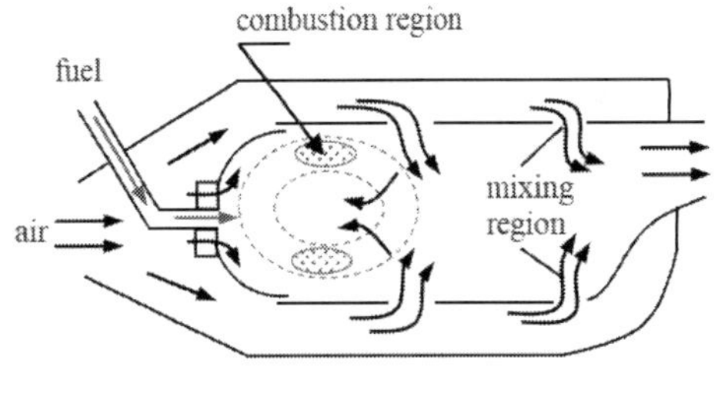

Fig. 1. Flow field model.

For the aircraft engine combustion chamber, the primary zone can be approximated with a premixed combustion, where the flame starts, propagating initially as a laminar flame and later developing into a turbulent flame. The experimental results presented in some publications [5] show a laminar flame speed of 0,5 m/s. The effect of turbulence consists in an enhancement of the energy transfer and an increase of the residence time in the flame tube volume. The performance of aircraft gas turbine engines depends extensively on the amount of energy release in the combustion chamber, as a result of air-fuel mixture burning, of the reaction rate, and of the time when combustion occurs, all of these being more or less determined by the combustion chamber's geometry and size.

* Corresponding author: rotaru.constantin@afahc.ro

Designing an aircraft engine involves knowledge in various academic areas such as fluid mechanics, thermodynamics, mathematics or materials science [6]. Controlling a system of this type implies an accurate understanding of each and every subsystem operation mode as well as the assembly of these sub-systems, regarded as a whole. The content of this study is based on the mathematical approach presented in reference [1].

2 Premixed flame model

The simplified conservation equations for laminar one-dimensional premixed flames are [1]

$$\begin{aligned}
&\frac{\partial \rho}{\partial t}+\frac{\partial(\rho u)}{\partial x}=0;\\
&\frac{\partial \rho Y_k}{\partial t}+\frac{\partial}{\partial x}\left[\rho(u+V_k)Y_k\right]=\dot{\omega}_k;\\
&\rho C_p\left(\frac{\partial T}{\partial t}+u\frac{\partial T}{\partial x}\right)=-\sum_{k=1}^{N}h_k\dot{\omega}_k+\frac{\partial}{\partial x}\left(\lambda\frac{\partial T}{\partial x}\right)-\\
&\qquad -\rho\frac{\partial T}{\partial x}\left(\sum_{k=1}^{N}C_{p,k}Y_kV_k\right),
\end{aligned} \tag{1}$$

where Y_k is the mass fraction of species k in the reaction mixture, V_k is the diffusion velocity, $\dot{\omega}_k$ is the reaction rate, C_p and $C_{p,k}$ are the heat capacities at constant pressure of the mixture and of the species k respectively, h_k the enthalpy, λ – heat diffusion coefficient, ρ – mixture density, T – temperature, u – axial velocity.
Typical inlet conditions are $u|_{x=0}=u_1$, $T|_{x=0}=T_1$ and also the flame must be ignited and the temperature must reach the adiabatic flame temperature at the outlet of the flow domain.

The basic model configuration for one-dimensional reactive flow is presented in Fig. 2.

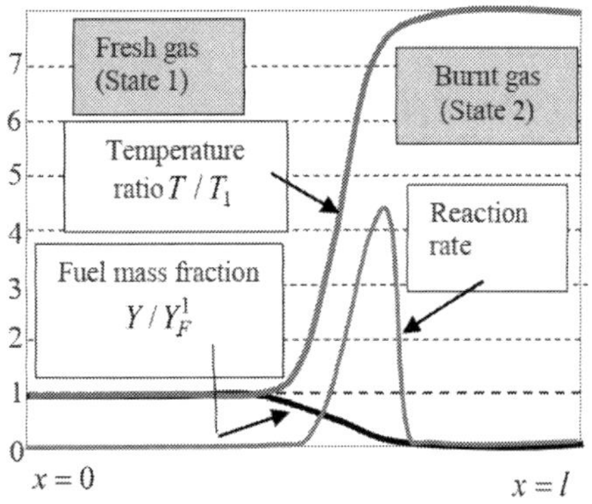

Fig. 2. Combustion computation model.

After some assumptions, the system of equations (1) can be transformed in the reference frame of the flame moving at speed s_L to the resolution of a single equation for the variable $\theta=(T-T_1)/(T_2-T_1)$. The link between the reduced temperature, θ, and reduced fuel mass fraction $Y=Y_F/Y_F^1$ is $\theta+Y=1$, so θ varies from 0 where $T=T_1$ to 1 in the burnt gases, and Y goes from 1 in the fresh gases to 0 in the burnt gases. The frequency of the collisions per unit volume contributes to the reaction rate in the form proposed by Arrhenius, $p^n f(T)e^{-E_a/(RT)}$, where p is the pressure, T is the reactant's absolute temperature, E_a is the activation energy and R is the universal gas constant [7].

In the aircraft engine combustion chamber, the pressure is approximately constant along the flame tube, so gas density can be expressed by the equation $p=\rho RT$ (the law of perfect gases) as follows:

$$\rho=\rho_1\frac{T_1}{T}=\rho_1\frac{1}{1+\alpha\theta/(1-\alpha)}, \tag{2}$$

where $\alpha=(T_2-T_1)/T_2$.

The exponential expression of the reaction rate, $e^{-E_a/(RT)}$ can be transformed into $e^{-\frac{\beta}{\alpha}}e^{-\frac{\beta(1-\theta)}{1-\alpha(1-\theta)}}$ with $\beta=\alpha T_{a1}/T_2$, where T_{a1} is the activation temperature; therefore the reduced reaction rate can be expressed as

$$\frac{\dot{\omega}_F}{\rho_1 Y_F^1 B_1}=T^{\beta_1}e^{-\frac{\beta}{\alpha}}\frac{(1-\theta)}{1+\dfrac{\alpha\theta}{1-\alpha}}e^{-\frac{\beta(1-\theta)}{1-\alpha(1-\theta)}}, \tag{3}$$

where subscript F refers to the fuel, B_1 is the reaction rate pre-exponential constant and β_1 is the temperature exponent. Fig. 3 presents four lines corresponding to $\alpha=0{,}75$, $\beta_1=0$ and different β exponents for a fuel inlet temperature of 300 K. The reduced reaction rate, expressed by $\dot{\omega}_F/(\rho_1 Y_F^1 B_1)$, has a maximum value for $\theta=1-1/(\alpha+\beta)$. The four curves correspond to $\beta=6$, $\beta=6{,}5$; $\beta=7$ and $\beta=7{,}5$.

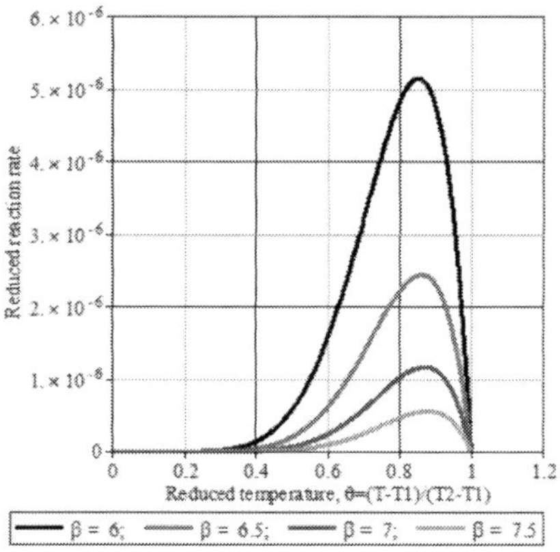

Fig. 3. The reduced reaction rate.

The combustion process of hydrocarbon fuels in air involves the collision of molecules (oxygen and fuel) and near stoichiometric combustion, and the reaction time takes a minimum of about 0,3 milliseconds. The reaction time scale may be approximated as

$$t_{reaction}=p^{-n}T^{-m}e^{\frac{E_a}{RT}}. \tag{4}$$

The residence timescale in an aircraft gas turbine combustor is longer compared with the reaction timescale.

3 Diffusion flame model

In a gas turbine engine, the fuel and oxidizer enter the combustion chamber separately; therefore, to solve the problem of the diffusion flame it is necessary to decouple these two main processes, namely the mixing and the combustion. In addition, some assumptions are made: the heat capacities of chemical species are equal and independent of temperature; the pressure is constant; the chemical species have the same heat capacities [8, 9]. The balance equation for the case that involves only fuel and oxidizer has the expression:

$$\frac{\partial \rho z_k}{\partial t} + \frac{\partial}{\partial x_i}(\rho u_i z_k) = \frac{\partial}{\partial x_i}\left(\rho D \frac{\partial z_k}{\partial x_i}\right), \tag{5}$$

where z_k has the expressions $z_1 = sY_F - Y_O$, $z_2 = \frac{C_P T}{Q} + Y_F$ and $z_3 = s\frac{C_P T}{Q} + Y_O$, according to the conservation equations for fuel, F, and oxidizer, O, and temperature T, $s = \left(\frac{Y_O}{Y_F}\right)_{st}$, Y_F and Y_0 are the mass fractions of fuel and oxidizer, D – diffusion coefficient [m^2/s], Q – heat of reaction, [kJ/kg], C_p C_P – the mixture heat capacity [J/(kg.K)], x_i – spatial coordinates and u_i – velocities spatial projections.

The normalized variables z_i for the scalars Z_i are defined by the expression

$$z_i = \frac{Z_i - Z_i^0}{Z_i^F - Z_i^0} \tag{6}$$

for $i = 1, 2, 3$, and they have the same boundary conditions and follow the same balance equation, which means that the mixture fraction measures the local ratio between the fuel and oxidizer and has the same boundary. The expression of the mixture fraction is

$$z = \frac{1}{\varphi + 1}\left(\varphi \frac{Y_F}{Y_F^0} - \frac{Y_O}{Y_O^0} + 1\right), \tag{7}$$

where $\varphi = sY_F^0 / Y_O^0$.

Equation (5) can be simplified for the case that corresponds to the aircraft turbojet engine, where the pressure in the combustion chamber is approximately constant, having the form

$$\frac{\partial z}{\partial t} + \frac{\partial}{\partial x_i}(uz) = \frac{\partial}{\partial x}\left(D\frac{\partial z}{\partial x}\right) \tag{8}$$

with the initial conditions $z(x, t = 0) = 1 - H(x)$, $z(-\infty, t) = 1$ and $z(+\infty, t) = 0$, with negative x for pure fuel and positive x for pure oxidizer.

The above equation (8) is reduced to

$$\frac{\partial z}{\partial t} = D\frac{\partial^2 z}{\partial t^2}, \tag{9}$$

assuming a constant diffusion coefficient and not moving before ignition.

This equation can be transformed by replacing variables x and t with $\eta = x/\left(2\sqrt{Dt}\right)$, which leads to

$$\frac{\partial^2 z}{\partial \eta^2} + 2\eta\frac{\partial z}{\partial \eta} = 0. \tag{10}$$

The Maplesoft solution of Equation (10) is:

$z(\eta) = \frac{1}{2} - \frac{1}{2}erf(\eta)$, where the exponential error function $erf(\eta)$ is defined as

$$erf(\eta) = \frac{2}{\sqrt{\pi}}\int_0^\eta e^{-x^2}dx. \tag{11}$$

Fig. 4 shows the plotting of $z(\eta)$ and $erf(\eta)$ for η from –5 to 5.

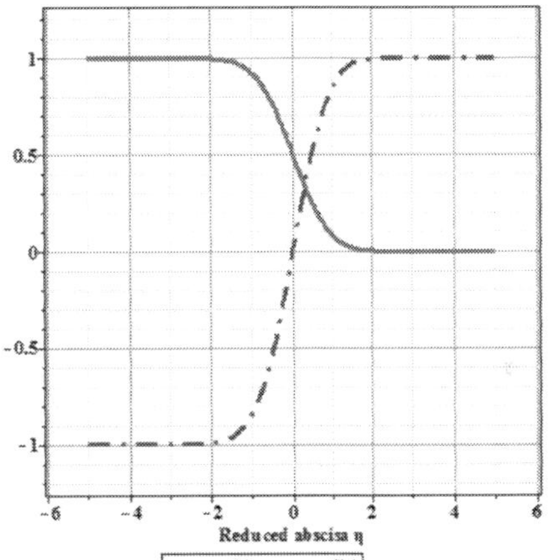

Fig. 4. The mixture fraction $z(\eta)$ and the exponential error function $erf(\eta)$.

For the assumption of infinitely fast chemistry, where the fuel and oxidizer cannot be found together, functions $T(z)$, $Y_F(z)$ and $Y_0(z)$ for a one-dimensional flame of jet fuel ($C_{12}H_{23}$) and air with $Y_F^0 = 1$ and $Y_O^0 = 0{,}21$ are presented in Fig. 5. Fuel and air temperatures are 300 K and 600 K and the fuel heat of the reaction is 43000 kJ/kg. The curve of the temperature is presented in Fig. 6. The atom-balance equation for an ideal mixture of fuel and air

$$C_x H_y + \left(x + \frac{y}{4}\right)\left[O_2 + \frac{79}{21}N_2\right] \rightarrow \sum_{i=1}^{NS} n_i A_i \tag{12}$$

gives the following mass stoichiometric and equivalence ratios for jet fuel: s = 3,4; φ = 16,19; z_{st} = 0,0581.

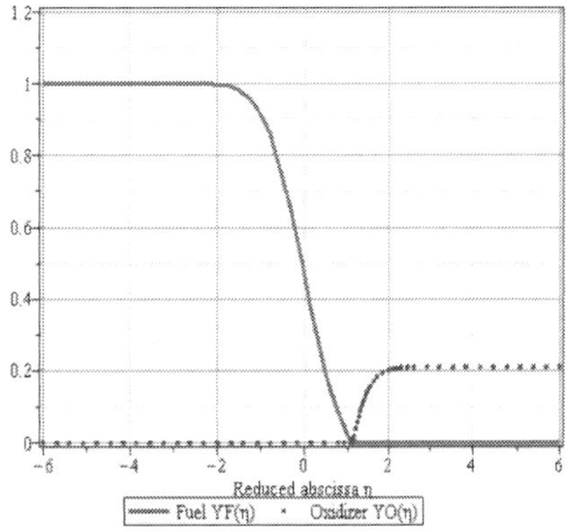

Fig. 5. Flame structure for $C_{12}H_{23}$ fuel.

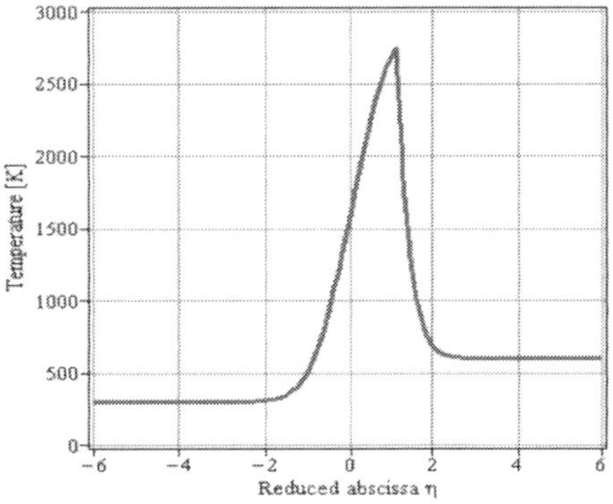

Fig. 6. The gas temperature.

The aircraft combustion chamber is fed by fuel mass flow rate $\dot{m}_F$ and oxidizer mass flow rate $\dot{m}_O$; therefore, the global equivalence ratio obtained if all fuel and oxidizer flow rates are perfectly mixed is $\varphi_g = \varphi \dot{m}_F / \dot{m}_O$. The mean temperatures after complete combustion, T_m, and maximum temperature, T_{ad}, are

$$T_m = \frac{T_F^0}{1+\frac{\varphi}{\varphi_g}} + \frac{T_O^0}{1+\frac{\varphi_g}{\varphi}} + \frac{QY_F^0}{C_P(\varphi+\varphi_g)}\min(\varphi_g, 1); \quad (13)$$

$$T_{ad} = \frac{T_F^0 + \varphi T_O^0}{1+\varphi} + \frac{QY_F^0}{C_P(\varphi+1)}. \quad (14)$$

The effect of the initial temperature on the adiabatic temperature is that only a half of the inlet temperature rise appears as rise in the final flame temperature.

Fig. 7 shows the variation of these temperatures versus the global equivalence ratio, φ_g for the aircraft fuel kerosene ($C_{12}H_{23}$) and air. Both temperatures are equal for φ_g = 1. Regarding aircraft engine pollution, the nitric oxide NO_x is mainly controlled by temperature and regardless of the $\dot{m}_F$, $\dot{m}_O$ and φ_g parameters the diffusion flame always reaches the same maximum temperature T_{ad} , and generates high NO_x pollutant in the stoichiometric regions, namely, in the vicinity of $z = z_{st}$.

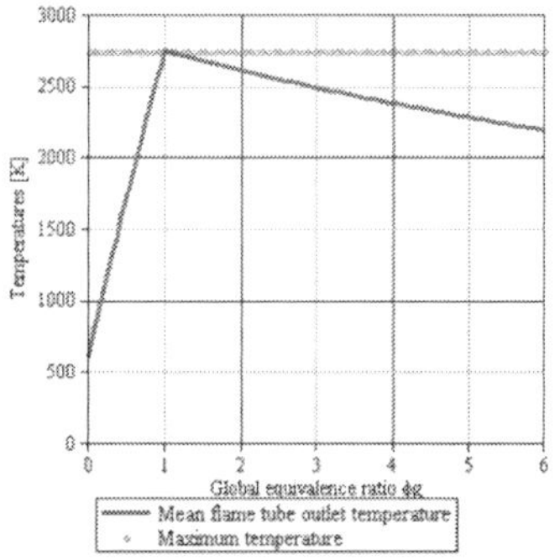

Fig. 7. Mean outlet and maximum temperatures.

The flame position x_f can be determined from the mixture fraction equation

$$z_{st} = \frac{1}{2}\left[1 - erf(\eta_f)\right] \Rightarrow \eta_f = erf^{-1}(1 - 2z_{st}) = 1.110$$

therefore, $x_f = 2\sqrt{D \cdot t}$; $\eta_{st} = 2.220\sqrt{D \cdot t}$.

For stoichiometric conditions, the flame velocity remains at x_f = 0 and is moving toward the fuel for φ < 1 and toward the oxidizer for φ > 1. Taking into account the conservation equation for fuel,

$$\frac{\partial \rho Y_F}{\partial t} + \frac{\partial}{\partial x_i}(\rho u_i Y_F) = \frac{\partial}{\partial x_i}\left(\rho D \frac{\partial Y_F}{\partial x_i}\right) + \dot{\omega}_F \quad (15)$$

in the case of constant density and zero velocity flow, the fuel reaction rate per unit flame area $\dot{\Omega}_F$, can be obtained by integrating $\dot{\omega}_f$ between two points located on both sides of the flame front, delimited by x_f on the right and the left hand sides [1]:

$$\dot{\Omega}_f = -\int_{x_f^-}^{x_f^+} \frac{\partial}{\partial x_i}\left(\rho D \frac{\partial Y_F}{\partial x_i}\right) dx = -\frac{\rho}{2} \frac{Y_F}{1 - z_{st}} \sqrt{\frac{D}{\pi t}} e^{-\eta_f^2}. \quad (16)$$

Fig. 8 displays $\dot{\Omega}_F$, where one can see that it goes to zero like $1/\sqrt{t}$, thus for keeping a diffusion flame burning, the reactants must be pushed against each other.

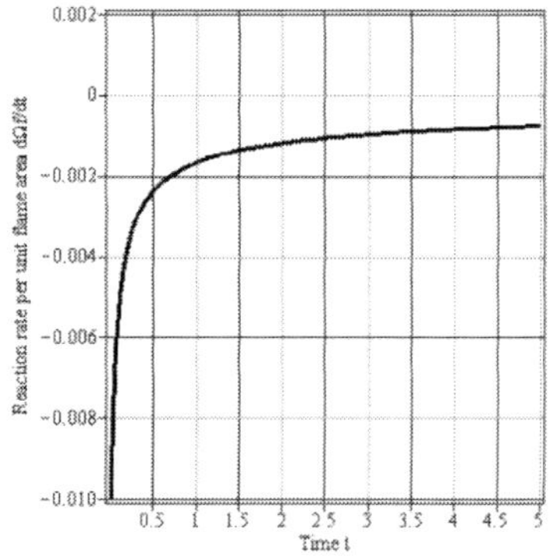

Fig. 8. The fuel reaction rate per unit flame area.

4 Numerical results

In order to simulate gas flow, a combustion chamber was built with a flame tube perforated with three rows of holes. The dimensions of the combustion chamber are close to those of a real combustion chamber, with 24 injectors. The range for the injectors' distribution is 0.35 m, whereas the supplied mass flow rate is 75 kg/s at maximum rev of the engine. Computational Fluid Dynamics calculations have been made for two cases, firstly for a flame tube with three rows of holes of the same diameter, and secondly for a flame tube with three rows of holes of various diameters. Some of the results and combustion chamber shapes are presented in the following images.

Fig. 9 illustrates the geometry of the flame tube and Fig. 10 shows the mesh domain. The first simulation of the flow through the flame tube was achieved for a uniform distribution of holes on the flame tube surface, in order to compare it with a flame tube with asymmetrically positoned holes.

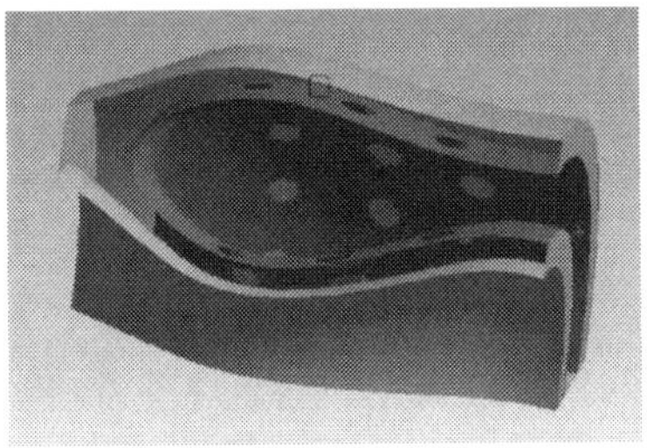

Fig. 9. The 3D view of the flame tube.

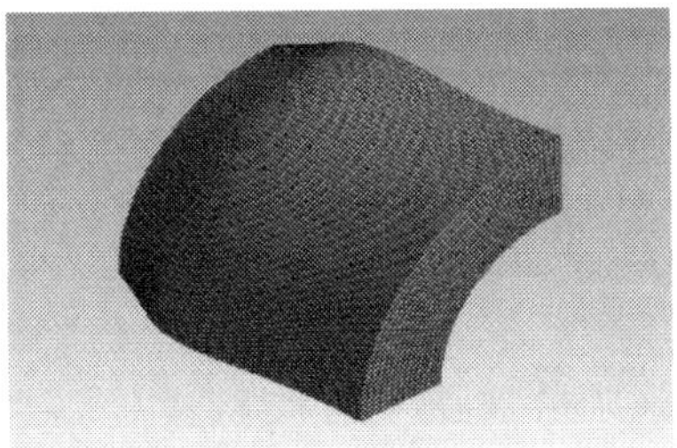

Fig. 10. The mesh domain.

Figures 11 and 12 show the current lines contour for the combustion chamber with six injectors in cases of symmetric and asymmetric distribution of holes. In the primary area of the flame tube the turbulence degree must be much higher.

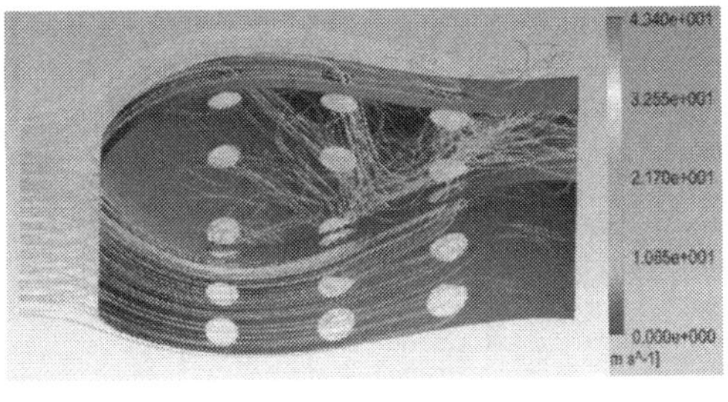

Fig. 11. The flow field lines contour.

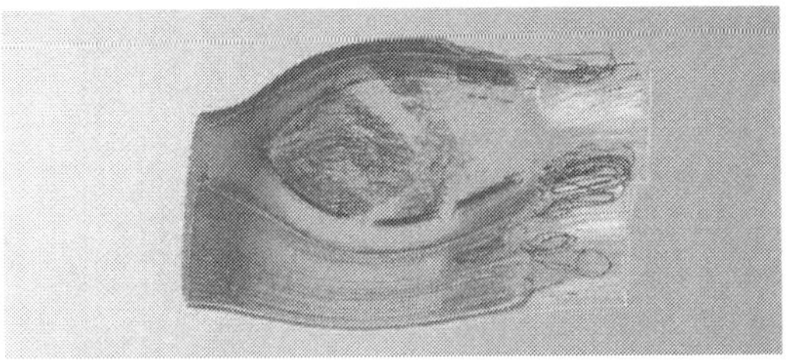

Fig. 12. The inner flow.

5 Conclusions

The prototype of the flame tube with asymmetric distribution of holes on its surface ensures a higher turbulence in the secondary area. Therefore, the residence time of the air–fuel mixture in the combustion chamber is higher, which allows for the complete burning of the mixture. The asymmetric distribution of the holes may lead to a helicoidal movement with a similar rotation direction to that of the turbine engine, which leads to a decrease in pressure losses in the stator of the first stage of the turbine.

The results obtained allow for the evaluation of the temperature field in the radial direction at the combustion chamber exit, in order that there are not high temperatures at the base of the rotor blades because they are subjected to mechanical stress in this root section. Similarly, the possibility of fuel injection through the stator blades can be further analysed, so that the maximum temperature in the flame tube can be reduced, and this difference in temperature be compensated by burning a certain amount of fuel through the stator wall of the first stage of the turbine.

References

1. T. Poinsot, D. Veynante, *Theoretical and Numerical Combustion*, Aquaprint, Bordeaux, France (2005).
2. J. Mattingly, *Elements of Propulsion. Gas Turbines and Rockets,* AIAA, Reston Virginia, USA (2006).
3. C. Rotaru, M. Mihăilă-Andres, P. Matei, International Journal of Turbo&Jet Engines, **31-3**, 229-237, DOI: 10.1515/tjj-2013-0048 (2014).
4. T. V. Chelaru, A. Chelaru, *Proceedings of RAST-2013*, 837-842, Istanbul, Turkey, 12-14 June, (2013)
5. S. Farokhi, *Aircraft Propulsion,* John Wiley & Sons Inc., USA, (2009).
6. H. Richter, *Advanced Control of Turbofan Engines,* Springer Science, USA, (2012).
7. A. Tudosie, *Proceedings of ICMT-2009*, ISBN: 978-80-7231-649-6, Brno, Czech Republic, 350-359, (2009).
8. C. Rotaru, I. Cîrciu, C. Aramă, C. Constantinescu, Review of the Air Force Academy, **3(30)**, 33-37, DOI: 10.19062/1842-9238.2015.13.3.5, (2015).
9. A. Tudosie, *Proceedings of ICATE-2012*, DOI: 10.1109/ICATE.2012.6403464, Craiova, 1-6, (2012).

Studying the wake contraction of the flow-field of a rotor in hover

Filip Panayotov[1], *Ivan* Dobrev[2], *Fawaz* Massouh[2], and *Michael* Todorov[1,*]

[1]Technical University – Sofia, Department of Air Transport, 8 Kliment Ohridski Blvd., 1000 Sofia, Bulgaria
[2]Arts et Métiers – ParisTech, DynFluid laboratory, 151 L'Hôpital Blvd., 75013 Paris, France

Abstract. This article presents a comparative study of three numerical algorithms for computation of the flow-field of a rotor in hover and the net static thrust that is produced. The flow-field is induced by a series of vortex rings, modelling the near wake of the hovering rotor and a single semi-infinite vortex cylinder, accounting for the velocity deficit in the far wake. All three models are based on the vortex theory and differ in the choice of the numerical scheme for the estimation of the exact position of the vortex rings, emitted at the tips of the blades of the rotor. Thus, the numerical models perform a real-time simulation of the propagation of the vortex rings in the downwash. The first model uses an Euler-predictor scheme, while the second and third models use respectively first and second order predictor-corrector schemes. The aim of the study is to assess the rapidity and accuracy of each algorithm. For that purpose, the numerical results are compared with the experimental data, obtained from a wind tunnel test of the model rotor. The best results in terms of computational speed and accuracy are obtained with the use of Adams-Bashforth predictor-corrector scheme of second order.

1 Introduction

Rotors of drones rarely perform long lasting hovers in zero wind conditions with constant performance parameters. This is due to atmospheric disturbances, which require continues thrust adjustments, in order to maintain a stable hover, while compensating for gusts and constant winds. Thus, of great interest is rotor performance during transient or unsteady operational regimes.

The study aims to develop a numerical model, capable of rapid and accurate computation of the thrust of a hovering rotor. Such a tool will make it possible to study rotor performance in unsteady flow conditions, resulting from the transition from one operational regime to another or from phenomenons, such as the tip vortex aperiodicity of a hovering rotor [1].

Previous research [2] performed by the authors, showed that the fastest numerical computation of the performance of a hovering rotor is achieved with the use of the vortex theory and more specifically with the model presented in section 3 "Numerical model". In this previous study [2], the numerical results showed good agreement when compared with the experimental data of a model rotor in hover. The main difference between the previous studies and the current study is that the latter features a free wake approach, while in the former the numerical models had a prescribed arrangement of the vortex elements, which modelled the flow-field of the rotor in hover. The vortex elements are allowed to move freely under the forces applied on them by the resulting induced velocity field, until the vortex system reaches a state of equilibrium. Thus, a greater accuracy is achieved, although the required computational time is significantly increased.

The Miller model is coupled with a BEM code, which allows for rapid determination of the forces acting on the blades of the rotor. This allows for rapid evaluation of the overall rotor performance by providing estimations for the produced net thrust and the required torque for stable hover. Those two can be later compared with experimental data measurements for the model rotor in hover.

2 State-of-the-art vortex models for a rotor in hover

According to [3], Joukovsky performed a series of theoretical studies on propellers, which set the mathematical foundations for the vortex theory such as applied to rotors. Those works were followed by numerous flow visualization experimental studies on small scale and full scale helicopter rotors in both forward flight and hover. They added more clarity on the structure of the flow-field around a helicopter rotor. These studies also provided the first semi-empirical numerical models for hovering rotors, derived from the obtained experimental performance data and the geometry of the trailing wake.

Jenny et al. [4] and Landgrebe [5] proposed vortex models for predicting the hovering performance of

* Corresponding author: Michael.Todorov@tu-sofia.bg

helicopter rotors. These models consist of a helical wake, composed of helical vortex sheets, trailing from each one of the rotor blades. Simplifying assumptions are introduced, which allow for the representation of the helical wake with a finite number of discrete vortex filaments, namely straight-line segmentation of the tip trailing helical vortex. The induced velocity field is computed with the use of the Biot-Savart law. These numerical models are capable of predicting adequately the performance of a hovering rotor, although the simplifying assumption for a rigid helical wake introduces a significant error when the rotor operates in too light or too heavy load conditions, corresponding to values of the coefficients of thrust C_T smaller than 0,0025 and bigger than 0,01. Another inconvenience is in order to have a higher degree of accuracy a greater number of straight-line segments per complete helical turn must be considered and a greater number of complete helical turns must be taken into account, which significantly increases the volume of the required numerical computations.

Landgrebe [6], Miller et al. [7, 8] and Reddy [9] proposed an alternative approach for modelling the hovering rotor by decomposing the trailing wake into several parts, namely: near, intermediate and far. Each of those parts is modelled with a different type and amount of vortex elements with the aim to achieve greater accuracy, while offering significant numerical simplification. The aforementioned papers demonstrate the possibility to obtain a significant degree of accuracy for the evaluation of the performance of hovering rotors with vortex systems, which comprise less than 30 vortex elements in total.

Karpatne et al. [1] demonstrated the capability of a vortex model, in which the wake of the hovering rotor is modelled entirely with a system of vortex rings. In their model the vortex rings are emitted with a specific time-step, depending on the angular velocity of the rotor and are allowed to move freely under the induced velocities resulting from the interaction with the remainder of the previously emitted vortex rings. Reference [1] showed that vortex models are capable of adequately simulating unsteady flow phenomena, thus allowing to study them.

3 Numerical model

Modelling the wake of a hovering rotor with a combination of different types of vortex elements for different segments of the wake is demonstrated in [1, 8, 9]. The present study is performed with the vortex model proposed by Miller et al. in [8]. In the rest of the article it will be referred to as the Miller model.

The wake of the Miller model consists of two parts: a near wake composed of 20 circular vortex rings arranged in series downstream of the rotor; and a far wake modelled by a single semi-infinite vortex cylinder, placed behind the last vortex ring of the near wake. Thus the model consists of only 21 vortex elements to be accounted for in the computation of the entire flow-field around the rotor.

The semi-infinite vortex cylinder allows the model to compensate for the velocity deficit formed downstream of a wake when modelled with a series of vortex rings. It allows to stabilize the numerical computation and ensures that the induced velocities in the far wake double those induced in the rotor plane, such as per the theory for the hovering rotor.

The crucial thing for the degree of accuracy and the adequate operation of the model is the proper setup of parameters d_0 and d_1, shown on Fig. 1.

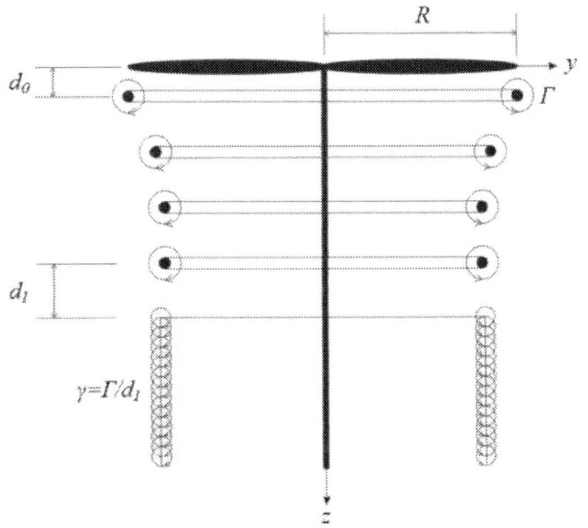

Fig. 1. Schematic representation of the arrangement of the vortex elements for the Miller model.

The distance d_0, at which the first vortex ring is placed behind the rotor, plays a key role in the distribution of the induced velocities along the blade. It is referred in [8] as the first blade-vortex encounter position, which represents the distance between the vortex ring emitted by the preceding blade and the following blade. This parameter can be varied, in order to study its influence on the induced velocity distribution on the blades and in the near wake downstream of the rotor.

Distance d_1 is the spacing between the last vortex ring of the near wake and the semi-infinite vortex cylinder, which models the far wake. The optimal values of parameters d_0 and d_1 are estimated for a comparison between numerical and experimental data for multiple operational regimes of the rotor.

The Miller model offers the potential for rapid numerical computation when compared to CFD and BEM models due to the necessity to compute for significantly smaller amount of vortex elements. An additional reduction of the volume of numerical computations is possible for hover and vertical flight regimes due to the presence of axis-symmetric flow condition for the rotor. For the studied rotor with two blades, only half of the flow-field can be computed in order to visualize the entire flow around the rotor.

4 Numerical study

The performed numerical study aims to evaluate the accuracy and computational rapidity of the free wake Miller model, coupled with three different algorithms for the precise estimation of the exact arrangement of the vortex rings in the wake.

Computations are made of the mutually induced and self-induced velocities between the vortex elements modelling the wake. The induced velocities on each vortex element are then used to compute its new position in the wake for the next computational time step Δt. Thus, the vortex rings propagate in the numerical domain and take the necessary geometry arrangement, in order to adapt to the latest flow conditions.

Once the solution has converged and the equilibrium arrangement of the vortex elements is obtained, then the distribution of the induced velocity V_z along the blades is computed. Thus, by knowing the angular velocity of the rotor Ω and the pitch angle of the blades θ, it is possible to determine the aerodynamic forces acting on the blades. Those forces are used for the computation of the net trust T produced by the rotor and the required mechanical power P for the rotor to perform hover. This computational cycle repeats itself for each time step Δt of the simulated time-period. The described algorithm is shown in Fig. 2.

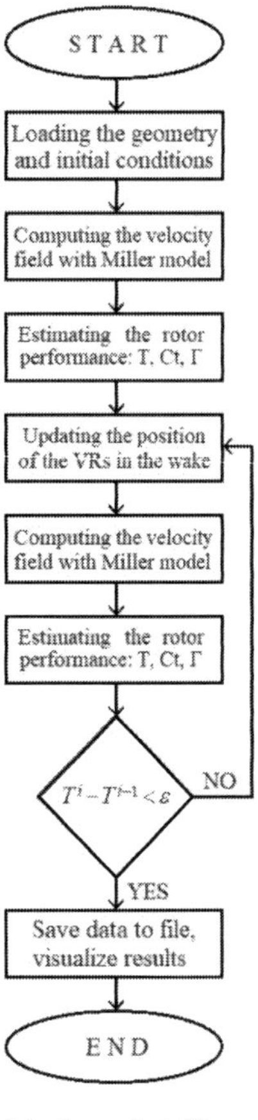

Fig. 2. Algorithm of the free wake Miller model.

By keeping the values of both the angular velocity of rotor Ω and the pitch angle of blades θ constant, the numerical model evaluates the rotor parameters in hover flight. However, the presented program module is also capable of studying the rotor performance in smooth transition from one operational regime to another, as well as for an abrupt step-change of the pitch angle of the blades θ.

For this present study assumed is a stable hover, which is a hover performed in zero wind conditions. The results from this study are presented in section 5 "Results".

The model rotor, which is used as reference geometry in those studies, has two rectangular untwisted blades of NACA0012 profile. The tip radius R of the blades is 288 mm, the root radius at the hub mounting r_0 is 65 mm and the chord length c is 25 mm. The pitch angle of the blades θ is adjustable and thus can be varied between wind tunnel experiments.

Fig. 3 shows the velocity triangle in a typical cross-section of the blade. The velocity of the axial flow through the plane of rotation, resulting from the work of the rotor, is designated by V_z. In the literature it is referred to as the axial induced velocity and is a function of the operational parameters of the rotor, namely angular velocity Ω and the pitch angle of blade θ. The speed of rotation of the rotor section in the plane is is refferred to as $U = \Omega y$, where y is the blade station radius, which varies from r_0 to R. For the tip of the blade $y = R$ and $U = \Omega R = V_{tip}$. The relative air speed for the cross-section of the blade W is computed according to (1):

$$W = \sqrt{U^2 + V_z^2} \,. \tag{1}$$

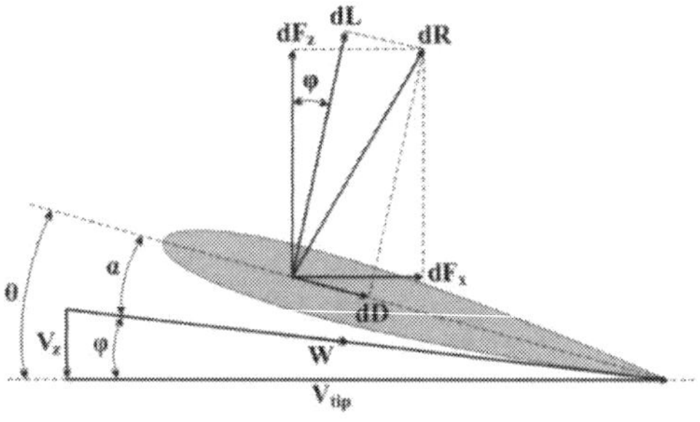

Fig. 3. Triangle of velocities in the cross-section of the blade.

4.1 Numerical approach

According to [1], by assuming uniform bound circulation and a single tip-trailing vortex per blade, the vortex strength Γ can be calculated as follows:

$$\Gamma = \frac{2T}{\rho N_b R V_{tip}} \,, \tag{2}$$

where T is the total thrust produced by the model rotor at a fixed angular velocity Ω and pitch angle of the blades θ. R is the tip radius of the blades, $V_{tip} = \Omega R$ is the velocity at the tip of the blades, N_b is the number of blades of the rotor and ρ is the density of the air.

For every change in either the angular velocity Ω or the pitch angle of the blades θ, the total thrust of the rotor T and the bound circulation Γ are also changed. Equation (2) is used for the computation of the bound circulation Γ.

In [10] the coefficient of thrust C_T is defined as:

$$C_T = \frac{T}{\rho A \Omega^2 R^2} \,, \tag{3}$$

 DOI: 10.1051/matecconf/201713301009

where T is the total thrust produced by the rotor; ρ is the density of the air; $\mathbf{\Omega}$ is the angular velocity of the rotor; R is the radius of the rotor and $A = \pi R^2$ is the area of the rotor disc. Equation (4) is used for the computation of the coefficient of thrust C_T.

From Fig. 3 it can be seen that with an increase of the axial induced velocity V_z there is a decrease of the angle of attack α. The smaller the angle of attack of the cross-sectional airfoil is, the smaller the local thrust increment ΔF_z is. Thus, the total thrust of the rotor T decreases and with it decreases the coefficient of thrust C_T.

Fig. 4 shows the key dimensions, used in the computation of the induced velocity at a point of the flow-field. The y_m, y_n, z_m and z_n dimensions are universal for both the cylindrical and contracting wake models. Those are the coordinates of two points, namely M and N. It is considered that the induction happens in point M by a vortex element, situated at point N.

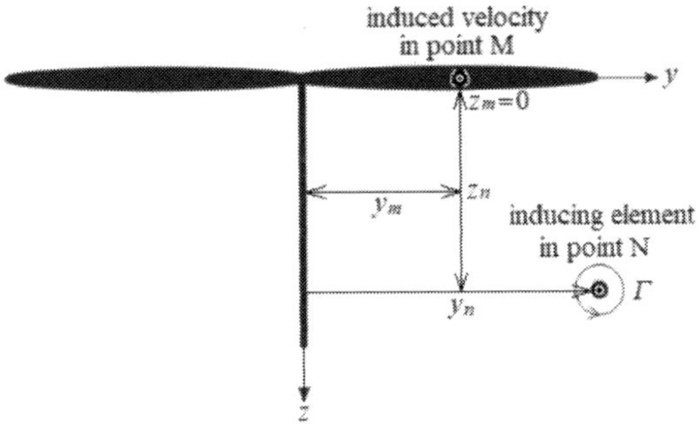

Fig. 4. Dimensions used in the computation of the induced velocity field of the rotor.

Lewis [11] introduces the dimensionless parameters z and y with (4) and (5):

$$z = \frac{z_m - z_n}{y_n}; \quad (4)$$

$$y = \frac{y_m}{y_n}. \quad (5)$$

Those are used for deriving the equations for the axial V_z and radial V_y induced velocities. in the case of the semi-infinite vortex cylinder, the axial velocity is computed with (8) and in the case of $y = 1$, (9) is used:

$$s = \frac{z}{\sqrt{z^2+(y+1)^2}}; \quad (6)$$

$$q = \frac{y-1}{y+1}; \quad (7)$$

$$V_{z,cyl} = \frac{\Gamma}{4\pi}\left\{\frac{\pi}{2} + s[K(k^2) - q\Pi(n,k^2)]\right\}; \quad (8)$$

$$V_{z,cyl_{(y=1)}} = \Gamma\left\{\frac{1}{4} + \frac{zK(k^2)}{2\pi\sqrt{z^2+4}}\right\}. \quad (9)$$

In the case of a semi-infinite vortex cylinder, the radial induced velocity is computed with (11):

$$B = \frac{2\Gamma}{\pi k^2\sqrt{z^2+(y+1)^2}}; \quad (10)$$

$$V_{y,cyl} = B\left[E(k^2) - \left(1 - \frac{k^2}{2}\right)K(k^2)\right]. \quad (11)$$

In the case of the system of vortex rings, the axial and radial induced velocities are respectively computed with (14) and (15):

$$C = \frac{\Gamma}{2\pi y_n\sqrt{z^2+(y+1)^2}}; \quad (12)$$

$$D = \frac{2}{z^2+(y-1)^2}; \quad (13)$$

$$V_{z,ring} = C\{K(k^2) - [1 + D(y-1)]E(k^2)\}; \quad (14)$$

$$V_{y,ring} = C\frac{z}{y}\{K(k^2) - (1 + Dy)E(k^2)\}. \quad (15)$$

Equations (8), (9), (11), (14) and (15) contain complete elliptic integrals of the first $E(k^2)$, second $K(k^2)$ and third kind $\Pi(n,k^2)$. The elliptic parameters n and k are introduced with (16) and (17):

$$k = \sqrt{\frac{4y}{z^2+(y+1)^2}}; \quad (16)$$

$$n = \frac{4y}{(y+1)^2}. \quad (17)$$

Equations (18) and (19) are used for the application of first-order Euler-predictor:

$$y_i^P = y_i^n + \Delta t V_{iy}(y_i^n, z_i^n); \quad (18)$$

$$z_i^P = z_i^n + \Delta t V_{iz}(y_i^n, z_i^n). \quad (19)$$

Equations (20) and (21) are used for the application of second-order Adams-Bashforth predictor:

$$y_i^P = y_i^n + \frac{\Delta t}{2}[3V_{iy}(y_i^n, z_i^n) - V_{iy}(y_i^{n-1}, z_i^{n-1})]; \quad (20)$$

$$z_i^P = z_i^n + \frac{\Delta t}{2}[3V_{iz}(y_i^n, z_i^n) - V_{iz}(y_i^{n-1}, z_i^{n-1})]. \quad (21)$$

Equations (22) and (23) are used in the corrector step:

$$y_i^{n+1} = y_i^n + \frac{\Delta t}{2}[V_{iy}(y_i^P, z_i^P) + V_{iy}(y_i^n, z_i^n)]; \quad (22)$$

$$z_i^{n+1} = z_i^n + \frac{\Delta t}{2}[V_{iz}(y_i^P, z_i^P) + V_{iz}(y_i^n, z_i^n)]. \quad (23)$$

The first model only updates coordinates y and z for each vortex element by applying only a first-order Euler-predictor without performing any correction step.

The second model updates coordinates y and z with a first-order Euler-predictor step, followed by a corrector step. This solution scheme is known as the Trapezoidal predictor-corrector scheme of first order.

The third model updates coordinates y and z with an Adams-Bashforth predictor step, followed by a corrector step. It is referred to as the Adams-Bashforth predictor-corrector scheme of second order.

Numerical instability issues are observed if the vortex rings in the near wake are not being initially slightly contacted. Thus, a simple arrangement with linear contraction higher than 10% is found to be enough to ensure stable computation.

5 Results

Fig. 5 shows three computational cases for the flow field of the model rotor in hover, computed with the Miller model with the second-order predictor-corrector scheme. The numerical computations are performed for 2000 RPM and $\theta = 10$ deg for three different values of parameter d_1.

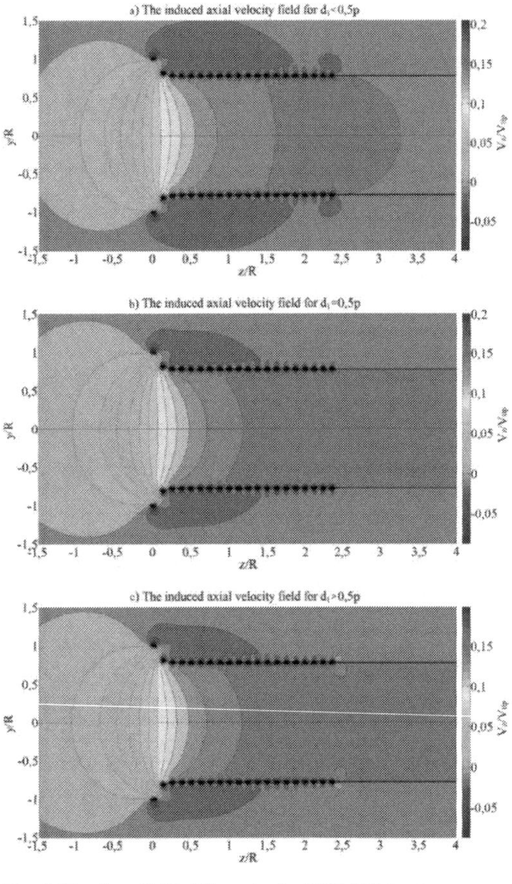

Fig. 5. The flow-field of the rotor for 2000 RPM and $\theta = 10$ deg, computed with the Miller model with the second-order predictor-corrector scheme.

The axial induced velocity V_z/V_{tip} in the far wake at $z/R > 10$ doubles its value, when compared with the induced velocity in the plane of the rotor at $z/R = 0$, only when the semi-infinite vortex cylinder is positioned at a specific distance behind the last vortex ring from the near wake. This distance was found to be equal to half of the averaged distance between the vortex rings in the near wake. Thus $d_1 = 0{,}5p$, where p is the average spacing between the vortex rings in the near wake.

Table 1 compares the coefficient of thrust C_T, computed with all three models, against the wind tunnel experiment for 2000 RPM and $\theta = 6$ deg. It presents also the percentage error $|\Delta|$ and the required computational time t_{comp} for the simulations. The spacing parameters for both computations are set as follows: $d_0 = 0{,}25p$ and $d_1 = 0{,}5p$.

Table 1. Comparison of the results obtained with all three models, against the wind-tunnel experiment data.

Models	Free wake Euler P	Free wake P-C 1 order	Free wake P-C 2 order	Wind tunnel experiment
C_T, 10^{-3}	2,977	2,781	2,717	2,569
$\|\Delta\|$, %	15,88	8,24	5,76	-
t_{comp}, s	0,94	1,17	1,23	-

6 Conclusion and Future work

The accuracy and rapidity of the numerical approach for the computation of the performance of a model rotor in hover was evaluated with three free wake vortex based models. On the basis of the results from the performed studies, it can be concluded that the free wake Miller model with the second-order predictor-corrector is only slightly slower than the model using the first-order predictor-corrector in terms of computational speed, while providing significantly higher degree of accuracy.

The existing percentage of error between the computational results and the experimental data is acceptable. Therefore, the proposed model can be used for preliminary studies and simulations of the expected thrust and overall performance of newly designed rotors.

The results from the conducted studies, presented in this paper, encouraged the authors to use a free wake Miller model with a predictor-corrector scheme of second order, in order to study the thrust of small rotors in unsteady operational condition. The intent is to conduct a series of studies on rotor performance during transient operational regimes, produced by both big and small rates of change in the angular velocity of the rotor Ω; and for both positive and negative rates of change.

It is of special interest to study the changes in the thrust, resulting from a rapid change of the pitch angle of the blades θ for a constant rotor angular velocity of Ω. Such a study will provide greater insights into rotor performance during the transition period, in which the flow-field adjusts to the new operational conditions of the rotor.

References

1. A. Karpatne, J. Sirohi, S. Mula, C. Tinney, J. of Fluids Engineering, **136** (2014)
2. F. Panayotov, I. Dobrev, F. Massouh, M. Todorov, *ICMT-2017*, 570-577, (2017) DOI :10.1109/MILTECHS.2017.7988822
3. V. Okulov, J. Soerensen, D. Wood, *Progress in aerospace sciences*, **73**, 19-46, (2015)
4. D. Jenney, J. Olson, A. Landgrebe, J. of the Amer. Helicopter Society, **13**(2), 1-26, (1968)
5. A. Landgrebe, USAAMRDL, Technical Report 7114, (1971)
6. A. Landgrebe, J. of the American Helicopter Society, **14**(4), 20-32, (1969)
7. R. Miller, J. of Aircrafts, **20** (1983)

DOI: 10.1051/matecconf/201713301009

8. R. Miller, T. Roberts, E. Murman, Comp. and Math. with Applic., **12A**, (1986)
9. K. Reddy, *7th Australasian hydraulics and fluid mechanics conference* (1980)
10. J. Leishman, *Principles of helicopter aerodynamics*, 43-44 (2000)
11. R. Lewis, *Vortex element methods for fluid dynamic analysis of engineering systems*, 147-170 (1991)

DOI: 10.1051/matecconf/201713302001

Vehicle propulsion systems design methods

Zlatina Dimitrova[1,2,*]

[1]PSA Groupe, Research and Innovation Department, 78943 Vélizy- Villacoublay, France
[2]Ecole Polytechnique Fédérale de Lausanne, IPESE, 1100, Lausanne, Switzerland

Abstract. To meet the targets of sustainable development and greenhouse emission reduction of the future vehicles fleet, the automotive industry needs to deploy cost-competitive and efficient advanced energy conversion systems for the future commercial personal vehicles. The efficiency improvement needs induce to search new structured methodologies allowing the integration of the efficiency/cost vision for different vehicle energy technologies, in the earlier design stage of the new vehicles and their propulsion systems. This article proposes a method to compare systematically different vehicles design options under different economic and environmental scenarios. The proposed methodology combines flowsheeting vehicles models, energy integration techniques, economic evaluation and life cycle assessment in a computational platform. The methodology is applied on electric vehicles and hybrid electric vehicles. This article presents the application of a systematic environomic optimization for vehicle energy systems. The efficiency, economic and environmental performances are assessed for different energy technology options and their integration in advanced vehicles powertrains. The performances indicators are compared and the trade-off are assessed to support decision making and to identify optimal energy systems configurations.

1 Introduction

The mobility is important for the economic and social development of the population, but also has to respect the environment. Many efforts from the research and the industry came to improve the efficiency of the vehicles, their comfort and security. The modern vehicles are characterized by low emissions and diversification of the powertrain for the energy integration adaptation. The new energy systems are more complex – additional electric motors, storage devices, torque converters and etc. are added with the intention to improve the system behavior [1]. For such complex system the heuristic design approaches fail. In this context the aim of the article is to propose and employ mathematical models and to use these models in a systematic model-based approach for powertrain design. In the traditional energy scenario by 2040, 90% of the global transportation will run on liquid petroleum based fuels [2]. The proliferation of hybrid and other advanced vehicles, along with the improvement of the conventional vehicle efficiency, will result in a flattening demand in petrol for personal transportation even as the number of personal vehicles in the world doubles [2].

By 2035, sales of conventional vehicles fall to a quarter of total sales, while hybrids dominate (full hybrids 23%, mild hybrids 44%). Plug-in vehicles, including full battery electric vehicles, are forecast to make up 7% of sales in 2035. Alternative fuels should be prospected for reducing the CO_2 impact of transport but also others environmental impacts have to be taken into account through an integrated approach based on the Well-to-Wheels analysis and Life Cycle Assessment (LCA) tools.

The scarcity of not only fuel resources but also the adverse effects of the operation of energy intensive systems on the environment (pollution, degradation) have to be taken into consideration, not only qualitatively but also quantitatively. Thus, the system can be properly designed and operated. The systematic consideration of thermodynamic, economic and environmental aspects for this purpose is called environomics [3]. During the 70s and 80s, the depletion of energy resources has been one of the primary concerns. Terms like thermoeconomics, exergo economics etc., have been coined to imply the attempts to save energy (exergy) by proper analysis and design of thermal and chemical plants. Environomic analysis is an extension of thermo-economics [4]. Many researches are performed on the energy conversion balance on the vehicle board. They are based on analytical methods. Katrasnik proposes in [5] analytically based method to calculate corrected fuel consumption of parallel and series hybrid electric vehicles (HEVs) at balanced energy content of the electric storage devices. The energy conversion phenomena are explained in [6]. Energy flows and energy conversion efficiencies of commercial plug-in hybrid-electric vehicles (PHEV) are analyzed for parallel and series PHEV topologies. The analysis is performed by a combined analytical and simulation approach. Various type models and algorithms derived from simulation and experiment are explained in details in [7]. Frequently they are based on quasi-static and or dynamic models of the vehicles. The

* Corresponding author: zlatina.dimitrova@mpsa.com

DOI: 10.1051/matecconf/201713302001

design of the converters and the stockers is optimized for global best tank-to-wheel efficiency. Genetic algorithms are mostly used for the design optimizations. Eren et al. in [8] deal with optimal sizing of HEV and propose a methodology for the optimization of HEV components using the multi-objective approach considering the minimization of operating cost, weight and volume simultaneously. In this work, it is proposed to use an optimization methodology to design environomic vehicle energy systems. This integrated methodology is needed to guide the evaluation of different energy vectors and vehicles propulsion systems combinations.

2 Methodology

The developed methodology combines flowsheeting models of technologies, energy integration techniques, economic evaluation and life cycle assessment in a computational platform. Multi- objective optimization techniques are used to explore the superstructure possibilities and to find the optimal design solution. The master optimization is done by using a genetic algorithm and mixed integer nonlinear programming. Such an integrated approach is innovative in comparison of the traditional heuristic design engineering method, based on iterations of designs and their cost evaluation. Fig. 1 illustrates the generic computational framework for environomic design of a vehicle energy system.

The superstructure contains a physical vehicle simulation model, with dynamic and thermal layouts. The cost equations are written in the economic model. The optimizer is based on a genetic algorithm. The optimization is decomposed into four major parts – a master multi objective optimization (MOO), a thermo economic simulation, a slave optimization (energy integration), where the energy integration occurs. The last part is the techno-economic evaluation. **Master Optimization**: The set of decision variables includes the types and the size of the equipment. The master optimization is solved by an evolutionary algorithm with 3 objectives: the minimization of the fuel consumption, the minimization of the investment cost and the environmental impacts for the technologies (Fig. 2). **Thermo-economic simulation:** in a second step the thermodynamic and economic state of the selected equipment is calculated by using a thermo-economic simulation models. **Economic evaluation**: the selected superstructure in the master level and the result of the slave optimization are used in the techno-economic evaluation phase to calculate objective functions of the master optimization – cost, and CO_2 emission. The multi-objective optimization results converged to the Pareto frontier curve.

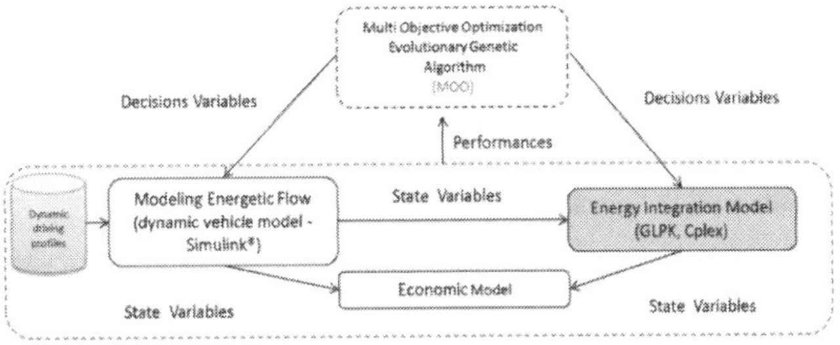

Fig. 1. Computational framework of environomic optimization.

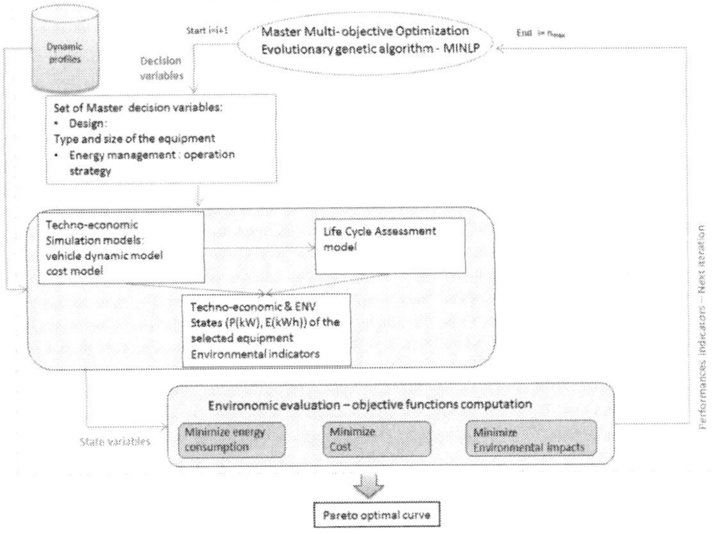

Fig. 2. Structure for multi-objective environomic optimization.

This article considers only the environomic optimization. The energy integration could be also activated in the superstructure. The energy integration method is presented in additional articles [9], [10], [11], and uses principles of energy and exergy analysis described in [12] for waste heat recovery from the cooling system and exhaust gases of the thermal engine.

2.1 Vehicles simulation models

The model components are modeled using differential equations and constitute the powertrain architecture. On the powertrain architecture a backwards approach is applied to estimate the energy consumption (Fig. 3). The energy flow is computed from the wheels to the energy sources.

Driving cycle: the input of the model is a predefined speed profile. The profile also defines the gear engaged along the cycle. **Vehicle model:** the vehicle model uses Newton's second law of motion to compute the traction force at the wheels F_t at every time step of the speed profile in order to find the power demand. It takes into account the rolling and the aerodynamic resistance and the uphill force if driving a slope. **Transmission model:** A transmission is used between the wheels and the energy converter to adapt the speed and the torque levels. The major part of the vehicle transmissions are 5 and 6 speed manual gear boxes. The gear ratio is imposed just in the New European Driving cycle (NEDC). More complex transmission such as continuously variable transmission (CVT) is also modeled. The advantage of the CVT is to choose the optimal fuel consumption gear ratio for every given drive profile. This is very suitable for customer's drives profiles. **Energy converters:** The energy converter transforms the energy (chemical or electrical) from the energy storage unit into mechanical power. As the dynamics of these converters can be complex the modeling is simplified using efficiency maps. These maps are obtained by measurement on test bench and usually represent the conversion efficiency as a function of the load and the speed demand of the primary shaft.

2.2 Vehicles cost model

The cost of the vehicle is computed for each run as a function of the size and efficiency of the energy converters and energy storage devices. The cost of the equipment comes from the literature and is related to the size of the components.

The cost of the electric motor includes the cost of the power unit. The battery cost is sensitive to the battery type and the energy storage capability of the material. The nominal cost represents the vehicle shell cost, without the powertrain components. This linear correlation (5), (Table 1) takes into account the price of the parts and the manufacturing cost of the vehicle shell and includes the margin of the carmaker. The correlation is built using the official customer prices for different vehicle classes and illustrates the link between the increasing cost and the increasing size and mass of the vehicle [14]. The car shell is defined as a completely equipped vehicle (body, interior equipment, wheels), except the powertrain. All symbols in the equations are explained in the nomenclature.

Table 1. Equations for the economic model.

Components	Cost [€]
Electric motor [13]	30 [€/kW] × P_{EM} [kW] (1)
Thermal engine [13]	15 [€/kW] × P_{TE} [kW] (2)
Battery [13]	600[€/kWh]×0,2477log($battery_{specific\ mass}$ (bat_type)+0,5126q_{bat}) [kWh] (3)
Supercapacitor [13]	15 [€/kW] × P_{SC} [kW] (4)
Nominal cost (car shell)	17,3×car_shell_mass[kg]–3905,4 [€] (5)
Electricity household [14]	0,14269 [€/kWh]
Diesel [14]	1,451 [€/L]

A simplified vehicle objective cost function is constructed (7), taking into account the vehicle powertrain cost (production) (6) and vehicle nominal cost (5).

$$Cost_{powertrain} = Cost_{ICE} + Cost_{EM} + Cost_{battery} + Cost_{supercapacitors} \quad (6)$$

$$Cost_{vehicule} = Cost_{powertrain} + Cost_{car\_shell} \quad (7)$$

2.3 Environmental model

In this work the Life Cycle Assessment (LCA) is applied as an indicator for the vehicle energy system design. The results of LCA are used to communicate on ecological labels and to demonstrate the CO_2 print improvement due to the introduction of new technologies: the powertrain efficiency progress or the light weighting of the vehicles, using new materials and innovation production processes. The literature shows that the functional unit for LCA vehicle study is to transport persons on 150000 km for 10 years [16] and this functional unit is also used for the study.

This study refers to one category from the CML short impact [17], used from the most part of the automotive industry, the Global Warming Potential (GWP) 100 years. The vehicle unitary processes and flow diagram are defined in Fig. 4 and includes three distinct successive phases: the production phase, the use phase

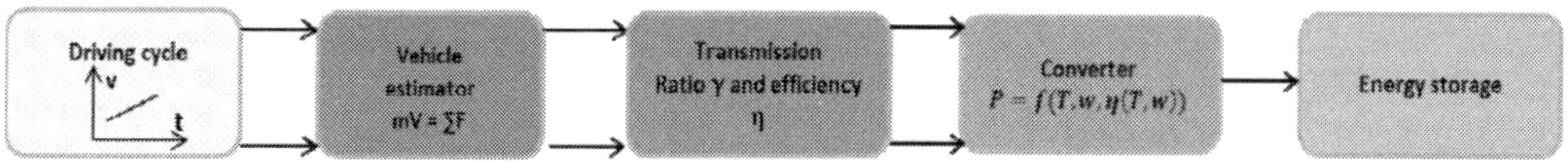

Fig. 3. Backwards simulation approach.

DOI: 10.1051/matecconf/201713302001

and the end-of-life phase. The life cycle inventory of the unitary processes is considered from the EcoInvent® database and from industrial data, concerning the vehicles production plants.

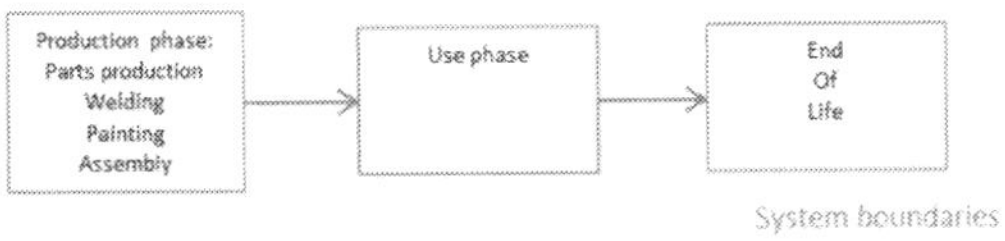

Fig. 4. Vehicle unitary processes flow chart – system definition.

3 Results—Application on electric and hybrid electric vehicles

3.1 Problem definition – environomic design of electric vehicles

In this paragraph a multi- objective optimization is performed for electric cars. The environmental impacts are chosen as optimization functions, in addition to the cost and the efficiency. The environomic optimization is applied and discussed for electric vehicles. The environmental impact category GWP (Global Warming Potential) is coming from the LCA model as performance indicator in the superstructure, and is used as an objective function. The decision variables are defined in Table 2:

Table 2. Decisions variables for design – multi-objective optimization problem.

Electric propulsion system components	Range	Unit
Electric machine	[15-50]	kW
Battery Ni-MH, 30 Wh/kg	[10-50]	kWh
Number of super capacitors in a parallel structure	[0-5]	[-]
Max Power of super capacitors	87,5	kW

The electric propulsion components are defined as decision variables for the design of the system. The optimization function p is defined in the equation (8):

$$p = \begin{pmatrix} \min(Cost(x), Env\_impacts(x), -Autonomy(x)), \\ x \in X_{Decision\_Variables} \end{pmatrix} \quad (8)$$

The Pareto curve is displayed in Fig. 5.

There is a correlation between the autonomy, the cost and the battery energy capacity. The major impact of the cost comes from the battery size, proportional to the battery energy. The coefficient of proportionality between the cost and the energy in the battery is 600 €/kWh. The cost-GWP plane allows visualizing the dependency between the cost and the GWP. There is a linear correlation with a positive slope. The cost optimum is reached with the GWP one. It enables to consider the cost or the GWP as the second objective for the multi-objective optimization, the first objective being the autonomy to maximize. Furthermore one can reduce the dimension of the optimization problem from 3 to 2.

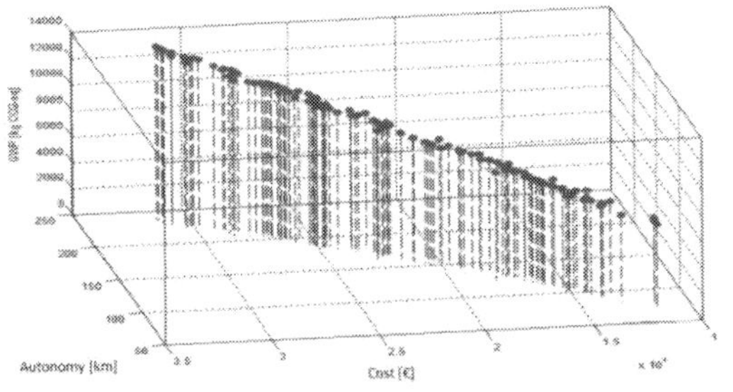

Fig. 5. 3D MOO Pareto frontier curve- electric vehicle.

The optimal solutions are detailed for two points with a characteristic autonomy:

- urban car adapted to commuting mobility (ID 2, 80 km)
- car with normal autonomy for electric propulsion (ID 4, 250 km)

The autonomy is related to the battery's energy capacity. The cost structure is shown in Fig. 6. A conventional vehicle with an internal combustion engine is given as a reference for comparison. The powertrain cost is related to the size of the powertrain components – power of the electric machine and the energy capacity of the battery. When increasing the autonomy, the customer investment cost for the powertrain increases despite the environmental bonus delivered by the government. The operating cost is related to the electricity consumption of the vehicle and to the price of electricity during the vehicle use phase and depends on the place of use. The calculation here is done with the hypothesis that the vehicle is driven in France with an electricity price of 0.14 €/kWh.

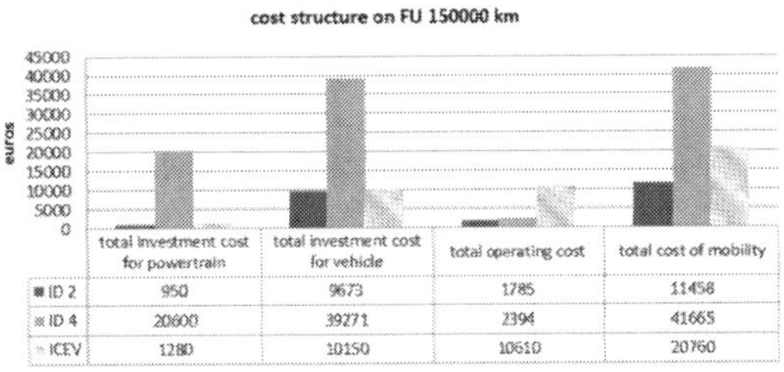

Fig. 6: Cost structure for the optimal solutions, FU – functional unit.

The operating cost is related to the electricity consumption of the vehicle and to the price of electricity during the vehicle use phase and depends on the place of use. The calculation here is done with the hypothesis that the vehicle is driven in France with an electricity price of 0.14 €/kWh. The internal combustion engine vehicle (ICEV) is characterized with equivalent investment and operating cost. The total mobility cost is of around 20000 €. Because of the high conversion efficiency of the electric powertrain and the price of the electricity, the operating cost of electric vehicles, based on the functional unit of 150000 km, is very competitive in comparison with vehicles with a conventional thermal powertrain.

The life cycle model is used to calculate the environmental impacts for the life cycle of the electric vehicle and especially the influence of the electricity production mix in the use phase (Fig. 7).

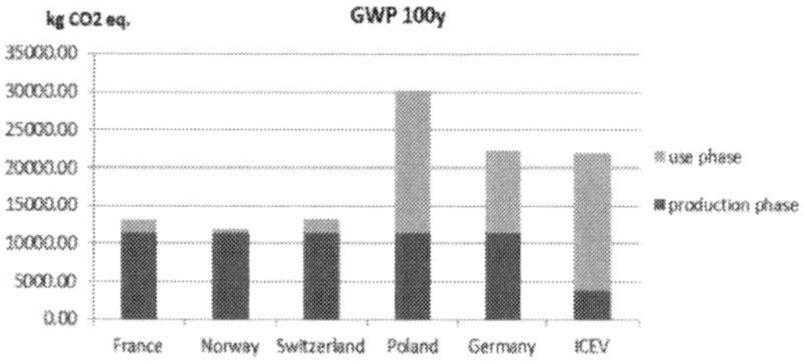

Fig. 7: GWP results for an electric car during the use phase in different countries.

In the electric powertrain production phase, the high voltage battery represents the major environmental impact - around 60%. An internal combustion engine vehicle from the same class is indicated for reference with 22000 kg CO_2 equivalent emissions. The ICEV is characterized with 17% of the equivalent emissions due to the production phase and the major part 83% of the emissions is due to its operation phase.

3.2 Problem definition – environomic design of electric vehicles

The vehicle model used in this case represents a commercial D-class [18] vehicle with a diesel electric powertrain (Fig. 8).

In the case of hybrid electric vehicles the use phase includes the GWP due of the CO_2 tank-to-wheels emissions emitted by the internal combustion engine during the vehicle operation over 150000 km. To add the wheel-to-wheels aspect the use phase contains also the GWP impact of the production of the energy vectors for charging the vehicles storage tanks – the diesel for the fuel tank and the electricity for the charging of the high voltage battery, over 150000 km.

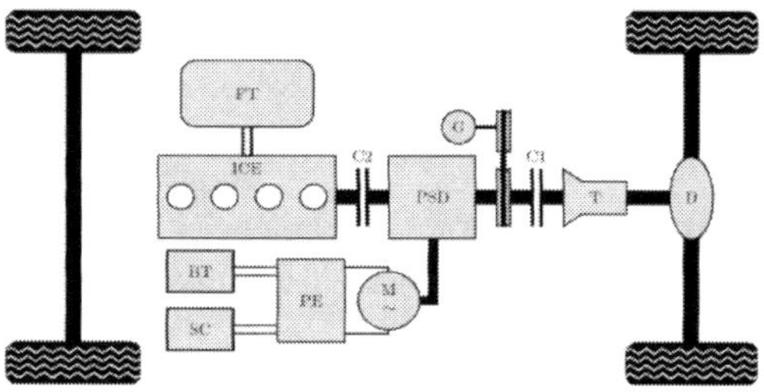

Fig. 8: Parallel hybrid electric architecture: FT – fuel tank, ICE – internal combustion engine, BT – high voltage battery, SC – super capacitor, PE – power electronics, M- electric motor, PSD – power split device, G – electric generator, C1- clutch 1, C2- clutch 2, T- Transmission, D- Differential.

The impact of electricity is considered only for the plug-in hybrid electric vehicles and the range extender vehicles, this means for vehicles equipped with a high voltage battery capacity superior to 3 kWh. The GWP is measured in kg CO_2 equivalent, and the total GWP is defined from well-to-wheels perspective in (9).

$$GWP_{well-to-wheels} = GWP_{total} = GWP_{vehicle\_production} + GWP_{tank-to-wheels\_CO2} + GWP_{diesel\_production} + GWP_{electricity\_production} \tag{9}$$

Thus the environomic optimization for hybrid electric vehicles is defined as:

$$\min(-\eta_{powertrain}(x))Investment\_Cost(x), GWP_{total}(x)), \quad x \in X_{decision\_variables}. \tag{10}$$

The decision variables for the powertrain design are defined in Table 3.

Table 3. Decision variables for design.

Decision variable for design	Range
ICE displacement volume [L]	[0,8-1-1,4-1,6-2,2]
Electric motor rated power [kW]	[1-150]
Battery energy [kWh]	[5-50]
Number of super capacitors [-]	[1-10]

The solutions of the three objective environomic optimization converged on a Pareto Frontier optimal curve (Fig. 9), representing the trade-off between the energy consumption and the cost and the total GWP impact of the vehicles on normalized driving cycle.

The vehicle use phase (including the operation CO_2 emissions and the emissions due to the energy vectors production) is clearly the major contributor to the total equivalent CO_2 emissions, in comparison of the equivalent CO_2 emissions for the vehicle production phase, for powertrain efficiencies between 25% and 35%. With the increasing of the powertrain efficiency over 35% and respectively the hybridization ratio (heavy Plug-in HEV and REX) and the size of the electric part of the powertrain, the impact of the vehicles production phase increases (Fig. 10). This is due to the increasing of the mass of the materials needed for production of the high voltage battery, the electric machine and the electronics.

Orders of magnitude for the total GWP evolution and the reparation of the impact of the different life cycles phases are given in Fig. 11, for different sizes of high voltage battery –this means for different hybridization ratio. The vehicles are considered to be operated in France with European diesel and French electricity mix production. This means that the emissions due to the energy vectors are thus estimated for an optimistic scenario. The operation of the Plug –In vehicles in countries with high carbon percentage use in the electricity generation (Germany, Poland, and China) will increase the contribution of the equivalent CO_2 emissions, coming from the electricity generation. The functional unit is 150000 km.

DOI: 10.1051/matecconf/201713302001

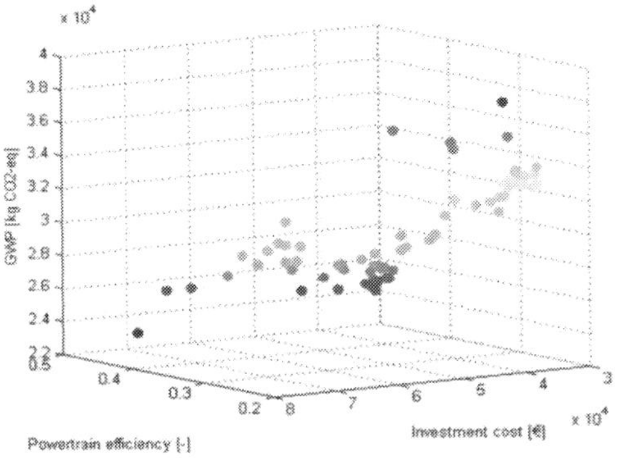

Fig. 9. 3D Pareto frontier curve: environomic optimization, NEDC cycle.

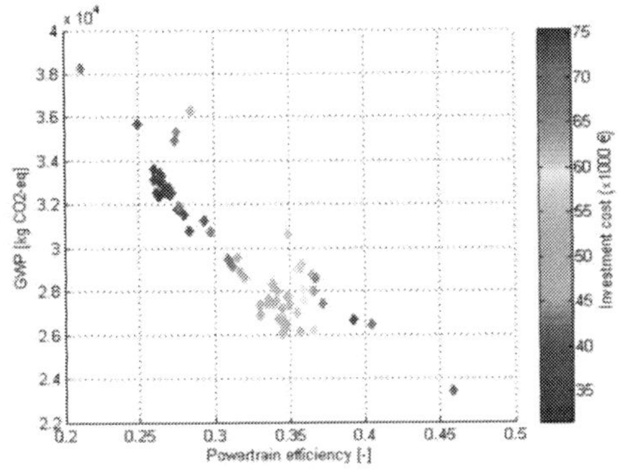

Fig. 10. Pareto curve – total GWP to powertrain efficiency, investment cost in color bar, NEDC.

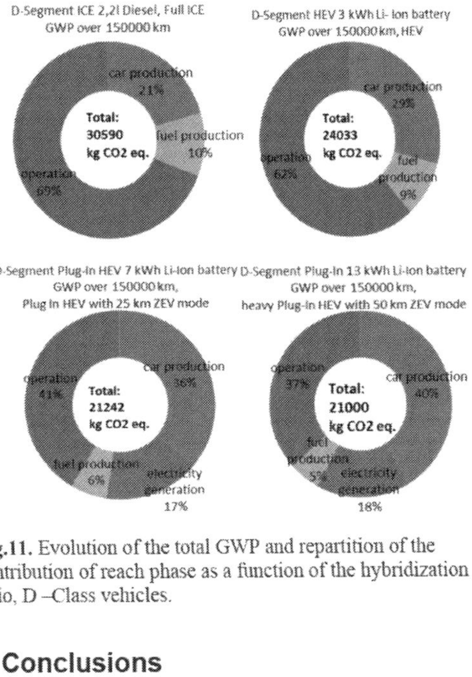

Fig.11. Evolution of the total GWP and repartition of the contribution of reach phase as a function of the hybridization ratio, D –Class vehicles.

4 Conclusions

The design methodology is applied on electric vehicles and hybrid vehicles. The competitiveness and the trade-off of different advanced powertrains are assessed and optimal design configurations are pointed out (Fig. 12).

The electric vehicles (EV) are suitable for urban mobility. The battery represents the biggest impact on the powertrain cost 75%-90% depending on the targeted autonomy of the vehicle. The energy density of the battery conditions the autonomy. The investment cost is high because of the important cost of the electric powertrain. The operating cost is particularly low 0.0119 €/km (order of magnitude for France). The environmental impact of the electric vehicle during the use phase is related directly to the place of use and the electricity production mix of the country. The hybrid electric vehicles appear to be a balanced solution between autonomy, cost and environmental impact. They combine the advantages of the low CO_2 emissions and low operating cost of the electric vehicles and overcome their limited autonomy problems. There are different categories of hybrid electric vehicles– non plug-in HEV, plug-in HEV and range extenders (REX). The use phase of the plug –in hybrid electric vehicles (PHEV) and REX is sensitive to the place of use of the vehicles and the electricity production mix. The PHEV present advantages in term of total GWP. PHEV with battery capacity of around 7 kWh seems to be the optimal environomic solution.

The cost/efficiency vision of the vehicles propulsions systems studied with the environomic design method is given in the Fig. 12. The equivalent CO_2 emissions are considered integrating the life cycle perspective and the well-to-wheels energy conversion chain.

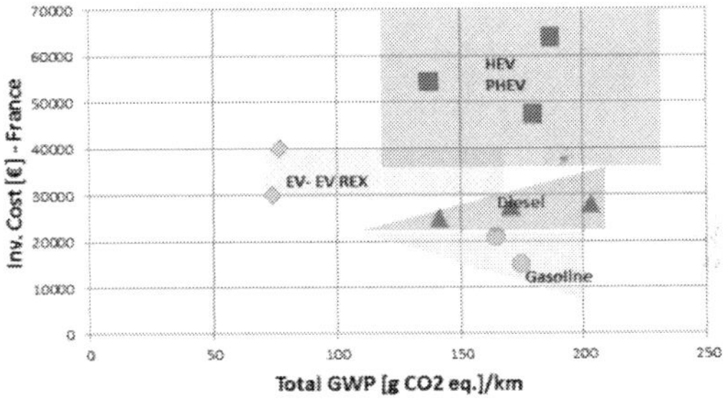

Fig. 12. Cost/ efficiency vision of the vehicles propulsion systems, well-to-wheels CO_2 emissions.

The methodology of environomic design can support automotive engineers to emerge advanced energy conversion systems concepts, taking into account economic, efficiency and environmental criteria on a holistic way. The proposed methodology combines flowsheeting vehicles models, energy integration techniques, economic evaluation and life cycle assessment in a computational platform. Multi- objective optimization techniques are used to explore the superstructure possibilities and to find the optimal design solutions, corresponding to the most efficient energy system definition for a vehicle. Such an integrated approach is innovative in comparison of the traditional heuristic design engineering method, based on iterations of designs and their cost evaluation.

Nomenclature

CML – Impact assessment method
EM – Electric motor
EV – Electric vehicle
F – Force in [N]
FU – Functional unit
GLPK, Cplex – Solvers
GWP – Global Warming Potential
HEV- Hybrid Electric Vehicle
m – Vehicle mass in [kg]
MOO – Multi Objective Optimization
MINLP– Mixed Integer Non Linear Programming
NEDC – New European Driving cycle
P – Power of the drive shaft in [kW]
$P_{battery}$ – Power delivered from the battery in [kW]
P_{EM} – Power of the electric motor in [kW]
P_{SC} – Power of the supercapacitors in [kW]
P_{TE} – Power of the thermal engine in [kW]
PHEV – Plug-in hybrid electric vehicle
q_{bat} – Capacity of the battery in [kWh]
REX – Range extender vehicle
T – Torque on the drive shaft in [Nm]
V – Vehicle speed in [m/s]
w – Rotation speed [rpm]
ZEV – Zero emissions vehicles
γ – Gear ratio [-]
η – Efficiency [-]
$\eta_{powertrain}$ – Powertrain efficiency in [-]

References

1. L. Guzzella, A. Sciarretta, Vehicle Propulsions System: introduction to modelling and optimization, third edition, (2013)
2. ExxonMobil, 2012 the Outlook of Energy- A view to 2040, www.exxonmobil.com/energyoutlook, accessed on 30.04.2015
3. C. Frangopoulos, "Thermoeconomic Functional Analysis: A Method for Optimal Design or improvement of Complex Thermal Systems, "Ph.D. Thesis, Georgia Institute of Technology, Atlanta, USA, (1983)
4. Y. El-Sayed, R. Evans, Journal of Engineering for Power, **92**, 27-35, (1970)
5. T. Katrašnik, Applied Energy, **87**, 3330-3339, (2010)
6. T. Katrašnik, Energy Conversion and Management, **52**, 2637-2650, (2011)
7. M. Hannan, F. Azidin, A. Mohamed, Renewable and Sustainable Energy Reviews, **29**, 135-150, (2014)
8. Y. Eren, H. Gorgun, International Journal of Hydrogen Energy, **40**, 2312-2319, (2015)
9. Z. Dimitrova, F. Maréchal, Energy, **83**, 539-550, (2015)
10. Z. Dimitrova, P. Lourdais, F. Maréchal, Energy, **86**, 574-588, (2015)
11. Z. Dimitrova, F. Maréchal, Applied Thermal Engineering, **91**, 834-847, (2015)
12. P. Punov, T. Evtimov, R. Chiriac, A. Clenchi, Q. Daniel, G. Descombes, International journal of energy research, 14, 1229-1241, (2017)
13. S. Ebbessen, P. Elbert, L. Guzzella, Oil & Gas Science and Technology – Rev. IFP Energies nouvelles, **68**, 109-116, (2013)
14. French Government. Statistiques de d_eveloppement durable. 2013. www.statistiques.developpement-durable.gouv.fr/energie-climat/s/prix-energies.html [accessed 18.09.13].
15. Z. Dimitrova, F. Maréchal, Energy, 76, 1019-1028, (2014)
16. S. Richet, P. Tonnelier, S. Martin. L'Analyse de cycle de vie (LCA) - demarche et applications chez PSA. PSA Peugeot Citroën; 2012, unpublished internal report.
17. J Guinée,. R. Heijungs. Environmental Toxicology and Chemistry **14**, 917–925, (1995)
18. La centrale, fiche technique de la Peugeot 508 SW, http://www.lacentrale.fr/fiche-technique-voiture-peugeot-508-swþ2.2þhdiþ204þfapþgtþbva6-2014.html [accessed 17.09.14].

DOI: 10.1051/matecconf/201713302002

Predictive and holistic energy distribution for hybrid electric vehicles

Zlatina Dimitrova[1,2,*]

[1]PSA Groupe, Research and Innovation Department, 78943 Vélizy- Villacoublay, France
[2]Ecole Polytechnique Fédérale de Lausanne, IPESE, 1100 Lausanne, Switzerland

Abstract. The improvement of the efficiency of vehicle energy systems stimulate active search to find innovative solutions during the design process. Engineers can use computer-aided processes to find automatically the best design solutions. This kind of approach named "multi-objective optimization" is based on genetic algorithms. The idea is to obtain simultaneously a population of possible design solutions corresponding to the most efficient energy system definition for a vehicle. These solutions will be optimal from technical and economic point of view .This paper presents a systematic optimization methodology for vehicle energy systems that delivers the designs of optimal vehicle powertrain solutions and their optimal operating strategies in a holistic way. The methodology is applied on D class hybrid electric vehicles, in order to define the powertrain configurations, to estimate the cost of the powertrain equipment and the optimal operating parameters. The optimal designs and operating strategies are researched for the normalized European driving cycle. The optimization can be done on-line and consider the next step of the driving cycle, which allow for predictive energy distribution strategies.

1 Introduction

1.1 Energy management strategies

Energy management strategies studies are done most of the time on the local level so that the energy consumption and performance of vehicles are improved. Zhang et al. did in [1] an analysis of the energy management strategies for hybrid electric vehicles: driving cycle recognition/prediction algorithms, integrated multi-objective, coordinated optimization energy management strategies. Most of the strategies are heuristics based and Trovao et Antunes proposed in [2] an analysis of meta-heuristic methods for power management of a dual energy storage system for electric vehicles. Chen and al. proposed in [3] an energy management strategy for a series of plug-in hybrid electric vehicle. They applied a number of quadratic equations to determine the engine fuel-rate with respect to battery power. The dynamic programming used for the solution of the optimal control problems is illustrated in [4], where Xu et al. study a multi-objective component sizing based on optimal energy management strategy of fuel cell electric vehicles. On-line predictive energy management is studied in [5].

This paper proposes a method for energy management of hybrid electric vehicles based on a holistic controller. A continuous energy distribution function is proposed. The main idea is to optimize in a simultaneous way the continuous energy distribution function. The optimization is done with a genetic algorithm. The optimization can be done on-line and considers the next step of the driving cycle, which is allowing predictive energy distribution strategies. The optimal sizing of the powertrain based on techno economic criteria is defined.

2 Methodology

2.1 Energy distribution function

The vehicle simulation tool is SIMULINK®. The vehicle model is based on mechanical and electrical flows. The level of the model is quasi-static. The vehicle is able to follow dynamic profiles generated from a library of driving cycles. The model has a loop energy management structure, linked to the required mechanical power, to follow the dynamic cycle. This energy management loop is called "back and forward" and allows, for a given design of the vehicle powertrain, to simulate the energy consumption of the vehicle, on the given driving profile. The energy flow is computed backwards from the wheels to the energy sources. The model is introduced and explained in details by the author in [6] (Fig. 1).

Hybrid- electric vehicles differ also according to the degree of hybridization of the powertrain and the battery capacity. In the optimization part the battery capacity and the hybridation ratio are part of the decision variables.

* Corresponding author: zlatina.dimitrova@mpsa.com

DOI: 10.1051/matecconf/201713302002

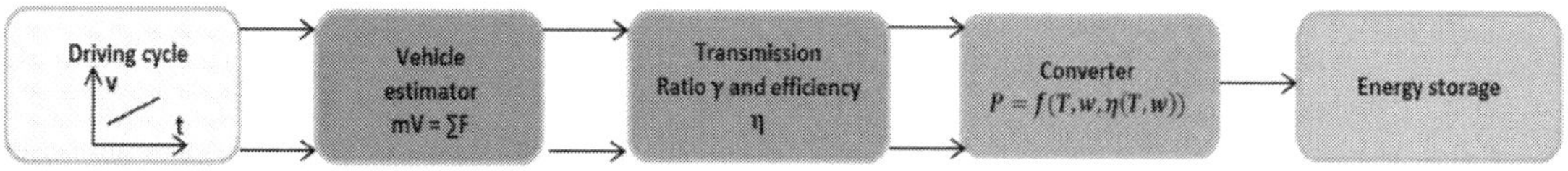

Fig. 1. Quasi-static model of the parallel thermal electric hybrid [6].

The solutions are classified according to the functional classification – mild, hybrid electric, plug-in hybrid electric and purely electric vehicles.

A power distribution strategy is put into place at the coupling of the electric motor and the combustion engine torques. This strategy defines the hybrid ratio, which represents the power contribution of the electric side of the powertrain. The model is based on a commercial D class diesel hybrid electric vehicle, with parallel hybrid architecture.

The degree of hybridization is defined in (1), as the ratio between the electric power and the total power. All symbols in the equations are explained in the nomenclature. The benefits, i.e. the reduction of the energy consumption and the additional cost associated with the hybridization, increase with the degree of hybridization. There are various possibilities to develop energy optimization strategies and most of them are heuristic based. Their guiding principles are that in the hybrid vehicle the engine should be used when its efficiency is relatively high and the battery charge and discharge should be regulated so that the state of charge remains within predefined limits. The operating modes of the powertrain are mapped, for example, the torque request is expressed as a function of the vehicle speed demand and the state of the charge of the battery [7].

$$H_r = \frac{P_{EM}}{P_{TOT}} = \frac{P_{EM,\max}}{P_{EM,\max} + P_{ICE,\max}}. \quad (1)$$

In this study, instead of having a discontinuous map with modes switching, a continuous function, relating to the hybridization ratio (the electric motor contribution), the state of charge of the battery and the speed demand is developed. Equation (2) gives the generic expression.

$$H_r = f(V, SoC). \quad (2)$$

Thus, the parameters relating to the state variables of the electric motor usage can be optimized for minimal energy consumption and minimal size of the powertrain components. In the multi-objective optimization problem these are the variables for operating strategies. The optimization problem is solved with an evolutionary genetic algorithm and the optimal operating strategies solution is solved in a holistic way. This is one of the novelties of this work. The batteries have low specific energy in comparison to fossil fuels. The energy management strategy makes use of the electric drive at low speeds and high state of charge (*SoC*) of the battery. The hybridization ratio decreases at high speeds and low *SoC* in order to save the battery from overload. A continuous S-curve function is suitable to archive these requirements (3).

$$H_r(SoC) = 3 + \left(\frac{c_1 \times SoC - (SoC_{\min} + SoC_{\max})}{SoC_{\min} + SoC_{\max}} \right)^3. \quad (3)$$

The behaviour of such shape is illustrated in Fig. 2, where $SoC_{\min} = 0{,}3$ and $SoC_{\max} = 0{,}8$.

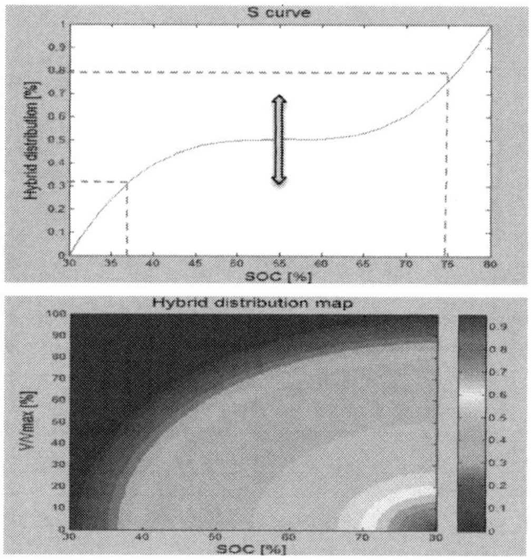

Fig 2. Hybridization ratio strategy.

During simulation, if the *SoC* is high, the energy management computes a high hybridization ratio and drains the battery, so its *SoC* will tend to the central plateau of the *S*- curve. On other hand, if the *SoC* is below the plateau, the strategy computes a low ratio of hybridization in order for the battery to get recharged during the regenerative braking. The strategy has a stabilizing effect on the state of charge. Coefficients c_1 and c_2 are responsible for the shape of the energy distribution. Parameter c_1 influences the height of the plateau. To extend the strategy to the vehicle speed, the *S*-curve is rotated in space, around the upper vertical axe. The speed axe is normalized using the maximal speed $V_{\max}$ of the driving cycle. Parameter c_2 is related to the stretch of the rotation. A simplified vehicle objective cost function is constructed (5), taking into account the vehicle powertrain cost (production) (4) and vehicle nominal cost.

$$Cost_{powertrain} = Cost_{ICE} + Cost_{EM} + Cost_{battery}; \quad (4)$$

$$Cost_{vehicle} = Cost_{powertrain} + Cost_{car\_shell}. \quad (5)$$

The multi-objective optimization tool used in this paper is presented in Fig. 3.

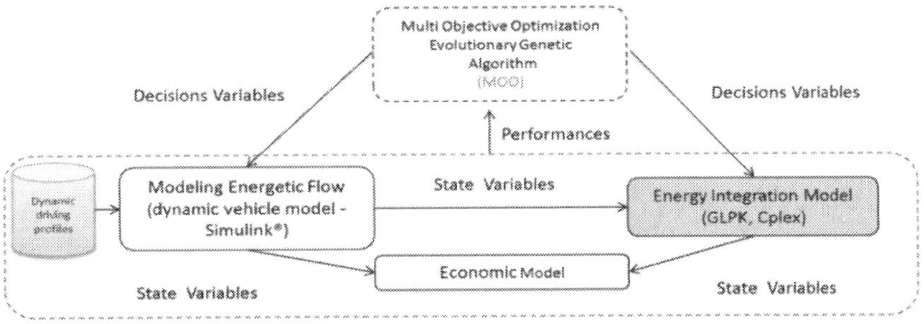

Fig. 3. Architecture of the multi-objective optimization tool.

The general computational approach has already been described in [6]. The superstructure contains a physical vehicle simulation model, with dynamic and thermal layouts. The cost equations are written in the economic model. The energy integration model uses the results from the dynamic and thermal flows calculations. The optimizer is based on a genetic algorithm.

2.2 Algorithm for holistic energy distribution

Fig. 4 presents a predictive energy management system for a hybrid vehicle, where the system uses vehicle input information, such as present location, time, 3-D maps, driving history, etc., to better determine the engine and motor power commands. The predictive energy management system includes a learning and predictions process block.

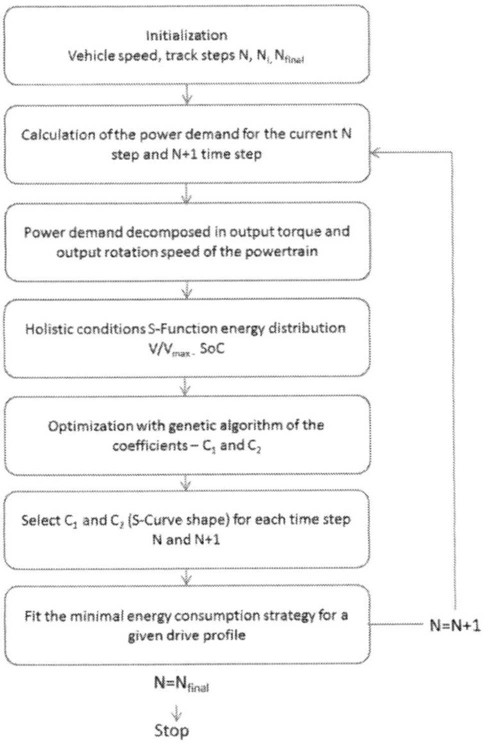

Fig. 4. Algorithm for a holistic energy distribution.

In this stage input information is obtained, a driver power demand for a series of N samples of future time horizon starting from the present is calculated. The calculated driver power demand is then applied to a predictive control strategy process block that generates the engine and motor power commands for each N sample. The predictive control strategy block employs optimization techniques based on genetic algorithms and is based on continuous S-curve shaped energy distribution curve to find the optimum sequence of power commands for the series of N samples (Fig. 4).

3 Results. Application in hybrid electric vehicles

3.1 Problem definition

A hybrid vehicle with multiple propulsion systems can be operated independently or together. The model contents are the electric machine, battery, supercapacitors, thermal engine and fuel tank, with diesel fuel. The vehicle model represents a commercial D- class [8] vehicle with a thermal electric powertrain.

In this study, instead of defining one vehicle with a set of parameters and then studying its performances over various driving cycles, the reverse is done. For each use the powertrain components and the energy management parameters are optimized. Thus, each usage has its optimal vehicle design. The objective is to size the components of the hybrid powertrain. The converters and the storage tanks are defined and the optimal operating strategy is estimated, regarding the energy consumption and the cost objectives. A two objective optimization is considered. The energy consumption and the powertrain cost are minimized. The optimization problem is defined as:

$$\min(-\eta_{powertrain}(x), Cost_{vehicle}(x)), s.t. x \in X_{decision\_variables} \quad (6)$$

The decision variables for the powertrain design and energy management are defined in Table 1.

After each iteration of the model, the mean powertrain efficiency in traction is calculated as:

$$\eta_{powertrain} = mean\left(\frac{P_{wheels}}{P_{fuel} + P_{battery}}\right). \quad (7)$$

Table 1. Decision variables for design and energy management.

Decision variable for design	Range
ICE displacement volume [L]	[0,8-1-1,4-1,6-2,2]
Electric motor rated power [kW]	[1-150]
Battery energy [kWh]	[5-50]
Energy strategy parameter c_1 [-]	[1,8-2,4]
Energy strategy parameter c_2 [-]	[0.1-10]

An additional step is to open the optimization of the parameters of the energy management in a predictive way. The trajectory is known and the best energy management should be used to minimize the fuel consumption of the vehicles during its trajectory. The possibility of the optimization structure for predictive energy management is illustrated by a simple example using the well-known New European Driving Cycle (NEDC). The cycle is divided into specific numbers and length of segments. Then, the multi-objective optimization is applied to maximize the powertrain efficiency. For each segment one can see the optimal energy management function.

3.2 Results and discussion

The cycle is divided into segments with a time division of 1 minute as time interval. The algorithm will look what is the trajectory 1 minute ahead and find the best parameters for the hybrid electric vehicle (HEV) (Fig. 5). By taking the value of the parameters of the closest point to the average efficiency and applying the same parameters into a complete NEDC run, one obtains the following results (Table 2). The optimization result is very close in terms of efficiency and cost to the average point, without decomposition of the cycle.

In the second stage a hybrid electric vehicle with 7 kWh of battery capacity is considered for the study. The powertrain parameters are listed in Table 3.

The optimal set of the S-Curve parameters is researched to minimize the fuel consumption of the vehicle for each cycle segment. Then the vehicles choose the optimal energy distribution as a function of the trajectory. The average value of the consumptions for the segmented NEDC is equal to 4,2 L/100 km.

The best fuel consumption with their associated S-Curve parameters, for each segment, are represented in Table 4.

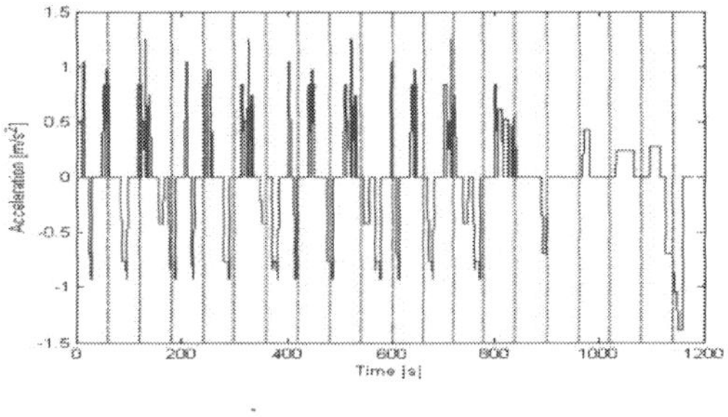

Fig. 5. Segmentation of the NEDC – one minute duration of the segments.

Table 2. Design and energy management parameters on a complete NEDC.

Parameters	Value
Powertrain efficiency [-]	0,325
Investment cost [€]	55428
Displacement volume [L]	1,9
Electric motor rated power [kW]	113
Battery size [kWh]	24
Energy strategy parameter 1	1,8
Energy strategy parameter 2	0,5
Consumption [L/ 100 km]	3
Vehicle mass [kg]	2187

Table 3. Design of the HEV for the NEDC segmentation.

Parameters	Value
ICE displacement volume [L]	2,2
Electric motor rated power [kW]	40
Battery capacity [kWh]	7

Table 4. Optimal fuel consumption on the NEDC segments.

ID	Fuel consumption [L/100 km]	Energy distribution c_1 parameter	Energy distribution c_2 parameter
1	5,5	1,80	0,49
2	3,5	1,84	0,46
3	4,3	2,02	0,27
4	2,4	2,00	0,39
5	3,9	1,89	0,20
6	5,5	2,16	0,38
7	2,0	2,02	0,17
8	4,6	1,80	0,33
9	6,1	1.93	0,40
10	1,5	1,80	0,41
11	7,9	2,37	0,18
12	3,3	1,80	0,51
13	4,0	2,30	0,14
14	6,6	2,15	0,15
15	4,2	2,21	0,36
16	2,7	1,80	0,44
17	4,2	1,87	0,31
18	4,9	2,14	0,38
19	5,9	2,20	0,34
20	1,6	2,22	0,51

As illustrated in Table 4, segment 10 demonstrates the lowest fuel consumption and segment 11 demonstrates the highest fuel consumption. The difference in the fuel consumption between these two segments can be explained by the different hybridation ratio that the holistic energy management controller defines. The energy distribution function for segment 10 shows in Fig. 6a, a dominant level of the electric traction that is used. This is the reason for the lowest fuel consumption results. The energy distribution function for segment 11 is shown in Fig. 6b, a dominant thermal engine mode. Thus, the highest fuel consumption is estimated.

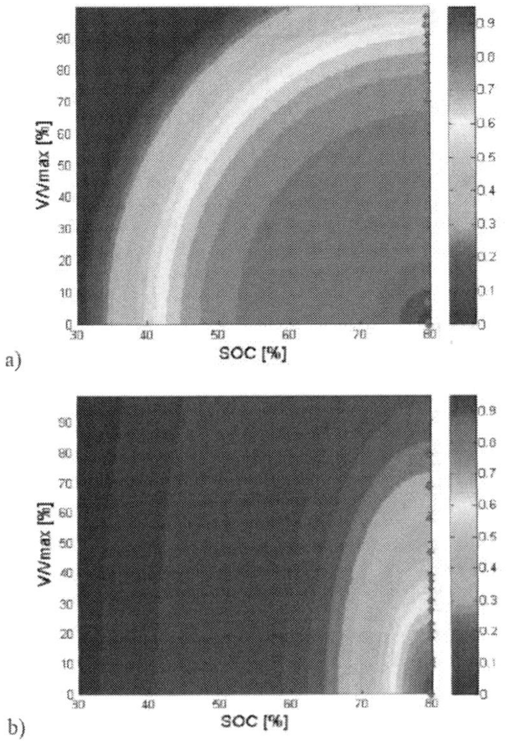

Fig. 6. *S*-Curve energy distribution function a) dominant electric motor mode b) dominant thermal engine mode.

4 Conclusions

This paper proposes a holistic method for control and design optimization of hybrid electric vehicles. It is based on a multi-objective optimization using a genetic algorithm. The optimal size of the propulsion system is based on an optimal energy function definition. A continuous energy distribution function is proposed for that. The main idea there is to optimize in a simultaneous way the continuous energy distribution function. The optimization can be done on-line and consider the next step of the driving cycle, which is allowing predictive energy distribution strategies. The optimal designs are given for every segment of the NEDC and they are compared on the average optimal design. The fuel consumption results are explained, related to the energy distribution function, applied for the segments. The method could be used for predictive control and for cost effective design, which will cover the power demands for the real drives.

Nomenclature

c_1, c_2 ,– energy management function coefficients in [-]
EM – electric motor
F – force in [N]
GLPK, Cplex – solvers
HEV- Hybrid Electric Vehicle
H_r – hybridization ratio in [-]
ICE – internal combustion engine
m – vehicle mass in [kg]
MOO – Multi Objective Optimization
NEDC – New European Driving cycle
P – power of the drive shaft in [kW]
$P_{battery}$ - power delivered from the battery in [kW]
P_{EM} - power of the electric motor in [kW]
P_{fuel} - power in the fuel in [kW]
P_{ICE} - power of the internal combustion engine in [kW]
P_{TOT} - total power of the vehicle powertrain in [kW]
P_{wheels} - power transmitted to the wheels in [kW]
T – torque on the drive shaft in [Nm]
SoC – state of charge of the battery in [%]
V – vehicle speed in [m/s]
w – rotation speed [rpm]
γ - gear ratio [-]
$\eta_{powertrain}$ - powertrain efficiency in [-]

References

1. P. Zhang, F. Yan, C. Du, Renewable & Sustainable Energy Reviews, **48**, 88-104, (2015)
2. J. Trovão, C. Antunes, Energy Conversion and Management, **95**, 281-296, (2015)
3. Y. Xu, D. Liu, Q. Wei, Energy Conversion and Management, **103**, 553-561, (2015)
4. Z. Chen, B. Xia, C.You, C. Mi, Applied Energy, **145**, 172-179, (2015)
5. L. Li, S. You, C. Yang, B. Yan, J. Song, Z. Chen, Applied Energy, **162**, 868-879, (2016)
6. Dimitrova, F. Maréchal, Energy, **91**, 630-644, (2015)
7. L. Guzzella. *Vehicle propulsion systems*, (2013)
8. La centrale, fiche technique de la Peugeot 508 SW, http://www.lacentrale.fr [accessed 17.06.17].

Comparison of real driving cycles and consumed braking power in suburban Slovakian driving

Tsvetomir Gechev[1,*], *Martin* Mruzek[2], and *Dalibor* Barta[3]

[1]Technical University – Sofia, Faculty of Transport, 8 Kliment Ohridski Blvd., Block 9, 1000 Sofia, Bulgaria
[2]University of Žilina, University Science Park, Univerzitná 1, 010 26 Žilina, Slovakia
[3]University of Žilina, Faculty of Mechanical Engineering, Univerzitná 1, 010 26 Žilina, Slovakia

Abstract. The paper compares the features of suburban real driving cycles performed with CORRSYS DATRON measurement equipment on routes in the region of Žilina, Slovakia. It observes differences in the maximum and average vehicle velocities and the amount of braking in relation to the elevation profile of each individual cycle. Consumed braking power was also calculated in the different cycles in order to review the potential electricity regeneration capabilities of hybrid electric vehicles, operating on the same routes. The change in braking energy depending on vehicle mass and presence of grade on the routes in the measured cycles was also assessed. The calculations and plotting were done by using Matlab software.

1 Introduction

Nowadays researchers and automobile producers come together in search of a way to increase the efficiency of vehicles by mastering or changing the technology and concepts incorporating the idea of modern transportation. This is due to the substantial usage of internal combustion engine (ICE) powered automobiles, which are the major cause of many environmental and health problems.

Electric vehicles (EVs) and hybrid electric vehicles (HEVs) are some of the main proposals that are to replace conventional vehicles and achieve a clean and high-efficient future transportation [1].

The adaptation of such vehicles would have a significant environmental benefit. It is because EVs and HEVs use electricity, generated by various sources, many of which are renewable. In France alone, where most of the electricity is produced from nuclear energy, the percentage of emissions would fall down considerably if EVs were to be used massively [2].

The efficiency of EVs is currently lower than that of conventional gasoline and diesel driven ICE automobiles but there are tremendous reserves in terms of future enhancement of the performance of such vehicles [3]. One such important feature of HEVs and EVs is their ability to recover kinetic energy from braking by using their electric motors as generators. For example, braking a 1500 kg vehicle from 100 km/h to 0 dissipates about 0,16 kWh of energy [1]. It could potentially be stored and reused thus making the vehicle more energy efficient.

Automobile energy consumption could be calculated by the use of driving cycles, therefore this principle is also suitable for estimation of the energy regeneration potential of HEVs and EVs.

The New European Driving Cycle (NEDC) acts as a simulation tool for the typical usage of a car in Europe. It is a standardized test applied in EU member states to enable comparability in results. However, the current NEDC is not representative of the present-day practical transportation conditions, but it is rather an old-dated pattern with low accelerations and many constant speed sections [4]. This paper presents real driving cycles performed on suburban roads by a contemporary conventional vehicle. The aim of the paper is to compare the features of the measured cycles and determine the amount of braking energy, potentially recuperated by the application of HEVs in short distanced suburban Slovakian driving.

2 Equipment

A typical commuter vehicle – Hyundai ix20 (Fig. 1), with the following characteristics (Table 1) was used for the measurement of the real driving cycles.

Fig. 1. Hyundai ix20 test vehicle.

* Corresponding author: tsv.gechev@gmail.com

Table 1. Vehicle characteristics.

Engine displacement, dm^3	1,591
Power, kW	91,5
Fuel type	gasoline
Air drag coefficient, -	0,3
Curb weight, kg	1255
Total permissible mass, kg	1710
Max speed km/h	185
Torque, Nm	156/4200 rpm
Width, m	1,765
Height, m	1,6

The measurement equipment consists of a DAS 3 unit with a contactless sensor based on the Doppler phenomenon used for the metering of time, velocity, acceleration and distance (Fig. 2), and a CORRSYS DATRON force pedal sensor (Fig. 3). The road grade is estimated with a GPS receiver (Fig. 3).

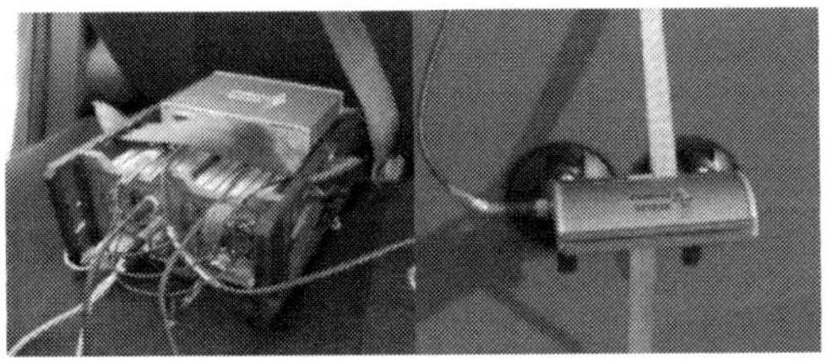

Fig. 2. DAS 3 unit (left) with a contactless sensor (right).

Fig. 3. CORRSYS DATRON force pedal sensor (left) and GPS receiver (right).

The total weight of the measurement equipment, passengers and their luggage is 241 kg.

3 Cycle measurement and comparison

3.1. Cycle overview

The city of Žilina is a big industrial and services center in northwestern Slovakia. It is situated at a crossroad and it offers many professional and educational opportunities. As a result there is a big number of commuters and students travelling each day or week from the neighboring small cities and villages in the region to Žilina.

On account of that, the vehicle was driven on two different suburban road sections: Žilina – Rajecke Teplice and Žilina – Terchova, in each respective direction, measuring four cycles with different characteristics (Fig. 4 – Fig. 7). The vehicle was driven on workdays (not in commuter trip peaks) according to the traffic requirements. The measured cycles were produced independently and could be used for simulation of various types of vehicles. The velocity values were filtered with a Savitzky – Golay filter in a Matlab environment in order to eliminate fluctuations and noise. Additionally, all plotting and calculations were done in Matlab.

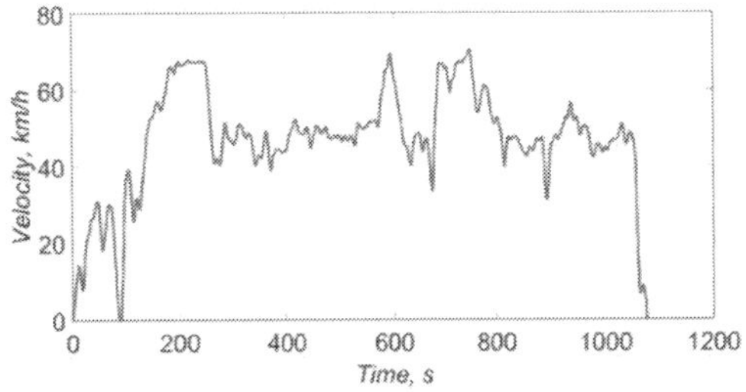

Fig. 4. Driving cycle Žilina – Rajecke Teplice.

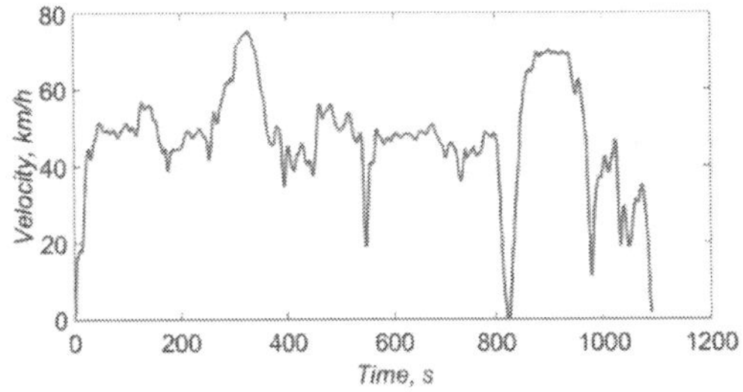

Fig. 5. Driving cycle Rajecke Teplice – Žilina.

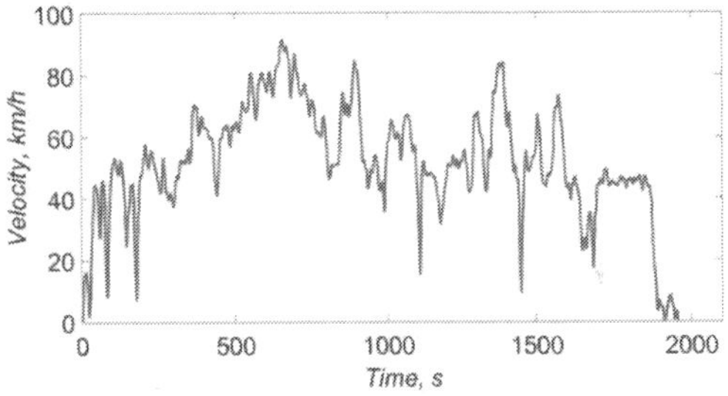

Fig. 6. Driving cycle Žilina – Terchova.

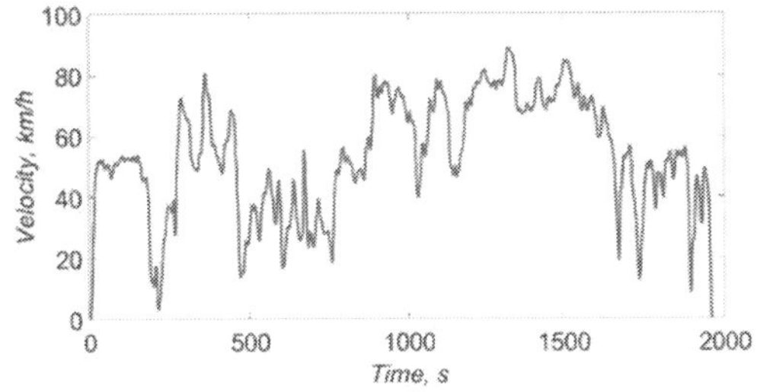

Fig. 7. Driving cycle Terchova – Žilina.

The cycles characteristics are displayed in Table 2. It can be seen that the average and maximum velocities are different. This is due to the fact that Rajecke Teplice is closer to the center of Žilina (14 km away as compared to 29 km to Terchova) and the traffic is denser. Additionally, the route to Rajecke Teplice goes through two big villages (compared to 3 of a moderate size on the way to Terchova) which results in more speed limitations. Despite the difference, the values are very close which shows that the conditions in which the cycles were measured are comparable.

Table 2. Cycle characteristics (F4 to F7 correspond to Fig.4 to Fig. 7).

Cycle	F4	F5	F6	F7
Max velocity, km/h	70,4	75,3	91,3	88,7
Average velocity, km/h	46,9	47,1	50,7	52,6
Distance, km	14	14,2	28,2	29,1
Duration, min	17,9	18,1	32,6	32,6
Denivelation, m	72,2		206	
Braking time, % of total	4,1	4,7	3,22	4,53
Idle time, % of total	0,46	0,36	0	0,25

Even though the cycles were not performed during traffic peaks, it is obvious from Fig. 4 and Fig. 5 that there could be rapid brake patterns dropping the vehicle speed to 0 in suburban traffic.

One of the most varying factors affecting the vehicle driving pattern and therefore the amount of braking is the road elevation profile (Fig. 8 and Fig. 9).

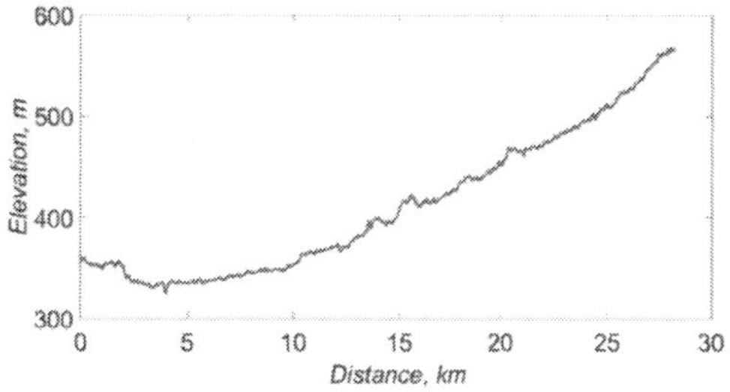

Fig. 8. Elevation profile Žilina – Terchova.

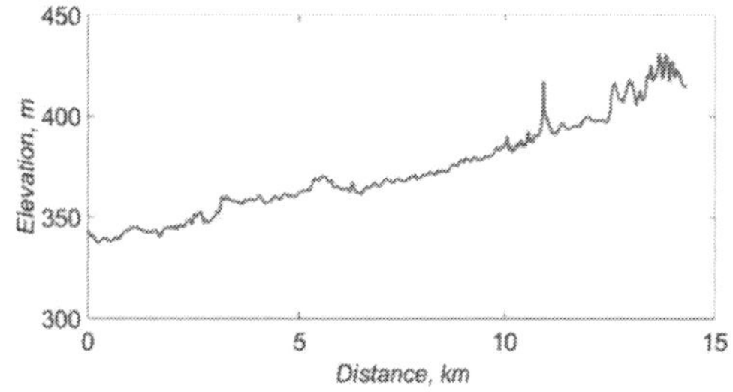

Fig. 9. Elevation profile Žilina – Rajecke Teplice.

The braking time data from Table 2, measured by the pedal force sensor, shows that cycle F7 has more braking time as a percentage of the whole driven time than the F6 cycle. This can be explained by the elevation profile (Fig. 8), which downgrades significantly in the direction Terchova – Žilina. The same applies for the comparison between cycles F4 and F5. The smaller change in the braking time percentage compared to the one in cycles F6 to F7 indicates smaller value differences in the braking energy between the uphill and downhill patterns.

3.2. Comparison of consumed braking power

The results of the driving cycles analysis show moderate height differences in the elevation profiles, which entails higher values of both tractive and braking energy consumption as compared to the ones measured on flat road. The required vehicle power, the braking and tractive energy dissipation are calculated based on the selected car parameters and the real driving cycles [5].

Vehicle force balance is given in (1) and presents the forces of aerodynamic drag resistance, inertia, grading resistance and rolling resistance, where ψ is the road resistance coefficient (2), f is the rolling resistance coefficient (3), G is the weight force, V is vehicle velocity, W is the coefficient of drag flow (4), a is vehicle acceleration and δ is the rotating parts' inertia coefficient (5) [6].

$$F = \psi G + WV^2 + \frac{\delta G}{g}a, \text{ N.} \quad (1)$$

$$\psi = f\cos\alpha \pm \sin\alpha. \quad (2)$$

$$f = f_0 + V^2.6{,}5.10^{-6}. \quad (3)$$

$$W = 0{,}5C_x\rho S, \text{ N.s}^2/\text{m}^2. \quad (4)$$

$$\delta = 1 + \sigma_1 + \sigma_2.i^2. \quad (5)$$

The rolling resistance coefficient at low speed f_0 = 0,011 corresponds to an asphalt road with moderate grip capabilities. Vehicle frontal area S = 0,8HB where H and B are vehicle height and width [6]. The rotating parts inertia coefficient δ is calculated on the basis of approximate estimation of the current shifted gear depending on the vehicle speed. Coefficients σ_1 and σ_2 are equal to 0,04 [6]. Table 3 shows the vehicle gear ratios and gear speed ranges. Idle time in all measured cycles is close to zero, therefore non-gear vehicle operation times can also be considered insignificant. The driver did not shift out of gear while driving in suburban traffic.

Table 3. Estimation of inertia from rotating parts.

Gear №	Gear ratio i	Est. speed	δ
1	3,615	0-15	1,56
2	1,955	15-30	1,19
3	1,370	30-40	1,11
4	1,036	40-50	1,08
5	0,893	50-70	1,07
6	0,733	70+	1,06

Multiplying (1) by the velocity gives the power output at driving wheels – (6) (Fig. 10) [6].

$$P = V\left(\psi G + WV^2 + \frac{\delta G}{g}a\right), \text{ W.} \quad (6)$$

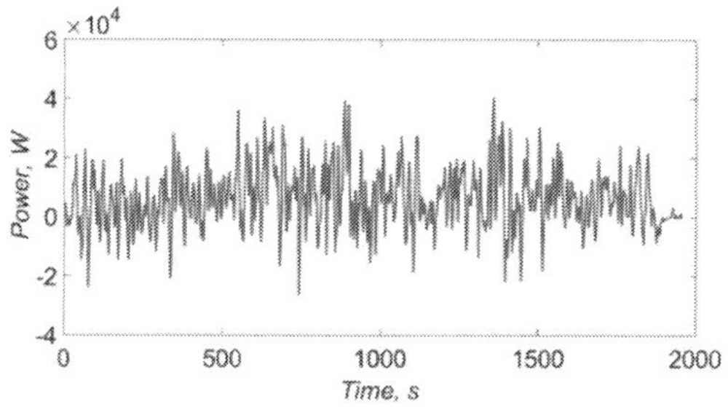

Fig. 10. Total power output, Žilina - Terchova.

For $P > 0$, the traction wheels accept power from the power plants and move the vehicle. In this situation, the braking power is zero. In contrast, when $P < 0$ vehicle kinetic energy is dissipated by the brake system. In this case, the driving power is zero. Integrating equation (6) over the driving time can give the values of both traction energy (TE) and braking energy (BE) (Fig. 11) [1].

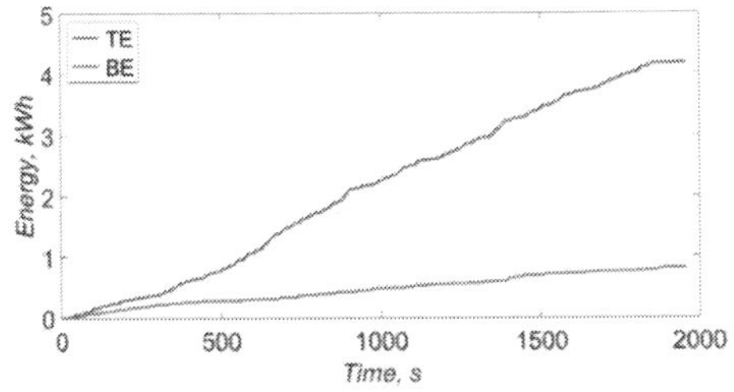

Fig. 11. Traction and braking energy dissipation over time, Žilina-Terchova (cycle F6).

The proper design and control of hybrid and electric vehicles, and more importantly, the design of automotive energy regeneration systems, depend on the amount of braking energy dissipated by a vehicle operating on certain road sections [1, 7]. Data from all measured cycles is given in Table 4.

Table 4. Tractive and braking energy dissipation for each cycle.

Cycle	F4	F5	F6	F7
Tractive energy, kWh				
• Per cycle	1,9	1,58	4,18	3,07
• Per km	0,14	0,11	0,15	0,11
Braking energy, kWh				
• Per cycle	0,68	0,73	0,81	1,33
• Per km	0,05	0,05	0,03	0,05
Percentage of BE to TE	35,8	46,2	19,4	43

When comparing cycles F4 and F5 it is important to note that cycle F4 has more consumed tractive energy due to the uphill characteristic of the road in the Žilina – Rajecke Teplice direction. More braking time in F5 (Table 2) and the downhill pattern indicate a slightly higher braking energy consumption value of F5 as compared to F4 cycle, given the brake force was approximately equal. Cycles F6 and F7 have the same comparative assessment with F6 having higher tractive energy and lower braking energy consumptions than F7.

Logically, routes with elevation profiles have significantly more energy regeneration potential in downgrade sections with percentage of braking energy amounting up to 46% and 43% of the tractive energy in the measured real driving cycles (F5 and F7).

3.2.1. Dependence of braking energy on vehicle mass and road elevation profile

It is important to estimate the braking energy in its correspondence to vehicle mass and the route elevation (Fig. 12). For the first comparison the following set of masses was used (including luggage), kg:

m_1 = 1255+70+5 = 1330 *just a driver*;
m_2 = 1255+70+65+10 = 1400 *2 passengers*;
m_3 = 1255+70+65+65+15 = 1470 *3 passengers*;
m_4 = 1255+241 = 1496 *actual mass in measured cycles*;
m_5 = 1255+70+65+65+65+20 = 1540 *4 passengers*;
m_6 = 1710 *total permissible mass*.

The bigger the vehicle mass is, the more braking power is put on the wheels (Table 5, Fig. 12). Consequently, there is more energy to be recuperated. Roads with change in the elevation also require more braking (Table 6).

Table 5. Change in kWh braking energy in relation to change of vehicle mass.

BE, kWh	m_1	m_2	m_3	m_4	m_5	m_6
F4	0,59	0,63	0,66	0,68	0,70	0,79
F5	0,64	0,68	0,72	0,73	0,76	0,85
F6	0,7	0,75	0,79	0,81	0,84	0,95
F7	1,15	1,22	1,3	1,33	1,37	1,56

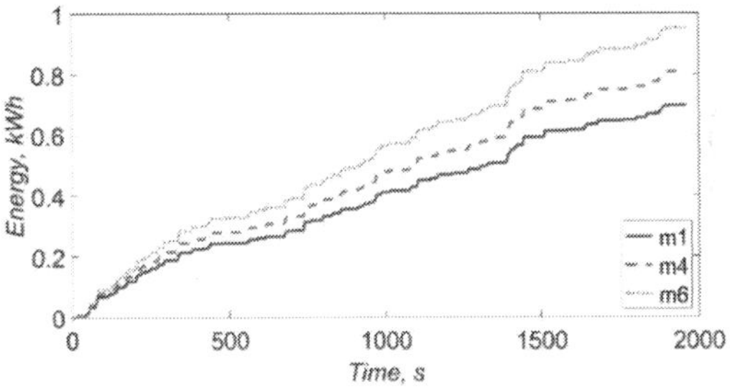

Fig. 12. Braking energy dependence on vehicle mass, Žilina – Terchova.

Table 6. Change in kWh braking energy in relation to change of road grade.

BE, kWh	F4	F5	F6	F7
Without grade	0,21	0,22	0,57	0,54
With grade	0,68	0,73	0,81	1,33

4 Conclusion

The results show that standard commuter vehicles driven on suburban mountainous roads in the region of Žilina spend high amounts of braking power which then dissipates as braking energy in the form of heat.

The braking energy for some of the cycles comprises as much as 43% of the total tractive energy used – cycle F7, elevation difference = 206 m, distance ≈ 28,5 km. In this cycle, by driving on a graded route, the braking energy values could rise by as much as 146% (Table 6), measuring up to 1,33 kWh in non-peak traffic situations, in comparison to non-graded routes of an equal distance. Additionally, the influence of the mass of the vehicle could increase the braking power by up to 35% at times when the vehicle operates at full mass load.

Conclusively, given the results, the usage of HEVs on mountainous suburban Slovakian roads by commuters is an appropriate measure to increase vehicle energy efficiency, to lower fuel costs and to ensure cleaner environment.

This paper is supported by the project VEGA 1/0927/15 "Research of the use of alternative fuels and hybrid drives on traction vehicles with aim to reduce fuel consumption and air pollutants production".

This paper is supported by the following project: University Science Park of the University of Žilina – II. phase (ITMS: 313011D13) supported by The Operational Program Research and Innovation funded by the European Regional Development Fund.

References

1. A. Emadi, Y. Gao, M. Ehsan, *Modern Electric, Hybrid Electric and Fuel Cell Vehicles*, 1-5 / 412-416, CRC Press, (2009)
2. C. Gribben, Debunking the Myth of EVs and Smokestacks, http://www.electroauto.com/info/pollmyth.shtml
3. I. Evtimov, R. Ivanov, *Proceedings of BulTrans-2009*, 94, (2009)
4. P. Mock, J. German, A. Bandivadekar, I. Riemersma, Discrepancies between type-approval and "real-world" fuel consumption and CO_2 values, *International Council on Clean Transportation*, (2012)
5. D. Barta, M. Mruzek, D. Kalincak, *Proceedings of BulTrans-2013*, 54, (2013)
6. B. Petkov, *Theory of the automobile, tractor and car*, 98-124, Tehnika, (1979)
7. D. Barta, M. Mruzek, *Proceedings of ICCSTE*-2015, No. 273, (2015)

DOI: 10.1051/matecconf/201713302004

Experimental and numerical study of the flow field around a small car

Ivan Dobrev[1,*], *Fawaz* Massouh[1], *Amélie* Danlos[2], *Michael* Todorov[3] and *Plamen* Punov[3]

[1]ENSAM, Dynfluid, 151 bd de L'Hôpital, 75013 Paris, France
[2]CNAM, Equipe Energétique, Turbomachines et Moteurs, Laboratoire Chimie Moléculaire, Génie des Procédés Chimiques et Energétique, CNAM, Saint Martin, 75003 Paris, France
[3]Technical University – Sofia, Faculty of Transport, 8 Kliment Ohridski Blvd., 1000 Sofia, Bulgaria

Abstract. This paper presents the aerodynamic study of a small car, which participated in Shell Eco-marathon Europe competition in the Urban Concept Hydrogen class. The goal is to understand the flow field around the vehicle. First, the flow is studied numerically using computational aerodynamics. The numerical simulation is carried out by means of CFD Fluent in order to obtain the drag force experienced by the vehicle and also the flow field. Then the flow field around the car is studied in a wind tunnel by means of particle image velocimetry (PIV). The comparison of the flow fields obtained numerically and experimentally shows good correspondence. The obtained results are very helpful for future car development and permit to improve the drag and to obtain a good stability.

1 Introduction

Vehicle aerodynamics is very important and has a significant impact on its general performance, especially on fuel consumption. A brief overview of the history of vehicle aerodynamics is presented in [1]. There exist different aerodynamic problems as drag reduction, vehicle stability at high speed, engine cooling, etc. However, the primary issue is drag reduction and many authors studied the drag sources and presented different methods for reducing it [1-3]. It should be noted that the field of vehicle aerodynamics is very large, has different tasks depending on car type and it is difficult to summarize all existing problems. In this paper, the authors focus their research on the aerodynamics of small car, which participated in Shell Eco-marathon Europe competition in the Urban Concept Hydrogen class. What is specific about this kind of vehicle is the low Reynolds number due to the low speed. The shape of the car is not optional but must observe the rules of the competition [4].

The aerodynamics of a small car to participate in the Shell Eco-marathon has been studied by a limited number of researchers. At first glance it seems that aerodynamic forces can be neglected due to the low vehicle speed. However, these forces are comparable with rolling resistance forces, because of the low vehicle weight. Therefore, the aerodynamic study is important for obtaining the best vehicle performance.

The authors of paper [5] present a fluid simulation of a solar prototype vehicle employing k-ε turbulence model. The results of simulations give satisfactory results for drag coefficients, which is by 8,2% less than the experimental value. The analysis of energy losses shows that the aerodynamic losses are 41% of all loses for this class of vehicle and the authors conclude that the losses depend primarily on rolling resistance and aerodynamic drag, even in the case when the speed does not exceed 38 km/h.

The authors in [6] simulated a flow around a small electric car for urban mobility at the stage of preliminary design. They studied the grid dependency of the solution and showed that the drag coefficient of 0,129 varies by less than 3% when the nodes increase from nearly 0,65 million to 2,4 million.

The paper [7] presents a study which is useful for the design of new vehicles for incoming races. The set of simulations was prepared for several Shell Eco Marathon models for a range of velocities from 20 to 45 km/h. The results of the simulations show a strong dependence of the aerodynamic drag on the Reynolds number. The authors demonstrate how flow simulation can contribute to the optimization of vehicle shape. Contrary to [5] they show that the aerodynamics drag is 50% of the rolling resistance but here the drag coefficient of 0,09 is much lower.

The author in [8] presents the development of the shape of a light road vehicle by means of flow simulation with Navier-Stokes and Boltzmann lattices solvers. The aerodynamic optimization is done by design evolution from low-drag body to the final vehicle design together with the aesthetic control of the shape after each change. The simulation is carried out on grid up to 11 million nodes with different turbulent models. The author reports that the lift coefficient shows a big change compared to the drag when the solver or turbulent model is changed.

* Corresponding author: ivan.dobrev@ensam.eu

The aim of this work is to simulate the flow around an existing model in order to obtain the aerodynamic drag and lift forces and also the flow field around the vehicle. Then this flow field will be compared with the real flow obtained in the wind tunnel by means of particle image velocimetry (PIV). It must be noted that the flow field around the vehicle and the pressure distribution are in close relationship with the aerodynamic drag. The velocity field shows the separation, which is very useful for understanding aerodynamic drag and makes it possible to study the impact of the vehicle shape on the drag. Hence this study will be useful for further amendments of vehicle shape for future races.

2 Numerical simulations

2.1. Numerical model

The flow around the vehicle with a length of 3,180 m, a width of 1,210 and a height of 1,010 m is modelled by means of commercial Navier-Stokes solver ANSYS Fluent 15. The computational domain is a box that includes the vehicle and has dimensions of 50 m × 12 m × 12 m (length × width × height). Such dimensions allow for the reduction of the boundary effects on the solution. On the upstream, top and side walls of this domain we apply the boundary condition of type "velocity inlet" because the flow is not perturbed by the vehicle. In this simulation the vehicle is fixed and in order to represent its movement, the velocity applied on all "velocity inlet" walls is equal to the desired vehicle speed.

As the vehicle wake perturbs the downstream so the boundary condition of the downstream wall must be of "pressure outlet" type. In order to represent the movement of the vehicle with respect to the ground and the wheels rotation, we apply Multi Reference Frame model (MRF). In this model the wall which represents the ground is moved respectively to the vehicle with the same velocity which is applied on walls with "velocity inlet" boundary condition. Also the wheels are supposed to be rotating around their axles with the angular velocity which corresponds to the desired vehicle speed.

The computational domain has a mesh size of approximately 20 million polyhedral nodes and 5 prism layers on the wall of the vehicle and also of the ground. The height of the first prism layer of 0,05 mm ensures y+ < 5, which is sufficient for this kind of computation. The calculation is unsteady, with time step of 0,01 s and 10 iterations per time step. The applied turbulence model is hybrid - Delayed Detached Eddy Simulation (DDES), [9]. This DDES model significantly improves the computation where flow separation occurs as in the tested case [9]. However, the calculation becomes power and memory consuming and requires a few days' calculation on a computer with 2 processors with 8 cores and a RAM memory of 64 Gb. In this study it is assumed that the solution converges if for a given averaging period, the drag and lift forces become relatively constant, which is achieved usually after 1500 time steps.

2.2 Numerical results and analysis

The first computation is carried out for vehicle speed of 10 m/s. The lift and drag coefficients are obtained by dividing the lift L and drag D forces by the upstream dynamic pressure and the vehicle cross section S:

$$C_D = \frac{D}{\frac{\rho V^2}{2} S}; \qquad (1)$$

$$C_L = \frac{L}{\frac{\rho V^2}{2} S}. \qquad (2)$$

Here ρ is air density of 1,225 kg/m^2, V is upstream velocity of 10 m/s and S is cross section of vehicle of 0,67283 m^2. The obtained drag coefficient is equal to 0,1614, which is typical for such kind of vehicle. The lift coefficient of –0,238 is negative which permits the increase of the vehicle stability. The flow field in the vehicle symmetry plane is presented in Fig. 1.

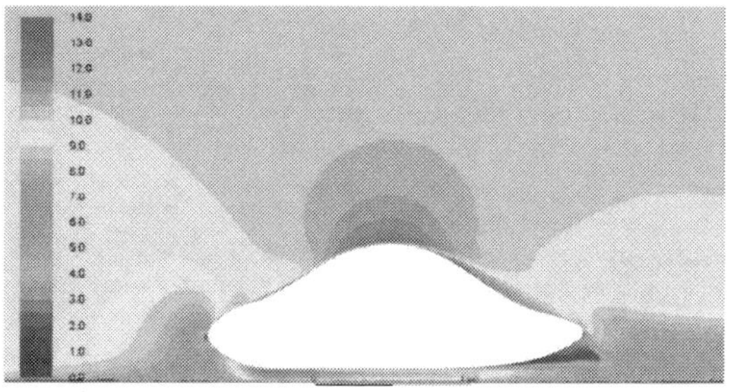

Fig. 1. Field of velocity V in m/s.

At the front end of the car, the flow between the car underside and the ground is accelerated. This acceleration is due to the high position of the front end. The rear end of the vehicle is also high, so the flow separates from underside where the recirculation zone occurs. However, the flow is unnecessarily accelerated at the upper side of the rear end, Fig. 2.

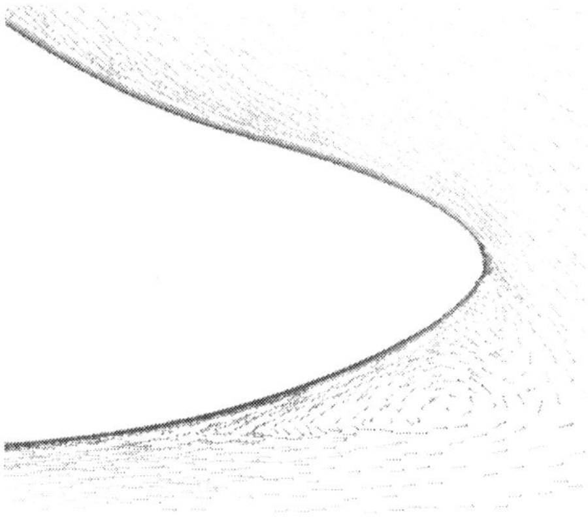

Fig. 2. Velocity vector at the rear end.

 DOI: 10.1051/matecconf/201713302004

The velocity field in the horizontal plane at a distance of 0,3 m from the ground is presented in Fig. 3. At the front end of the vehicle, the flow is unnecessarily accelerated due to the high curvature near the front wheel. On the contrary, the rear end angle of the vehicle is very important and the flow detaches from the lateral sides. The detachment is not steady and its position varies periodically with time similarly to wake behind a bluff body.

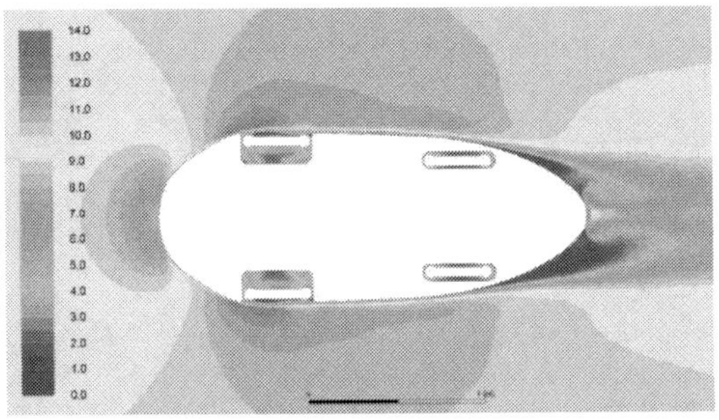

Fig. 3. Velocity vector at the rear end.

The wake behind the vehicle is represented by means of iso-vorticity surfaces, Fig. 4. The vortex wake shows also that the flow is detached from the rear lateral sides. In this image the vehicle surface is coloured by means of static pressure. The lowest pressure which is represented by deep blue colour corresponds to high velocity and permits to visualise the flow acceleration at the front wheels.

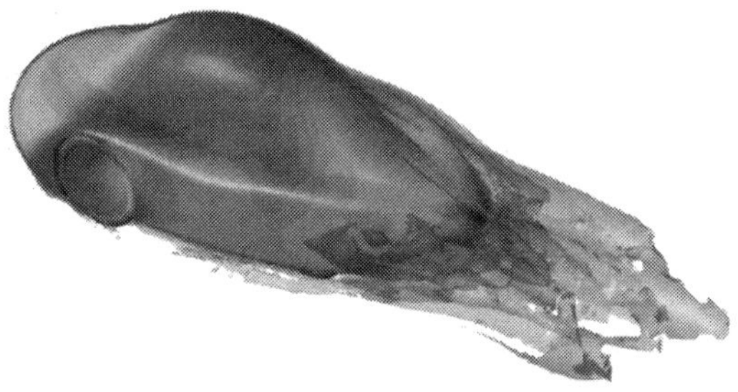

Fig. 4. Vortex wake behind the car.

In this study, for the forces and PIV measurements, we use a vehicle model with a length of 0,19 m, which is quite small compared with a real vehicle. In addition, the ground of the test bench is fixed, therefore the relative movement between the vehicle and the ground cannot be represented. For this reason a second simulation is needed with a Reynolds number corresponding to wind tunnel experiment. To simplify the solution, compared to previous case, only fluid viscosity is changed and the ground and the wheels are fixed. As can be expected the drag coefficient is by 24% bigger and equal to 0,2. The lift coefficient becomes now positive and equal to 0,038. Such a big change can be explained by the decrease of the flow velocity between the vehicle underside and the ground.

3 Experimental study

3.1 Force measurements

The experimental study is conducted in the wind tunnel of Dynfluid Laboratory at ENSAM-Paris. This wind tunnel is a Prandtl type (closed-circuit) and is equipped with a three-blade fan with a rotor diameter of 3 m. The flow is homogenized in a settling chamber by means of honeycomb strengtheners and wire mesh. The nozzle contraction ratio is 12,5, which permits to obtain uniform velocity profile with a turbulence level lower than 0,25%. The test section is semi-guided and the flow is not constrained laterally, which allows for an increase in the dimensions of the tested models. The test section has a cross-section of 1,35 m × 1,65 m and a length of 2 m. Its static pressure is equal to the atmospheric pressure and the upstream velocity depends only on the stagnation pressure in the settling chamber. During the experiments the tunnel velocity is measured by a differential pressure transducer connected with the settling chamber and the atmosphere. The fan is driven by a frequency-controlled asynchronous motor, which permits a change in the wind tunnel velocity from 1,8 m/s up to 40 m/s.

The wind tunnel is equipped with two superposed balances linked to each other. The first balance is a six-component and can measure up to 1500 N with a precision of 0,075 N. The second balance can measure only four components: lift, drag, pitch and roll. This balance has a precision of 0,01 N in the interval of 0 N to 50 N. The strain gauges of the balances are connected to a data acquisition system HBM MGCPlus, which is equipped with high quality strain gauges amplifiers and analog to digital converters. The acquisition system is controlled by means of PC with a specially developed Labview program.

The studied vehicle model with a length of 0,19 m is placed on the second balance by means of rectangular flat plate with dimensions of 0,5 m × 0,2 m. This plate represents the ground. The measurements for upstream velocity of 25 m/s are conducted during a time interval of 60 s with a sampling rating of 10 Hz. For such low forces as in the present case study, the drag coefficient can be determined with a precision of 5%. The averaged drag coefficient is 0,24, which is 20% bigger than the simulation value. Such discrepancy can be justified because the solver has difficulties in taking into account the viscosity effects for a low Reynolds number. Additionally, the boundary layer is completely turbulent without any laminar part as it exists in reality.

3.2 PIV measurements

The velocity around the vehicle model is obtained by means of particle image velocimetry (PIV). This method is non-intrusive and does not perturb the flow. The flow is seeded by special mist generator which permits to visualise the flow. The method of PIV is based on cross-correlation between two successive images which permits to calculate the particle displacement. As this displacement is low, the shutter speed is very high and the camera needs high intensity light which is ensured by the flash of impulse laser. Specially dedicated software ensures the synchronisation between the PIV camera, the frame grabber and the impulse laser: The principles of PIV are explained in details in [10].

In this exploration the PIV system is controlled by DynamicsStudio 2015a of Dantec. The images are taken with a Nd-Yag Litron Nano-L 200-15, with a power impulse of 200 mJ and a Dantec FlowSense 4M camera with a resolution of 2048×2048 px. The camera is equipped with AF Micro-Nikkor 60 mm f/2,8D lenses. The synchronization system is equipped with NI-1428 frame-grabber card to synchronize the images with laser flashes. The flow is seeded with micro-droplets of olive oil generated by a mist generator Dantec 10F03. The average droplet diameter is 2-5 μm.

The wind tunnel velocity during the exploration is 13,7 m/s. To obtain good averaging the totality of 500 double images are taken. A typical raw image is presented in Fig. 5 where it is possible to observe the small olive droplets illuminated by the laser. Additionally, it can be observed that the flow underside is not illuminated therefore the velocity vectors cannot be calculated for this area.

Fig. 5. PIV raw image.

To calculate the PIV velocity in this study we apply Adaptive PIV method, which permits to obtain a spatial resolution of 1,5 mm with good precision. The flow which is obtained by averaging the 300 snapshot is presented in Fig. 6.

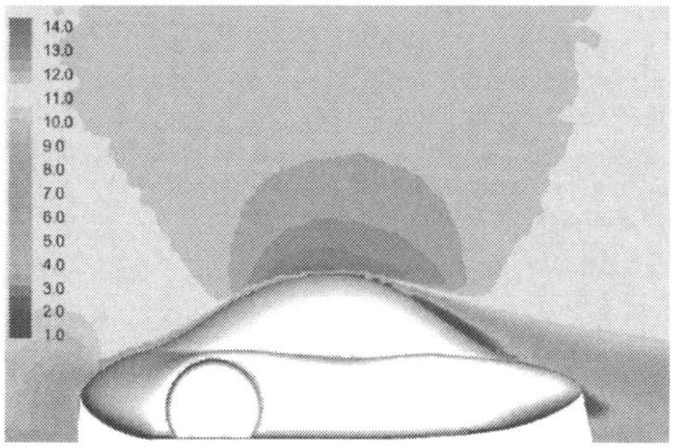

Fig. 6. Averaged velocity field obtained by means of PIV.

Generally, the simulation result presented in Fig. 1 is similar to the experimental velocity field apart from the flow behind the canopy, which is slightly separated.

4 Conclusions

This numerical and experimental study concerns the aerodynamics of a vehicle which participated in Shell Eco-marathon. The objective is to obtain drag and lift coefficients and also the flow field around the car. The presented numerical study makes it possible to obtain the aerodynamic performance of the real vehicle and also of the small vehicle model, which is used for wind tunnel exploration. The lift coefficient varies moderately due to the influence of Reynolds number while the lift exhibits strong variation due to the additional influence of ground fixing.

Finally, the small vehicle model is tested in the wind tunnel for obtaining the drag and the flow field in the plane of symmetry. In general, the results of the comparisons between the experiments and the simulation are satisfactory. The numerical simulations reveal the recirculation zones which should be avoided. The study shows that CFD modeling will be useful for future design of the vehicle.

References

1. W. Hucho, *Aerodynamics of road vehicles: from fluid mechanics to vehicle engineering* (Elsevier, 2013)
2. J. Katz, *Race car aerodynamics: designing for speed* (R. Bentley 2006)
3. S. McBeath, *Competition Car Aerodynamics: A Practical Handbook* (Veloce Publishing House 2015)
4. http://www.shell.com/energy-and-innovation/shell-ecomarathon.html
5. G. Galmarini, S. Dell'Agostino, M. Gobbi, G. Mastinu, Solar Prototype for Shell-Eco Marathon Race (No. 2017-01-1260) SAE Technical Paper (2017)
6. Abo-Serie, E. Oran, O. Utcu, Aerodynamics consideration for low drag Shell Eco-Marathon competition car, *ICAME-2016*, Istanbul, Turkey (2016)
7. A. Cieslinski, W. Prym, M. Stajuda, D. Witkowski, Investigation on Aerodynamics of Super–Effective Car for Drag Reduction Mechanics and Mechanical Eng., **20**(3), 295-308, (2016)
8. D. Cogan, The aerodynamic design and development of an urban concept vehicle through CFD analysis, Diss, Cape Peninsula Univ. of Tech. (2016)
9. P. Spalart, Detached-eddy simulation, Ann. Rev. of fluid mech., **41**, 181-202, (2009).
10. M. Raffel, C. Willert, S. Wereley, J. Kompenhans, *Particle image velocimetry: a practical guide*, Springer. (2013)

Improving ATV's stability by means of a gyroscopic mechanism

Iliyana Minkovska[1,*], *Georgi* Kadikyanov[1], *Rosen* Ivanov[1], and *Rusi* Rusev[1]

[1]University of Ruse, Department of Engines and Vehicles, Rousse, Bulgaria

Abstract. This paper presents research carried out to improve ATV stability by means of a built-in gyroscopic mechanism. For this purpose, a mathematical model was used. The negative gyroscopic effect of wheel rotation on ATV stability is considered. The theoretical results for ATV motion and the impact of the built-in gyroscopic mechanism on its stability under various conditions are presented.

1 Introduction

The growing interest in ATV driving increases the risks of accidents. Most frequently, the reason why they occur is stability loss when ATV moves in curves [1, 2]. Loss of stability has two aspects: lateral sliding and lateral overturn. Depending on the different performance (grip coefficient, radius of curve, center of gravity coordinates in ATV riding by different riders, etc.) when ATV moves in curves, one of the above possibilities to lose stability may arise. It is essential which one will come first. The stability loss for lateral overturn presents a greater risk to the life and health of the driver (Fig. 1).

Fig. 1. Stability loss at lateral overturn with a risk for the life and health of the driver.

To improve the stability at overturn, a built-in gyroscopic mechanism in the ATV structure is useful [3]. It is a metal solid disk with high angular rotation speed. The gyroscopic mechanism axis is parallel to the transverse horizontal of vehicle. In this way the gyroscopic moment is used when ATV makes turns [4]. This feature is may be negative to the ATV controllability. On the other hand, the appropriate axis allocation and proper rotation direction may be used to improve the stability of movement in curves.

The main difficulty in using such devices comes from the danger to break the rotor because of the forces occurring at the high angular rotation speed. To prevent the consequences of device breaking (for monolithic rotor), it should be placed in a hard-to-break container.

The aim of this study is to determine the possibility to improve the stability of ATV in curve movement by means of a built-in gyroscopic mechanism.

2 Methodology of mathematical calculations

Due to the kinematics of wheel vehicles, a gyroscopic moment arises when wheels rotate, which negatively influences controllability and stability. To calculate the gyroscopic moment of the four wheels equation (1) is used:

$$M_{Gk} = 4\, J_k\, \omega_k\, \Omega,\ \text{Nm}, \tag{1}$$

where

J_k, kg.m^2 is the wheels mass moment of inertia;

ω_k, rad/s – angular rotation speed of wheels;

Ω, rad/s – angular rotation speed of ATV around the center of curve.

The wheels mass moment of inertia is determined by the standard methods through an experimental device shown in Fig. 2.

According to [5] under the different operation modes the critical velocity, whichever is lower, against lateral sliding and lateral overturn is taken for calculating Ω, rad/s – angular rotation speed of ATV around the center of curve.

When determining the critical velocities in [5] a two-mass model is used for selecting the lower of velocities on the front or rear drive axles. Only cases for which the grip coefficient is 0,75 are considered.

* Corresponding author: iminkovska@uni-ruse.bg

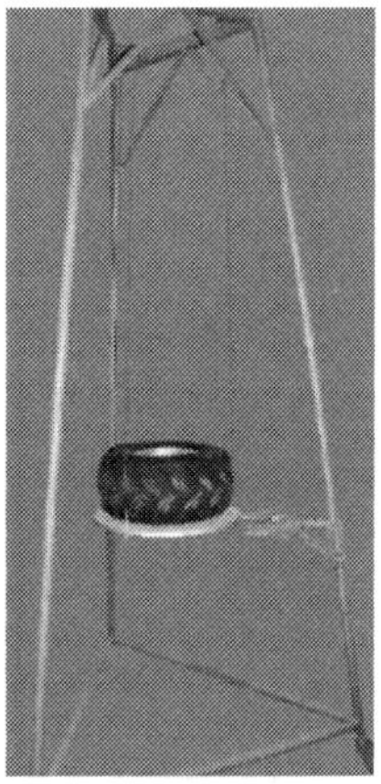

Fig. 2. Experimental device to determine the mass moment of inertia of the wheels.

In cases where the riders who are with higher BMI (body mass index [6]) the critical velocities against overturn are lower than the critical velocities against lateral sliding.

Figure 3 shows a schematic diagram of the forces acting on the ATV with no built-in gyroscopic mechanism when it moves in a curve.

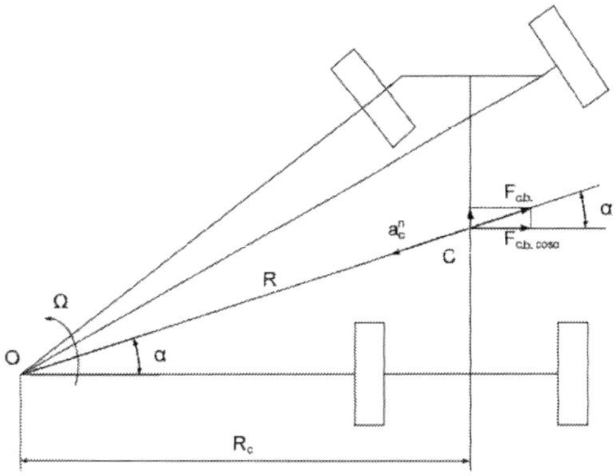

Fig. 3. Schematic diagram of the forces acting on the ATV with no built-in gyroscopic mechanism when it moves in a curve.

The angular rotation speed of ATV around the center of curve according to Fig. 3 is

$$\Omega = \frac{V_a}{R} \text{ rad/s}. \quad (2)$$

In theoretical calculations of the gyroscopic mechanism for the purposes of design and layout, the following values are assumed: disk diameter $d = 300$ mm and disk width $b = 30$ mm. For steel disk with permissible tension of $\sigma_T = 400$ MPa, density $\rho = 7850$ kg/m^3, radius $r = 0{,}15$ m, the maximum permissible rotation speed will be

$$\omega = \sqrt{\frac{\sigma_T}{\rho\, r^2}}, \text{ rad/s}; \quad (3)$$

$$n = \frac{30\,\omega}{\pi}, \text{ min}^{-1}. \quad (4)$$

The gyroscopic mechanism is positioned in the middle of the rear part of vehicle. Its minor influence on the change of center of gravity coordinates of ATV is ignored.

The gyroscopic stabilizing moment will be

$$M_{Gs} = J_d\,\omega_d\,\Omega, \text{ Nm}, \quad (5)$$

where
J_d, kg.m^2 is the disk mass moment of inertia;
ω_d, rad/s – disk angular rotation speed.

The centripetal acceleration is

$$a_c^n = R\,\Omega^2, \text{ m/s}^2. \quad (6)$$

The centrifugal force corresponding to the critical velocities V_a from [5] without embedded gyroscope is

$$F_{c.b.} = a_c^n\, m_A, \text{ N}, \quad (7)$$

where
m_A, kg is the total mass: of ATV mass and rider.

Then, under these conditions for the stabilizing moment the following is obtained

$$M_s = F_{c.b.}\, h_A \cos\alpha, \text{Nm}, \quad (8)$$

where
h_A, m is the vertical coordinate of the center of gravity;
α,° – the angle between the centrifugal force directrix and the axis of rear drive axle.

The total stabilizing moment regarding the gyroscopic effects is

$$M_o = M_{Gs} + M_s - M_{Gk}, \text{Nm}. \quad (9)$$

The centrifugal force corresponding to this moment is

$$F_{c.b.o} = \frac{M_o}{h_A \cos\alpha}, \text{Nm}. \quad (10)$$

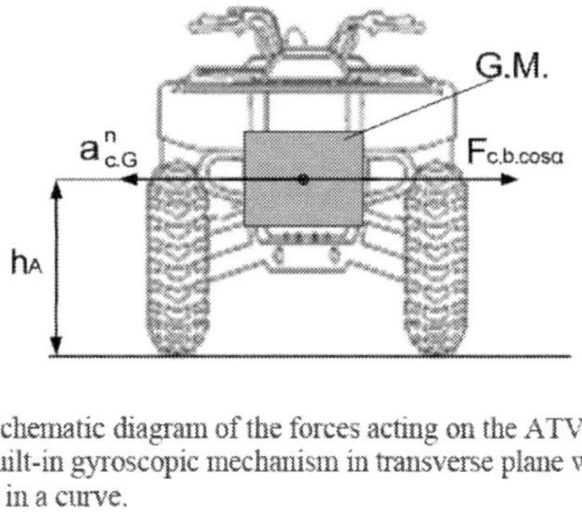

Fig. 4. Schematic diagram of the forces acting on the ATV with a built-in gyroscopic mechanism in transverse plane when it moves in a curve.

According to Fig. 4 the centripetal acceleration under the action of the total stabilizing moment is

$$a_{cG}^{n} = \frac{F_{cbo}}{m_A}, \text{m/s}^2. \quad (11)$$

Angular rotation speed of ATV around the center of the curve with centripetal acceleration according to (6) and (11) is

$$\Omega_o = \sqrt{\frac{a_{cG}^{n}}{R}}, \text{rad/s}. \quad (12)$$

Then the critical velocity against overturn taking into account the gyroscopic moments: the effect of wheels rotation and the built-in gyroscopic mechanism is

$$V_{A,G} = \Omega_o R, \text{m/s}. \quad (13)$$

The critical velocities obtained from equation (13) are compared to the critical velocities obtained in [5] for 0,75 value of the grip coefficient.

3 Results of the study

The critical velocities against lateral sliding and lateral overturn when taking into account the effect of the gyroscopic mechanism built in ATV for riders with different body mass index (BMI) is shown in Figures 5 to 9.

It can be seen from Figures 5 to 9 that the critical velocities against lateral sliding for radiuses of curves bigger than 10 m and 15 m do not depend on the BMI of different riders. This is so because these velocities are not affected by the mass of the vehicle or rider. For smaller 3 m and 6 m radiuses of curves, the critical velocities against lateral sliding differ for some of the riders as the loss of stability occurs at either rear or front axles.

The critical velocities against lateral overturn for riders with higher BMI (more than 30 for any radius of curves, and more than 25 for radius of curve 3 m) are lower than those against lateral sliding. This is very dangerous as the loss of stability with lateral overturn occurs all of the sudden. Particularly for riders with higher BMI it occurs because of the higher vertical coordinate of the center of gravity.

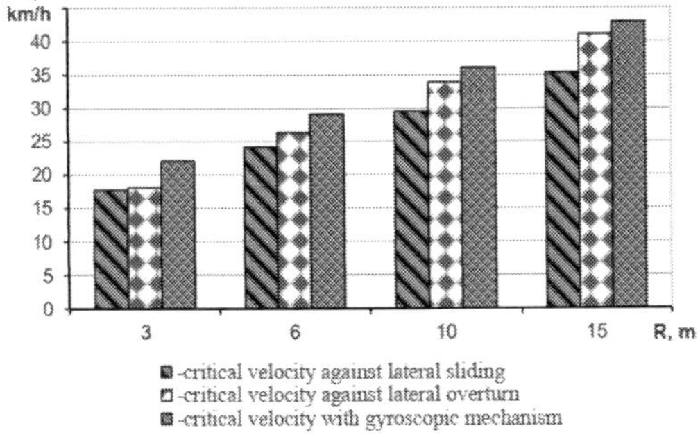

Fig. 5. Critical ATV driving velocities with a rider of BMI 21.

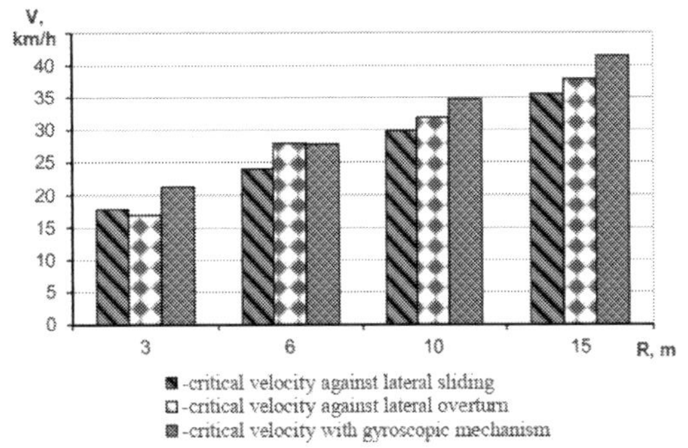

Fig. 6. Critical ATV driving velocities with a rider of BMI 25.

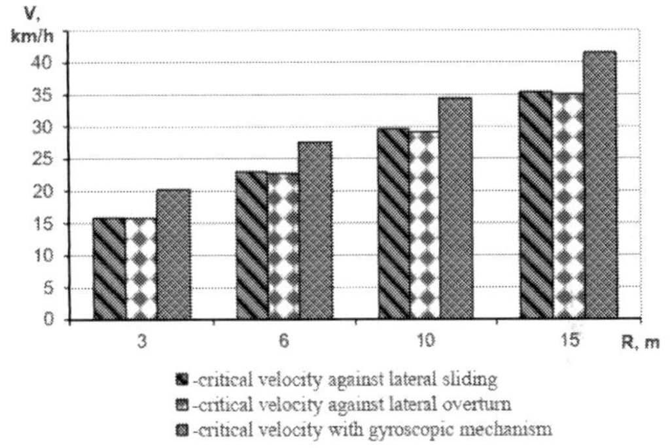

Fig. 7. Critical ATV driving velocities with a rider of BMI 30.

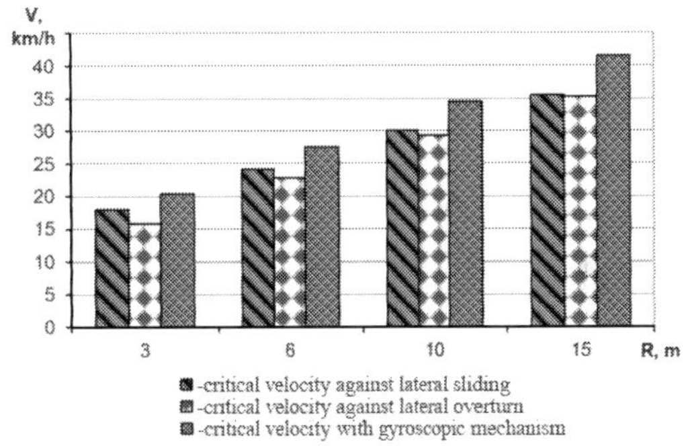

Fig. 8. Critical ATV driving velocities with a rider of BMI 31.

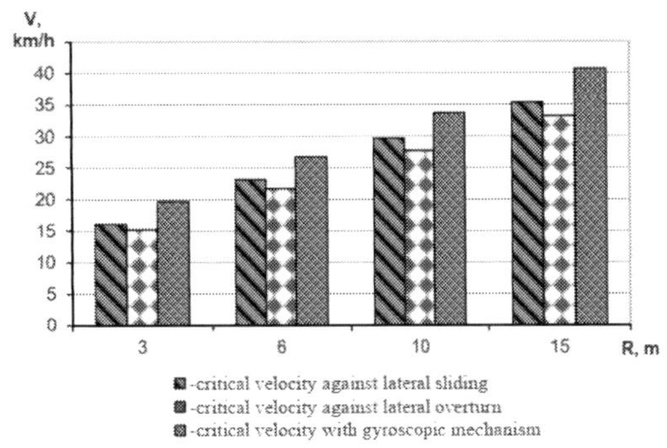

Fig. 9. Critical ATV driving velocities with a rider of BMI 40.

It can be seen from the graphics that the critical velocities against overturn increase with a buildt-in gyroscopic mechanism in ATV which prevents its occurrence. In such cases the critical velocity against lateral sliding remains lower. Then, for all riders, at first the lateral sliding will occur. This phenomenon is not so risky for the riders' health and life.

4 Conclusions

1. For riders with lower BMI of up to 25, the gyroscopic mechanism has no influence on the ATV stability. In such cases lateral sliding occurs before reaching lateral overturn velocities.

2. For riders with higher BMI, embedding a gyroscopic mechanism increases the critical velocity against lateral overturn by up to 5% for a rider with BMI 40.

3. Increasing the critical velocity against lateral overturn by embedding a gyroscopic mechanism in the cases considered guarantees that the less dangerous phenomenon of lateral sliding will appear at first.

The study was supported by contract BG05M2OP001-2.009-0011-C01 of University of Ruse "Angel Kanchev", "Support for the development of human resources for research and innovation at the University of Ruse "Angel Kanchev". The project is funded with support from the Operational Program "Science and Education for Smart Growth 2014 - 2020" financed by the European Social Fund of the European Union.

References

1. D. Blower, J. Woodrooffe, P. Green, A. Matteson, M. Shrank, Determination of events leading to sport utility vehicle rollovers. *Transport.* Res. Record 1908. 180–186, (2005).
2. S. Tanner, M. Aitken, J. Warnock, Analysis of All Terrain Vehicle Crash Mechanisms. Journal of undergraduate research bioengineering, 53-58 (2010)
3. R. Sharp, D. Limebeer, A motorcycle model for stability and control analysis, *Multibody System Dynamics*, **6**(2), 123–142, (2001)
4. T. Sung, S. Han, Y. Han, J. Lee, N. Jeong, S. Hwang, S. Choi, Designs and analyses of flywheel energy storage systems using high-Tc superconductor bearings. *Cryogenics,* **42**, 357–362, (2002)
5. G. Kadikyanov, I. Minkovska, G. Staneva, A Study of the Influence of Some Exploitation Factors on the ATV Stability, *Transport Problems*, Katowice, 262-269, (2017)
6. P. Deurenberg, M. Yap, W. VanStaveren, Body mass index and percent body fat: a meta analysis among different ethnic groups. Int. J. of Obesity, 1164-1171, (1998)

Influence of shock absorber temperature on vehicle ride comfort and road holding

Nikolay Pavlov[1,*]

[1]Technical University – Sofia, Faculty of Transport, 8 Kliment Ohridski Blvd., 1000 Sofia, Bulgaria

Abstract. In this work shock absorber damping characteristics at different temperature conditions are determined. The variation of damping properties of the shock absorber depending on the temperature is given. An assessment of ride comfort and road holding characteristics for different temperatures is made.

1 Introduction

Vehicle dynamic behaviour, ride comfort and tire-road contact are important parts of the design, construction and testing of a vehicle. Research in these areas of automotive engineering has been carried out by many authors in recent years. Some of the most comprehensive books and monographs are [1-4], and the main results and models described by their authors have been taken into account and used in this research work. Shock absorbers heat load and oil temperature are studied in the literature [5-7]. There are no studies how shock absorber temperatures directly influence the vehicle ride comfort. The conditions under which the vehicles are operated depend mainly on four factors: vehicle load, road irregularities character, vehicle speed and ambient temperature [8]. Oscillations of varying frequency and amplitude occur under these various conditions. For most cars, manufacturers set up non-adjustable shock absorbers for good roads, average load (70-80 % of max.), normal speed (moderate) and ambient temperature of 20 °C (corresponding to standard atmospheric conditions). With this setting, shock absorbers effectively dissipate the vibrations and provide the smallest vertical accelerations of the passenger traveling in vehicle. But in reality the conditions are varied and in practice many combinations are possible which do not meet those assumed by the car maker.

Oil viscosity in the dampers changes when temperature changes. Relatively small temperature changes cause large changes in viscosity. This property of oils can be graphically expressed by the viscosity-temperature characteristic. Fig. 1 illustrates how viscosity varies dramatically with temperature over the required operating temperature range [9].

From the viscosity-temperature characteristic it can be seen that when the temperature drops to negative values, the viscosity of the oil increases sharply. At very low winter temperatures, the increased viscosity can cause damage and complete failure of the shock absorber and other chassis components. When big bumps have to be overcome, damage can occur even when the car sets off.

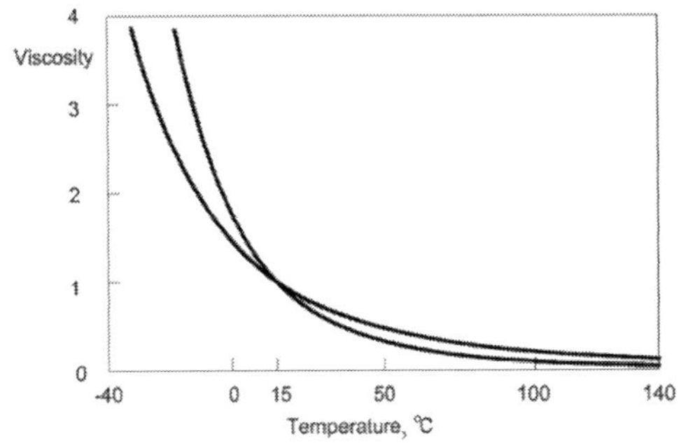

Fig. 1. Mineral oil viscosity relative to its room-temperature viscosity for two types of oil (temperature viscosity sensitivity coefficient $C = 2000$ and 3000 K) [9].

After hard damping work of the shock absorbers, resistive forces gradually decrease due to oil heating and reduced viscosity, but reaching optimum operating temperature is either too slow or not possible at all and depends mostly on ambient temperature, type of road roughness and speed of the car. At high ambient temperatures and intense damping, the viscosity of the oil greatly decreases, which reduces the effectiveness of the shock absorber as a suspension dissipative element.

The purpose of this work is to study the impact of the shock absorber working temperature on the vehicle ride comfort and tire-road contact.

2 Factors influencing the working temperature of shock absorbers

The minimum working temperature of the shock absorber is mainly affected by the air temperature. This temperature depends on the climate specificities of the latitudes in which the vehicles are operated and changes with the seasons. For example, in northern countries such as Canada, Sweden and Norway are characterized by very low temperatures in winter and from low to medium temperatures in the other seasons.

* Corresponding author: npavlov@tu-sofia.bg

For countries located in the equatorial part of the globe and the Mediterranean countries, high temperatures are typical throughout the year. In these two cases, account must be taken of the climate and the dampers are loaded with lower viscosity oils, which are for cold countries and with higher viscosity oils for the warmer countries. More serious is the problem with cars operated in continental climate (countries such as Bulgaria, Serbia, Hungary, etc.), which is characterized by high temperatures in summer and low temperatures in winter, i.e. a large range of annual temperatures (minimum / maximum). The minimum and maximum annual air temperatures in 2016 for the capital of Bulgaria - Sofia are shown in Fig. 2.

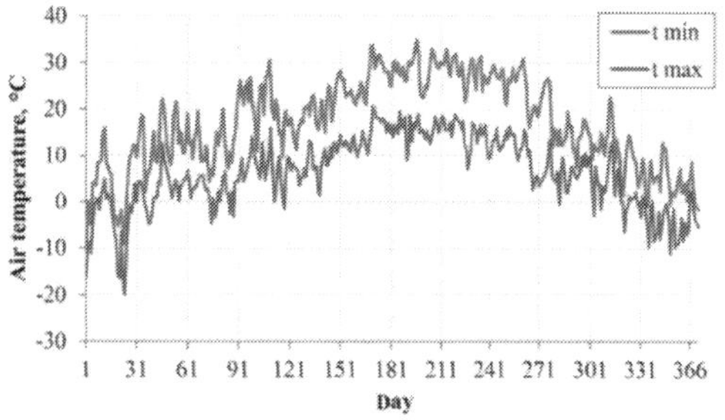

Fig. 2. Air temperature (°C) in 2016 in Sofia [16].

The graph shows that in the coldest month - January (2016) - the minimum temperatures reached –20°C. During the warmest month for 2016 - July, the maximum temperatures reached +35°C.

At maximum operating temperature, there are other factors that have an impact besides the environment. Depending on the type of road and the characteristics of its unevenness, the dampers work more or less intensively, as this influences their temperature rise. Due to the feature of their being mounted in the vehicle's suspension, they can be blown by the opposite wind more or less. Also, heating depends on the type of shock absorbers used - single-tube or double-tube. Single-tube gas-filled shock absorbers are less heated as the walls of the working cylinders are in direct contact with the air and are cooled better. Although modern hydraulic shock absorbers use hydraulic additives to improve the viscosity-temperature characteristics and synthetic oils developed for lower temperature sensitivity than mineral or vegetable oils, the ideal zero viscous variation cannot be achieved. There are also special designs of valves for temperature compensation, but in many modern shock absorbers they are not used [9].

3 Shock absorber temperature characteristics

The temperature characteristic is the dependence of the viscous damping force on the temperature. The temperature range in which the dampers must work is –40 to +130°C [9]. According to [10] the temperature characteristic is recorded for temperatures from –30 to +100°C, and according to [11] for temperatures from –50 to +100°C. For cars operating on the territory of a continental climate, it can be assumed that the minimum temperatures shown in Fig. 2 must be taken into account and the minimum temperature for testing is taken to be –10°C.

For the purpose of the present study shock absorbers from two different manufacturers were selected. They are designed to be mounted in the rear suspension of a medium-sized car.

To obtain the temperature characteristics of the shock absorbers we used specialized electro-mechanical shock tester model Shock Dyno - Hi Speed, make from Intercomp® (Fig. 3). The procedure is similar to the one described in [13, 14], but the difference is that the temperature varies.

Fig. 3. Shock absorber tester.

The force-displacement characteristics of the shock absorbers obtained during the test are shown in Figures 4 and 5.

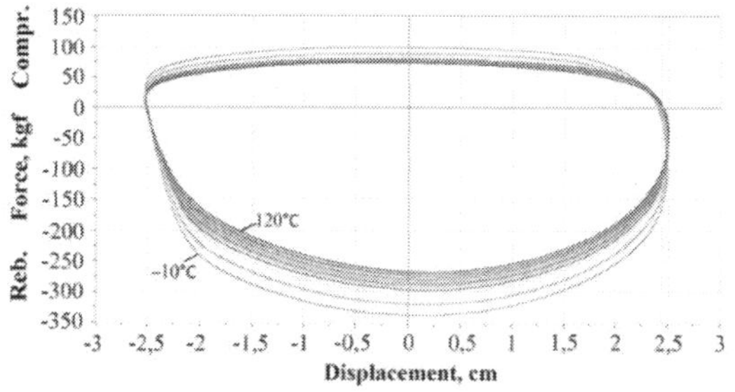

Fig. 4. Force-displacement characteristics of the shock absorber No.1.

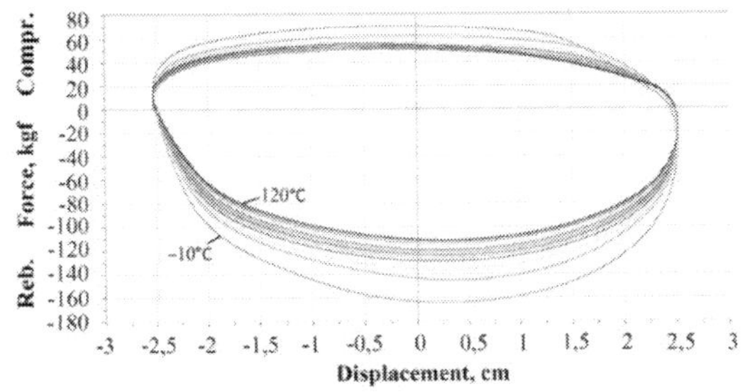

Fig. 5. Force-displacement characteristics of the shock absorber No.2.

From the force-displacement characteristics at different temperature values, the temperature characteristics of the shock absorbers in the range from –10 to +120°C are built, as shown in Figures 6 and 7.

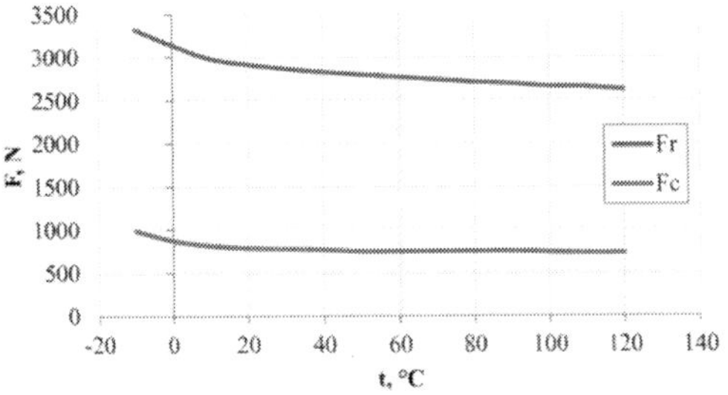

Fig. 6. Temperature characteristic of shock absorber No.1.

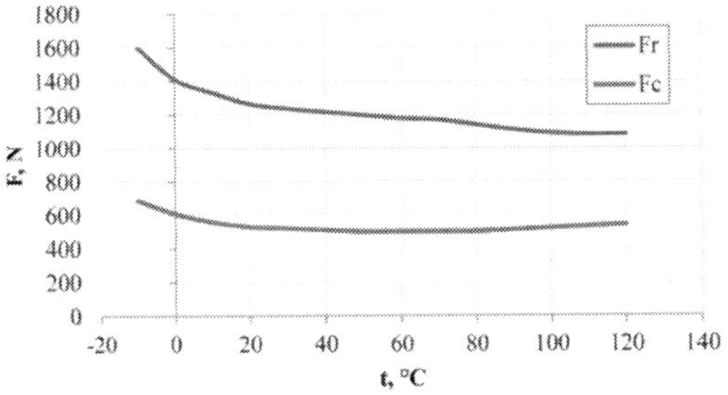

Fig. 7. Temperature characteristic of shock absorber No.2.

The calculation damping coefficient of shock absorber β, is obtained as the average of the resistive force of the shock absorber F_{av}, divided by the velocity of the piston that should be 0,52 m/s for the initial design calculations [12]:

$$\beta = \frac{F_{av}}{v};\tag{1}$$

$$F_{av} = \frac{F_c + F_r}{2},\tag{2}$$

where F_c and F_r are respectively the resistance forces of the shock absorber during compression and rebound, N.

The asymmetry ratio is:

$$k_a = \frac{F_r}{F_c}.\tag{3}$$

The calculated parameters of the tested shock absorbers for different temperatures are shown in Tables 1 and 2.

Table 1. Results for shock absorber No.1.

t, °C	F_{av}, N	β, (N.s)/m	k_a,-
-10	2148,39	4131,52	3,38
0	2006,15	3857,97	3,60
10	1898,24	3650,45	3,66
20	1849,19	3556,13	3,71
30	1819,76	3499,53	3,70
40	1795,23	3452,37	3,69
50	1770,71	3405,20	3,75
60	1755,99	3376,90	3,71
70	1741,28	3348,61	3,67
80	1726,56	3320,31	3,63
90	1716,75	3301,44	3,61
100	1697,13	3263,71	3,61
110	1687,32	3244,85	3,65
120	1672,61	3216,55	3,61

Table 2. Results for shock absorber No.2.

t, °C	F_{av}, N	β, (N.s)/m	k_a,-
-10	1142,87	2197,82	2,33
0	1010,43	1943,13	2,32
10	946,67	1820,51	2,39
20	897,62	1726,18	2,39
30	878,00	1688,45	2,38
40	863,28	1660,15	2,38
50	848,57	1631,86	2,39
60	838,76	1612,99	2,35
70	833,85	1603,56	2,33
80	819,14	1575,26	2,27
90	809,33	1556,39	2,17
100	804,42	1546,96	2,09
110	804,42	1546,96	2,04
120	809,33	1556,39	2,00

4 A numerical study of ride comfort and road holding for different shock absorber temperatures

1DOF and 2DOF models of the quarter car are used to determine the comfort and tire-road contact at different shock absorber operating temperatures. The simplest model is the one with 1 degree of freedom (Fig. 8, a). It takes into account the unsprung mass m acting on one wheel (1/4 of the car mass).

More complex is a dual mass model with two degrees of freedom (Fig. 8, b). It consists of two bodies – sprung mass m and unsprung mass m_t and two elastic elements c and c_t, representing respectively the elasticity of the spring and tire elasticity. Dissipative elements are β and β_t and represent the shock absorber and tire dissipation. Through this model we receive results for the characteristics of the sprung and unsprung masses.

The basis of the method is Lagrange equation of the second order:

$$\frac{d}{dt}\left(\frac{\partial T}{\partial \dot{q}_i}\right) - \left(\frac{\partial T}{\partial q_i}\right) + \left(\frac{\partial U}{\partial q_i}\right) + \left(\frac{\partial R}{\partial \dot{q}_i}\right) = Q,\tag{4}$$

where T and U are the kinetic and potential energy of the system; R - Rayleigh dissipation function; q_i, $\dot{q}_i$ - vector of generalized coordinates and velocities; Q - kinematic disturbance; t - time.

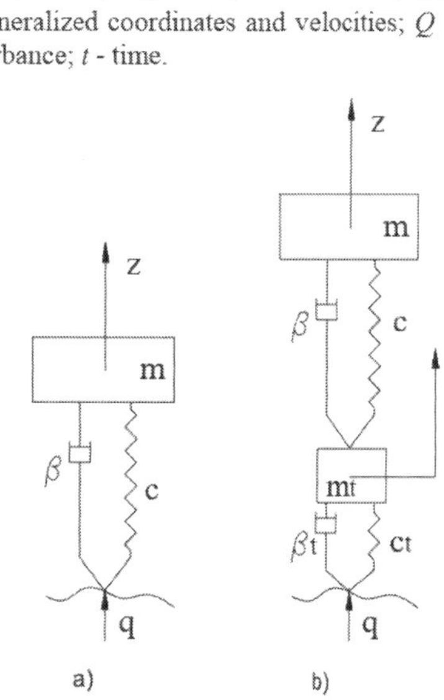

Fig. 8. 1DOF (a) and 2DOF (b) quarter car model.

The numerical simulations were carried out in MATLAB with sprung mass m = 350 kg, unsprung mass m_t = 30 kg, spring ratio of the spring is c = 20 kN/m, tire spring ratio c_t = 150 kN/m, shock absorbers damping parameters as given in Tables 1 and 2, tire damping β_t = 50 N.s/m.

The results from the numerical experiments for 1DOF dynamic model are shown in Figures 9 to 12, and for 2DOF dynamic model are shown in Figures 13 to 20.

The free oscillation step response of shock absorber No.1 and No.2 are shown in Figures 9 and 10.

The forced vibrations for the 1DOF model are shown in Figures 11 and 12.

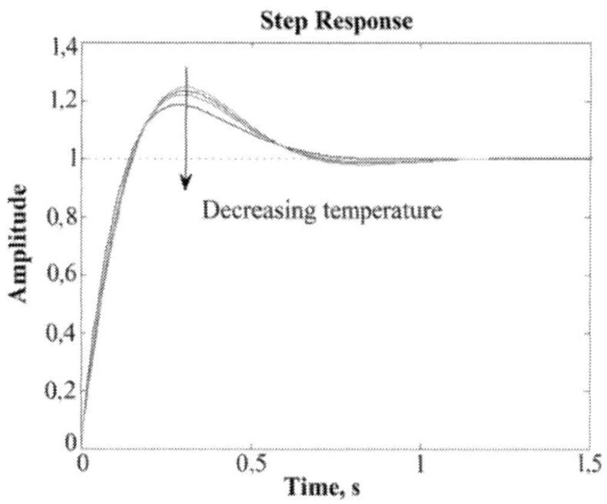

Fig. 9. Oscillogram of 1DOF model damping oscillations at different temperatures of shock absorber No.1.

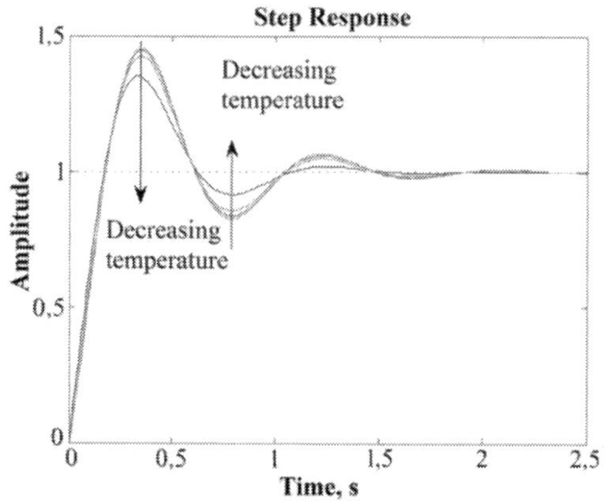

Fig. 10. Oscillogram of 1DOF model damping oscillations for different temperatures of shock absorber No.2.

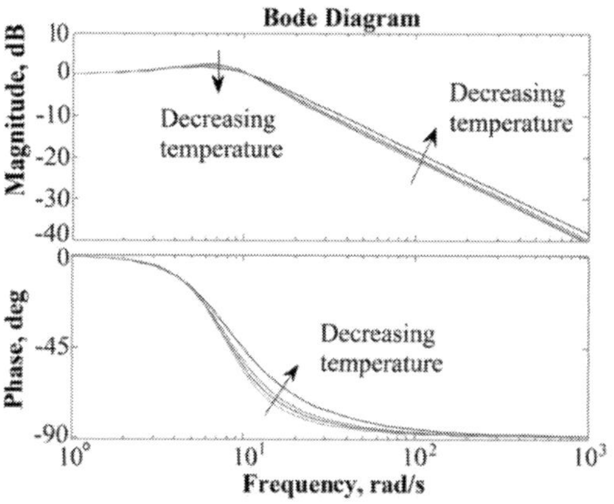

Fig. 11. Amplitude-frequency and phase-frequency response of 1DOF model at different temperatures of shock absorber No.1.

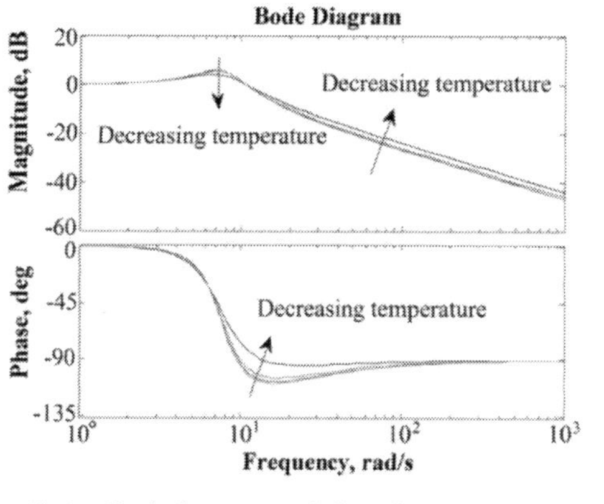

Fig. 12. Amplitude-frequency and phase-frequency response of 1DOF model at different temperatures of shock absorber No.2.

The results from the 2DOF model for random vibrations acceleration spectral density (ASD) are given in Figures 13 to 16. Excitation characteristics are typical of pavement in good condition.

A decrease in the temperature increases the shock absorber damping coefficient. As a result there is an

DOI: 10.1051/matecconf/201713302006

increase in the accelerations of the sprung mass in the resonant and the post-resonant region for the shock absorber No.1 (Fig. 13). The rise of spectral density from 20°C to –10°C is about 20%, and the difference between spectral densities in a temperature range of –10 to 120°C is about 40%. A reduction in the damping coefficient in high temperatures improves the ride comfort in post-resonance band for both shock absorbers.

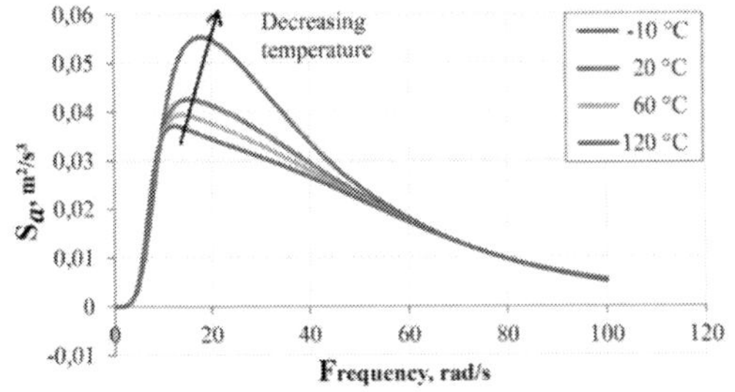

Fig. 13. Spectral density of acceleration of the sprung mass at different temperatures of shock absorber No. 1.

There is an increase of the spectral density in the post-resonant region for shock absorber No. 2, but reduction of acceleration in the resonant region (Fig. 14).

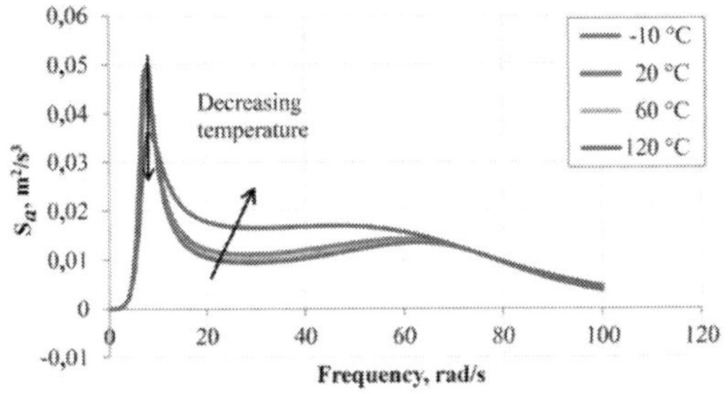

Fig. 14. Spectral density of acceleration of the sprung mass at different temperatures of shock absorber No. 2.

By accelerating the unsprung masses (Figures 15 and 16) we can notice a significant reduction (of over 40% from 120°C to –10°C) in the high rates of damping (low temperatures) to a given frequency range. The most obvious thing is the reduction of the accelerations at the resonant frequency of these masses for both shock absorbers. Small accelerations testify to small intensity of vibration of the wheels. That means better contact between tire and road.

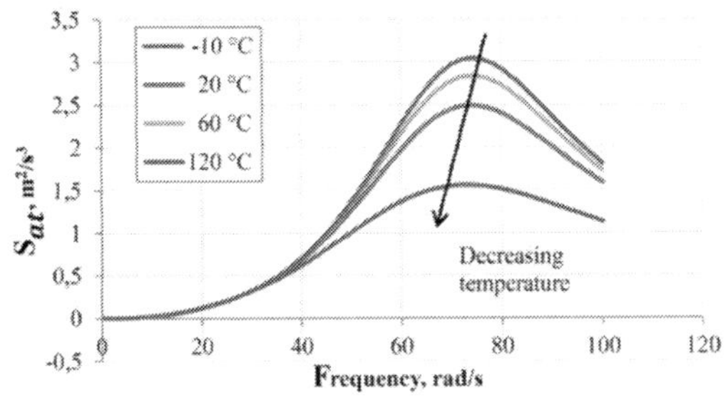

Fig. 15. Spectral density of acceleration of the unsprung mass at different temperatures of shock absorber No. 1.

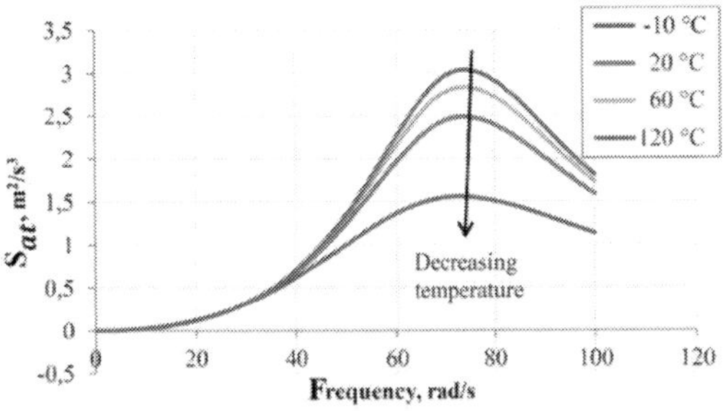

Fig. 16. Spectral density of acceleration of the unsprung mass at different temperatures of shock absorber No. 2.

This analysis demonstrated conflicting demands on damping and the need to reduce coefficient β for maximum comfort and increase it when we need maximum contact between tire and road.

As can be seen from the graph in Figures 17 and 18, there is a fall by about 10 % in the root mean square (RMS) of vertical acceleration of the sprung masses for both shock absorbers and a rise of unsprung mass RMS by about 25 % for shock absorber No. 1 and by 30 % for shock absorber No. 2.

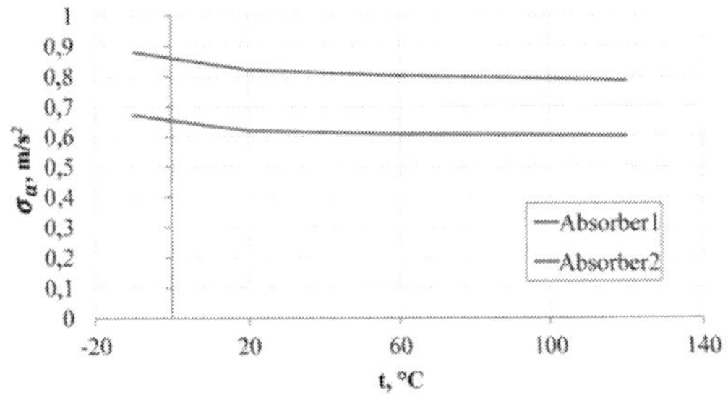

Fig. 17. RMS value of the vertical acceleration σ_a of the sprung mass of shock absorbers No. 1 and No. 2.

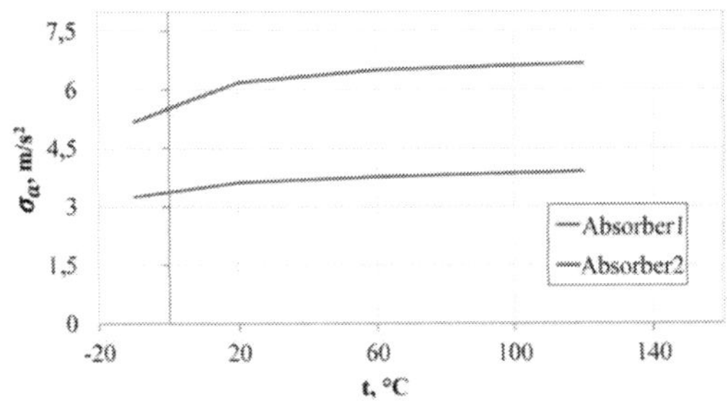

Fig. 18. RMS value of the vertical acceleration σ_a of the unsprung mass of shock absorbers No. 1 and No. 2.

An example of the assessment and comparison of vertical acceleration of sprung mass for different shock absorber temperatures, according to the ISO standard for whole-body vibration, is shown in Fig. 19. For the simulation was used sinus excitation with an amplitude equal to 1,1 cm, corresponding to asphalt with good quality, and frequency from 1 to 100 Hz. In the same frequency range was studied the vibration of the unsprung mass with damping parameters for shock absorbers No.1 and No.2 (Fig. 20). At higher amplitude of irregularities, more noticeable differences in RMS values can be achieved for the different temperature conditions.

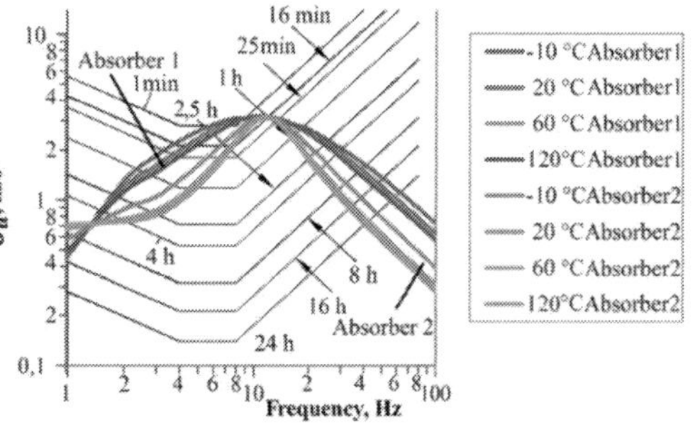

Fig. 19. RMS value of the vertical acceleration σ_a of the sprung mass of shock absorbers No.1 and No.2 according to ISO 2631-1 (1985) [15].

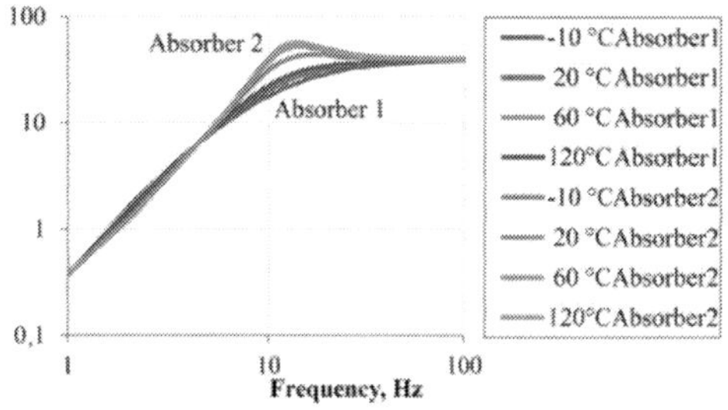

Fig. 20. RMS value of the vertical acceleration σ_a of the unsprung mass of shock absorbers No.1 and No.2.

5 Conclusion

The present study examines the impact of shock absorber temperature on vehicle ride comfort and road holding. For this purpose, shock absorbers temperature characteristics were obtained. The results were used for numerical simulations with 1DOF and 2DOF quarter car models. On the basis of the simulation, the conclusion can be drawn that a decrease in temperature of up to –10°C does influence the ride comfort of vehicles, expressed in growth in spectral density of vertical acceleration (ASD) in the resonant and the post-resonant region with 20% in comparison with the acceleration at 20°C. Similarly 20 % fall is observed between ASD at 120°C in comparison with the ASD at 20°C. For ASD of the unsprung masses there is over 40% reduction in values at –10°C in relation to values at 120°C. There is 25-30% rise in RMS of vertical acceleration of the unsprung masses for temperature interval between –10°C and 120°C. As well 10% fall for RMS of the sprung masses in the same temperature interval. Assessment of vertical acceleration RMS value of sprung mass according to the ISO standard tolerance curves at different temperatures is made.

References

1. G. Genta, L. Morello, *The automotive chassis. Vol. 2: System design*, Springer, (2009)
2. G. Genta, *Vibration dynamics and control*, Springer, (2009)
3. M. Mitschke, H. Wallentowitz, *Dynamik der Kraftfahrzeuge*, Springer, (2014)
4. M. Demić, *Optimizacija oscilatornih sistema motornih vozila*, Mašinski Fakultet, Kragujevac, (1997)
5. M. Demić, D. Diligenski, Numerical simulation of shock absorbers heat load for semi-active vehicle suspension system. Thermal Science, **20(5),** 1725-1739, (2016)
6. M. Demić, D. Diligenski, M. Milovanović, A contribution to research of the influence of degradation of vehicle vibration parameters on thermal load of shock absorbers. Journal of Applied Engineering Science, **11(1),** 23-30, (2013)
7. K. Reif, K. Schmidt, F. Schimmack, F. Niedermeier, N. Kennes, Estimation of the oil temperature in adjustable vibration dampers, ATZ Worldwide, **112(3),** 22-27, (2010)
8. K. Kosev, *Hydraulic shock-absorbers for motorcar and motorcycles*. Technica Publishing, Sofia, in Bulgarian, (1976)
9. J. Dixon, *The shock absorber handbook*. Wiley,(2007)
10. *Telescopic Hydraulic shock absorbers for automobiles - bench test methods*, Standardization Publishing, Sofia, in Bulgarian, (1983)
11. GOST R 53816-2010, Vehicles. Hydraulic telescopic shock absorbers. Technical requirements and test methods, Standardinform, Moskow, in Russian, (2010)
12. B. Heißing, *Chassis handbook: fundamentals, driving dynamics, components, mechatronics, perspectives*, Vieweg+Teubner Verlag, (2011)
13. N. Pavlov, A method and test equipment for obtaining characteristics of controlled hydraulic shock absorbers, *EKO Varna*, 41-47, in Bulgarian, (2015)
14. N. Pavlov, E. Sokolov. Vibration characteristics of quarter car semi-active suspension model - numerical simulations and indoor testing, Trans Motauto World, **2(1),** 11-16, (2017)
15. ISO 2631-1. Evaluation of human exposure to whole-body vibration – part 1: general requirements. Geneva: International Organization for Standardization, (1985)
16. http://www.stringmeteo.com/ (10 August 2017)

MATEC Web of Conferences **133**, 02007 (2017)
BulTrans-2017
DOI: 10.1051/matecconf/201713302007

Study of the wheel loader vibration with a developed multibody dynamic model

Nikolay Pavlov[1,*], *Evgeni* Sokolov[1], *Mihail* Dodov[1], and *Stoyan* Stoyanov[1]

[1]Technical University – Sofia, Faculty of Transport, 8 Kliment Ohridski Blvd., 1000 Sofia, Bulgaria

Abstract. In this paper a multibody dynamic model of a compact wheel loader is created. This model includes the sprung and the unsprung masses of the tractor and the mass of the bucket. Results for the natural frequencies and the accelerations of the dynamic system are obtained. The dynamic model is verified by comparing the calculated results with the ground test results.

1 Introduction

Aggregating tractors with different agricultural implements (ploughs, tillers, sprayers, fertilizer spreaders) or construction machinery (excavators, hydraulic hammers, front-end loaders, etc.), results in an increase in their mass and moments of inertia. Their most negative impact is the increase of the mass moment of inertia around the lateral axis. It leads to a decreased frequency of oscillation around this axis and an increase in their amplitudes when the tractor is moving. As a result decreases ride comfort, which leads to fatigue in the driver-operator and a decrease in his ability to work.

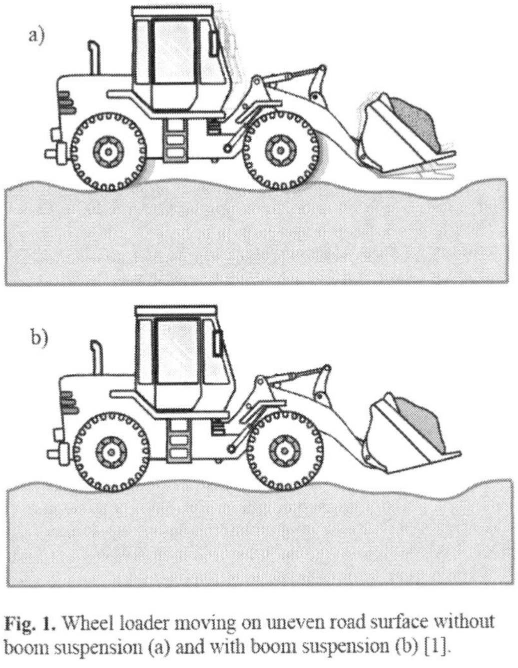

Fig. 1. Wheel loader moving on uneven road surface without boom suspension (a) and with boom suspension (b) [1].

In addition, there are major changes in the dynamic force of the front axle wheels, sometimes resulting in loss of contact with the road [1]. This leads to deterioration in handling the vehicle and traffic safety. If the movement is carried out with a load in the front loader bucket, angular oscillations with large amplitudes can result in the loss of part of the load (Fig. 1). Often, the operator has to decrease the travel speed to reduce the oscillations [2, 3].

Without an effective system for reducing the vibrations the efficiency, safety, and life of the machine can all be reduced, but most wheel loaders adopt rigid or semi-rigid suspension [4]. In order to improve the operation comfort of the wheel loader and reduce material falling out of the bucket, a type of vibration absorber system is created. To solve the problem, dynamic systems with hydraulic accumulators, operating as elastic elements and damping in hydraulic cylinders were proposed. The system is passive and requires a compromise solution when choosing constant coefficients of elasticity (the mass of the gas in accumulators is constant) and damping for different loads in the bucket of the loader. In order to fully solve the problem for all types of loading, more sophisticated mechatronic valve systems and pressure regulators in the hydraulic cylinders are used.

The purpose of this paper is to present five degrees of freedom (5DOF) dynamic model of a tractor front-end loader system. The model can be used for studying wheel loader vibration, ride comfort and road holding.

2 Multibody model

A multibody dynamic model of the system of a compact wheeled tractor with front-end loader was developed. Both axles of the tractor are elastic suspended. The model has five degrees of freedom. They are determinate of the vector of the generalized coordinates:

$$q = \left[z_0, \theta_0, \theta_{fl}, z_1, z_2\right]^T. \qquad (1)$$

* Corresponding author: npavlov@tu-sofia.bg

A scheme of the model is shown in Fig. 2. This model includes the sprung mass (tractor body), the front and the rear unsprung masses (wheels and axles of the tractor) and the mass of the loader bucket.

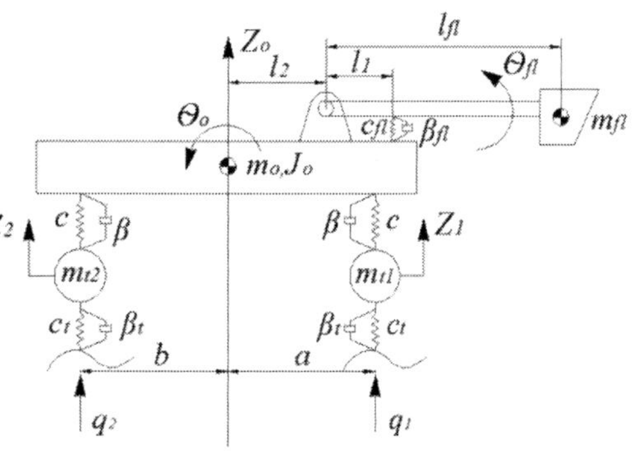

Fig. 2. Multibody dynamic model with five degrees of freedom.

The theoretical basis of the method is Lagrange equation of motion of the second kind:

$$\frac{d}{dt}\left(\frac{\partial T}{\partial \dot{q}_i}\right) - \left(\frac{\partial T}{\partial q}\right) + \left(\frac{\partial U}{\partial q}\right) + \left(\frac{\partial R}{\partial \dot{q}}\right) = Q(t), \quad (2)$$

where T is the kinetic energy of the system; U is the potential energy of the system; R is Rayleigh dissipation function; q_i are generalized coordinates; $Q(t)$ is function of kinematic disturbance.

The kinetic energy of the system is:

$$T = \frac{1}{2}m_0\dot{z}_0^2 + \frac{1}{2}J_0\dot{\theta}_0^2 + \frac{1}{2}m_{fl}l_{fl}^2\dot{\theta}_{fl}^2 + \frac{1}{2}m_{t1}\dot{z}_1^2 + \frac{1}{2}m_{t2}\dot{z}_2^2. \quad (3)$$

The potential energy of the system is:

$$U = \frac{1}{2}c_{fl}\left(\theta_{fl}l_1 - \theta_0 l_2\right)^2 + \frac{1}{2}c(z_0 + \theta_0 a - z_1)^2 + \frac{1}{2}c(z_0 - \theta_0 b - z_2)^2 + \frac{1}{2}c_t(z_1 - q_1)^2 + \frac{1}{2}c_t(z_2 - q_2)^2. \quad (4)$$

The Rayleigh dissipative function is:

$$R = \frac{1}{2}\beta_{fl}\left(\dot{\theta}_{fl}l_1 - \dot{\theta}_0 l_2\right)^2 + \frac{1}{2}\beta\left(\dot{z}_0 + \dot{\theta}_0 a - \dot{z}_1\right)^2 + \frac{1}{2}\beta\left(\dot{z}_0 - \dot{\theta}_0 b - \dot{z}_2\right)^2 + \frac{1}{2}\beta_t(\dot{z}_1 - \dot{q}_1)^2 + \frac{1}{2}\beta_t(\dot{z}_2 - \dot{q}_2)^2. \quad (5)$$

Their derivatives are:

$$\frac{d}{dt}\left(\frac{\partial T}{\partial z_0}\right) = m_0\ddot{z}_0; \quad (6)$$

$$\frac{d}{dt}\left(\frac{\partial T}{\partial \theta_0}\right) = J_0\ddot{\theta}_0; \quad (7)$$

$$\frac{d}{dt}\left(\frac{\partial T}{\partial \theta_{fl}}\right) = J_{fl}\ddot{\theta}_{fl} = m_{fl}l_{fl}^2\ddot{\theta}_{fl}; \quad (8)$$

$$\frac{d}{dt}\left(\frac{\partial T}{\partial z_1}\right) = m_{t1}\ddot{z}_1; \quad (9)$$

$$\frac{d}{dt}\left(\frac{\partial T}{\partial z_2}\right) = m_{t2}\ddot{z}_2; \quad (10)$$

$$\frac{\partial U}{\partial z_0} = 2cz_0 + c(a-b)\theta_0 - cz_1 - cz_2; \quad (11)$$

$$\frac{\partial U}{\partial \theta_0} = c(a-b)z_0 + \left(c_{fl}l_2^2 + ca^2 + cb^2\right)\theta_0 - c_{fl}l_1 l_2\theta_{fl} - caz_1 + cbz_2; \quad (12)$$

$$\frac{\partial U}{\partial \theta_{fl}} = l_1^2\theta_{fl} - c_{fl}l_1 l_2\theta_0; \quad (13)$$

$$\frac{\partial U}{\partial z_1} = -cz_0 - ca\theta_0 + (c + c_t)z_1 - c_t q_1; \quad (14)$$

$$\frac{\partial U}{\partial z_2} = -cz_0 + cb\theta_0 + (c + c_t)z_2 - c_t q_2; \quad (15)$$

$$\frac{\partial R}{\partial \dot{z}_0} = 2\beta\dot{z}_0 + \beta(a-b)\dot{\theta}_0 - \beta\dot{z}_1 - \beta\dot{z}_2; \quad (16)$$

$$\frac{\partial R}{\partial \dot{\theta}_0} = \beta(a-b)\dot{z}_0 + \left(\beta_{fl}l_2^2 + \beta a^2 + \beta b^2\right)\dot{\theta}_0 - \beta_{fl}l_1 l_2\dot{\theta}_{fl} - \beta a\dot{z}_1 + \beta b\dot{z}_2; \quad (17)$$

$$\frac{\partial R}{\partial \dot{\theta}_{fl}} = \beta_{fl}l_1^2\dot{\theta}_{fl} - \beta_{fl}l_1 l_2\dot{\theta}_0; \quad (18)$$

$$\frac{\partial R}{\partial \dot{z}_1} = -\beta\dot{z}_0 - \beta a\dot{\theta}_0 + (\beta + \beta_t)\dot{z}_1 - \beta_t\dot{q}_1; \quad (19)$$

$$\frac{\partial R}{\partial \dot{z}_2} = -\beta\dot{z}_0 + \beta b\dot{\theta}_0 + (\beta + \beta_t)\dot{z}_2 - \beta_t\dot{q}_2. \quad (20)$$

After applying the Lagrange equation (2) for equations describing the laws of motion of the system is valid:

$$[M]\ddot{q} + [R]\dot{q} + [C]q = [Q], \quad (21)$$

wherein $[M]$ is the matrix of inertia; $[C]$ is the stiffness matrix; $[R]$ is the Rayleigh's dissipative damping matrix.

They have the following form:

$$M = \begin{bmatrix} m_0 & 0 & 0 & 0 & 0 \\ 0 & J_0 & 0 & 0 & 0 \\ 0 & 0 & m_{fl}l_{fl}^2 & 0 & 0 \\ 0 & 0 & 0 & m_{t1} & 0 \\ 0 & 0 & 0 & 0 & m_{t2} \end{bmatrix}; \quad (22)$$

$$C = \begin{bmatrix} 2c & c(a-b) & 0 & -c & -c \\ c(a-b) & (c_{fl}l_2^2 + ca^2 + cb^2) & -c_{fl}l_1 l_2 & -ca & cb \\ 0 & -c_{fl}l_1 l_2 & c_{fl}l_1^2 & 0 & 0 \\ -c & -ca & 0 & (c + c_t) & 0 \\ -c & cb & 0 & 0 & (c + c_t) \end{bmatrix} \quad (23)$$

$$R = \begin{bmatrix} 2\beta & \beta(a-b) & 0 & -\beta & -\beta \\ \beta(a-b) & (\beta_{fl}l_2^2 + \beta a^2 + \beta b^2) & -\beta_{fl}l_1 l_2 & -\beta a & \beta b \\ 0 & -\beta_{fl}l_1 l_2 & \beta_{fl}l_1^2 & 0 & 0 \\ -\beta & -\beta a & 0 & (\beta + \beta_t) & 0 \\ -\beta & \beta b & 0 & 0 & (\beta + \beta_t) \end{bmatrix}. \quad (24)$$

$[Q]$ is the kinematic excitation vector.

3 Numerical simulations

The numerical simulations were performed in MATLAB with the parameters given in Table 1.

Table 1. Model parameters.

Symbol	Value	Unit
m_0	1 000	kg
J_0	1 000	Nm^2
m_{fl}	100	kg
m_{t1}	120	kg
m_{t2}	120	kg
c	34 000	N/m
c_t	160 000	N/m
c_{fl}	42 000	N/m
β	3 000	Ns/m
β_t	100	Ns/m
β_{fl}	2 000	Ns/m
a	0,77	m
b	0,43	m
l_1	0,33	m
l_2	1,07	m
l_{fl}	0,9	m

The symbols used in Table 1 are:
m_0, J_0 – Sprung mass and moment of inertia;
m_{fl} – Front loader bucket mass;
m_{t1}, m_{t2} – Front/rear unsprung mass;
c, β – Suspension stiffness and damping;
c_t, β_t – Tyre stiffness and damping;
c_{fl}, β_{fl} – Boom suspension stiffness and damping;
a – Distance between front axle and tractor mass centre;
b – Distance between rear axle and tractor mass centre;
l_1 – Distance between boom joint and boom suspension;
l_2 – Distance between boom joint and tractor mass centre;
l_{fl} – Boom lenght.

To find the natural frequencies of the system, the equations are presented in Cauchy normal form [5]:

$$\dot{y} + Ly = 0, \tag{25}$$

where

$$y = [y_1, y_2 \dots y_{10}]^T = [q_1, q_2, \dots q_5, \dot{q}_1, \dots \dot{q}_5]^T; \tag{26}$$

$$L = \begin{bmatrix} M^{-1}O & M^{-1}C \\ I & 0 \end{bmatrix}, \tag{27}$$

where $[I]$ is identity matrix and $[O]$ is zero matrix.

The natural frequencies with given in Table 1 parameters are:
1,19 Hz – tractor vertical vibration (z_0);
0,5 Hz – tractor angular vibration (θ_0);
1,72 Hz – front loader boom angular vibration (θ_{fl});
6,4 Hz – wheel vertical vibration (z_1, z_2).

To obtain the results for vibration of the system under sinusoidal disturbance, the equations are presented like:

$$\dot{y} + Ly = Y, \tag{28}$$

where

$$L = \begin{bmatrix} M^{-1}R & M^{-1}C \\ -I & 0 \end{bmatrix}; \tag{29}$$

$$Y = \begin{bmatrix} M^{-1}Q(t) \\ 0 \end{bmatrix}. \tag{30}$$

After integration of the equations using the Runge-Kutta method (ode45 solver in MATLAB), results for displacement, velocity and acceleration in a given time interval are obtained. Disturbing actions are applied in the contact patch centre of the tire with the road. They have the following form:

$$q = q_0(1 - cos(vt)), \tag{31}$$

where $q_0 = 0{,}04$ m – amplitude of unevenness; v – frequency of the disturbing action, rad/s.

A front-end loader with special construction was designed and made for verification of the dynamic model (Fig. 3).

Fig. 3. Specially designed wheel loader.

The loader has pneumatic suspension to realise various stiffness ratios in the boom suspension and was attached to a compact wheeled tractor with suspended axles. Fig. 4 shows the ground test on special track with characteristics similar to the disturbances from the numerical experiment. An accelerometer for measuring acceleration was mounted on the driver seat (Fig. 5).

Fig. 4. Proving ground test on special track.

Fig. 5. Position of the accelerometer on the driver seat.

DOI: 10.1051/matecconf/201713302007

A comparison of the calculated results and the ground test results for vertical acceleration on the seat can be made using Fig. 6 and Fig. 7.

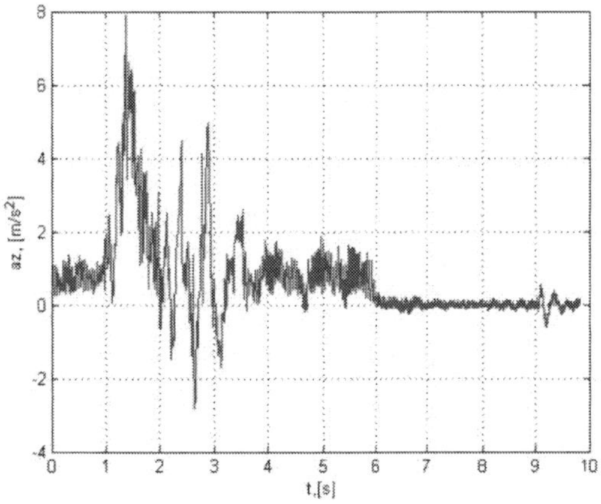

Fig. 6. Measured sprung mass vertical acceleration.

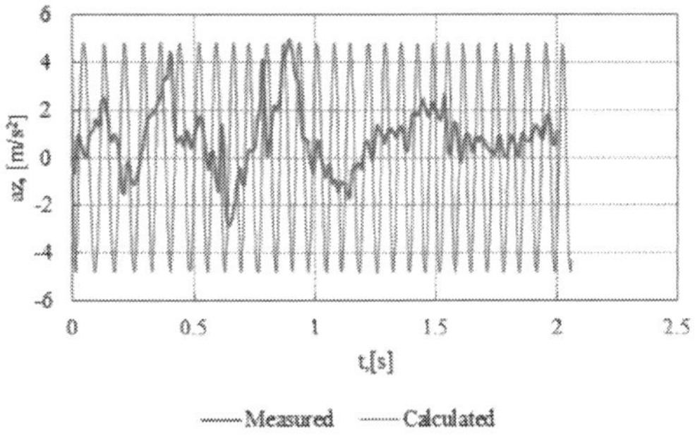

Fig. 7. Measured in time range of 2 to 4 s (the tractor moves on the track) and calculated sprung mass vertical acceleration.

The movement speed during the test was 8 km/h. The maximum values of the accelerations in the range of 2 to 4 seconds in Fig. 6 are commensurable with those obtained in the numerical experiments (Fig. 7). The asymmetry in measured acceleration comes probably from asymmetric characteristics of suspension shock absorbers which are not taken into account in the model. Before and after this time interval, the tractor practically does not move on the special test track. The maximum acceleration observed shortly after the 1-th second corresponds to the clutch engagement when the tractor begins to move. The acceleration between 4th and 6th second is obtained when the tractor moves on a meadow and after the 6-th second it is just from the engine force disturbance.

4 Conclusion

The developed multibody model makes it possible to study the vibration of sprung and unsprung masses, as well as the vibration of the loader boom. The model allows to study the vibrations at different stiffness and damping ratios of the boom suspension. The results may be used for estimation of dynamic properties related to wheel loader vibration, ride comfort and road holding. The specially designed front loader allows for the realization of different stiffness ratios of the boom suspension. The front-end loader has been made and a ground test was conducted, which shows good comparability with the results from the numerical simulations.

The research presented in this paper was carried out with financial support from the Scientific and Research Sector of the Technical University of Sofia – Internal Funding Session 2017. Grant No. 171ПР0020-04.

References

1. http://www.hydraforce.com/Applications/WheelLoader_solutions.html (Accessed 12 August 2017)
2. T. Langer, Comfort evaluation criteria for pitching vibration damping of agricultural tractors, *Proceedings of LAND.TECHNIK*, Germany, (2016)
3. R. Rahmfelda, M. Ivantysynova, An overview about active oscillation damping of mobile machine structure. Int. Journal of Fluid Power, **5(2),** (2004)
4. Z. Wang, The simulation analysis and modelling of the coupling vibration absorber system for the wheel loader. Applied Mechanics and Materials, **121-126,** 2358-2362, (2012)
5. M. Todorov, Study the vibration of forklift trucks and optimization of their parameters, PhD Thesis, Technical University – Sofia, in Bulgarian (1996)

Design solutions for structures with optimal internal stress distribution

Yurii Kitov[1], *Maryna* Verevicheva[1], *Glib* Vatulia[1*], *Yevhen* Orel[1], and *Svitlana* Deryzemlia[1]

[1]Ukrainian State University of Railway Transport, Structural Mechanics and Hydraulics Department, 7 Feuerbach Sq., 61050 Kharkiv, Ukraine

Abstract. The development of transport infrastructure, reconstruction of roads and railways, together with growing urban traffic necessitate the construction of new bridges and/or the reconstruction of existing ones. The primary objective of design is obtaining a reasonable system in compliance with the requirements for stress-strain state: equal strength or equal torque, maximum rigidity or minimum weight of the structure. In the paper, the object of optimization is the design of a three-span girder bridge of a specified length under the constant load. The sets of both statically determinate and statically indeterminate systems are considered, namely: three-span hinged girders in which the hinge joints are located in the end spans or middle span, and three-span continuous girders on unilateral supports or on resilient supports. Two variable design parameters are taken into account: the length of each end span z_1 and the position of the hinges x for the statically determinate systems, and the displacement of support assemblies joints Δ or the value that affects the displacement, i.e. flexibility of yielding supports c for the statically indeterminate ones. The authors proved that there are optimal solutions in the set of statically indeterminate systems both with unilateral and with elastic constraints.

1 Introduction

Along with the development of new constructive solutions and methods for assessment of the stress-strain state and bearing capacity of building structures, the issue of optimizing their parameters is also addressed now.

In [1] three principles are described that underlie the design of steel constructions: metal saving, reduction of labour intensity of manufacture and installation time. In different periods, these principles were taken into account in different ways, depending on the state-of-the-art in the industry and the urgent needs of society, as well as the existing structural engineering methods. Before methods of optimization appeared, methods of determining the weight depending on the design parameters of structures were developed. Afterwards, methods of determining the labour intensity of manufacture and the installation of structures were also developed.

The beginning of the modern epoch of optimization of building constructions dates back to more than 70 years ago when the expediency of application of developed mathematical programming methods in this area was substantiated [1-7].

Along with the obtained applied results, these works contain the analysis of the principal quality criteria, restriction types, as well as problem-solving techniques for optimal design.

Initially, the only tasks in using the mathematical programming methods with problems of steel structure optimization were to minimize the girder and the plane frame weight. Further, problem statements got more complicated, in particular, including the transition to the economic optimality criteria, adjustment of the stress-strain state of construction, unification and typification of elements, which finally allowed bringing the optimization theory closer to the design practice in engineering, transport and construction [8-17].

2 General provisions

We assume that the optimal internal force distribution in a structure is the set of internal forces in which the system becomes optimal, for example, a system of minimum volume or cost.

One of the optimal design types is systems that have strength conditions as limitations. The construction is optimal, if the maximum stresses in all its elements are equal to the estimated ones, that is, a system has equal strengths. If the stress-strain state of all elements is the same, and their cross-sections are equal, then the system can be considered to have equal strength if the forces in the nominal cross-sections are equal. To obtain the optimal system with equal strengths, equal forces of the elements should be obtained by changing the geometric characteristics of those elements.

* Corresponding author: glebvatulya@gmail.com

All constructions can be conditionally divided into two sets: the set of statically determinate systems and the set of statically indeterminate ones.

In the set of statically determinate systems, the forces are determined by means of the static equilibrium equations; their distribution at the given stress depends only on the linear dimensions of elements and their mutual arrangement, but not on the displacement of the support constraints.

In the set of statically indeterminate systems, those forces are also determined by means of equilibrium equations, as well as by equations of strain compatibility. Their distribution depends both on the linear dimensions of the elements and on the displacement of the support constraints. Therefore, in the set of statically indeterminate systems, an optimal system can be obtained not only by changing the linear dimensions of elements, but also by displacing the support constraints. For this purpose, the system – for example, a continuous girder – must have unilateral support constraints or resilient supports.

To determine forces in the set of statically indeterminate systems, additional equations are required along with equilibrium equations, namely, strain compatibility equations. Hence, the force distribution depends both on the linear dimensions of the elements and on their deformations. Therefore, equal strength of all elements cannot be achieved only by changing their linear dimensions.

Let us use an example to show that in the set of statistically indeterminate systems with unilateral as well as elastic constraints, optimal solutions exist also under constant load. They completely coincide with the solutions in the set of statically determinate systems.

3 Optimization of the design of the three-span girder bridge

3.1 Problem statement

An optimal design of a three-span girder bridge of a given length l under constant load $q = \text{const}$ should be obtained. The cross-section of the girder is constant and equal in all spans, the layout is symmetrical.

Let us consider four design options: two of them belong to the statically determinate systems and the other two – to the statically indeterminate ones (Fig. 1).

The first two are three-span statically determinate hinge-connected girders in which hinge joints are located in the end spans (Fig. 1, a) or in the middle span (Fig. 1, b).

The second two are three-span continuous girders on unilateral supports (Fig.1, c) or on resilient supports (Fig. 1, d).

There are two design parameters. The first one which effects the force distribution is the length of each end span z_1, which is a common parameter for all types of girders. The second parameter is specific for each set: in the statically determinate systems, it determines the position of hinges x; in the statically indeterminate systems, it is displacement of the support joints Δ or the value effecting the displacement, i.e. flexibility of resilient supports c.

3.2 Optimization conditions

Since the load bearing elements are girders of constant cross-section, they are in the equal strain-strain states under load. Therefore, the optimal bridge design is equally strong girders with equal breakdown torques in the characteristic cross-sections: in the first girder, above the support and in the second girder (Fig. 1).

There are three characteristic cross-sections with breakdown torques; therefore, two optimality conditions are required, namely:

$$M_{1\max} = \left|M_{opt}\right|, \tag{1}$$

$$\left|M_{opt}\right| = M_{2\max}. \tag{2}$$

On the basis of the above conditions, optimality equations are derived; there are two of them for this problem. We can determine the optimal design parameters for different types of girders by solving these equations together.

3.3 Continuous girder with hinge joints in the end spans

For such a structure (Fig. 1, a), let us write:

$$M_{1\max} = \frac{qx^2}{8},\ M_{opt} = \frac{q\,z_1(x - z_1)}{2}, \quad M_{2\max} = q\left[\frac{x\,z_1}{2} + \frac{l(l - 4z_1)}{8}\right]. \tag{3}$$

From (1), we can obtain the expression for x:

$$x = 2z_1\left(\sqrt{2} - 1\right). \tag{4}$$

Parameter z_1 can be numerically determined by exhaustive search with 0,001 increment until the condition (2) is met.

3.4 Continuous girder with hinge joints in the middle span

We can write the equations for such a structure (Fig.1,b):

$$M_{1\max} = \frac{q}{8z_1^2}\left[(x + z_1)^2 - lx\right], \quad M_{opt} = -\frac{q\,x(l - 2z_1 - x)}{2},\ M_{2\max} = q\frac{\left[l - 2(x + z_1)\right]^2}{8}. \tag{5}$$

From (2), we can obtain the expression for x:

$$x = \frac{\sqrt{2} - 1}{2\sqrt{2}}(l - 2z_1). \tag{6}$$

Parameter z_1 can be numerically determined by exhaustive search with 0,001 increment until the condition (1) is met.

3.5 Continuous girder with unilateral constraints

For this structure (Fig.1, c) the equations can be written as:

$$M_{opt} = \frac{-(l-2z_1)^3 - z_1^3 + (l^3 - 2lz_1^2 + z_1^3)k}{4(3l-4z_1)}q, \tag{7}$$

$$M_{1\max} = \left[\frac{xk(l^3 - 2lz_1^2 - z_1^3)}{4z_1(3l-4z_1)}\right]q + \left[\frac{x(-l^3 - 6lz_1^2 + 6l^2z_1 - z_1^3)}{4z_1(3l-4z_1)} - \frac{x^2}{2}\right]q, \tag{8}$$

at

$$x = \frac{(l^3 - 2lz_1^2 + z_1^3)k - l^3 - 6lz_1^2 + 6l^2z_1 - z_1^3}{4z_1(3l-4z_1)}, \tag{9}$$

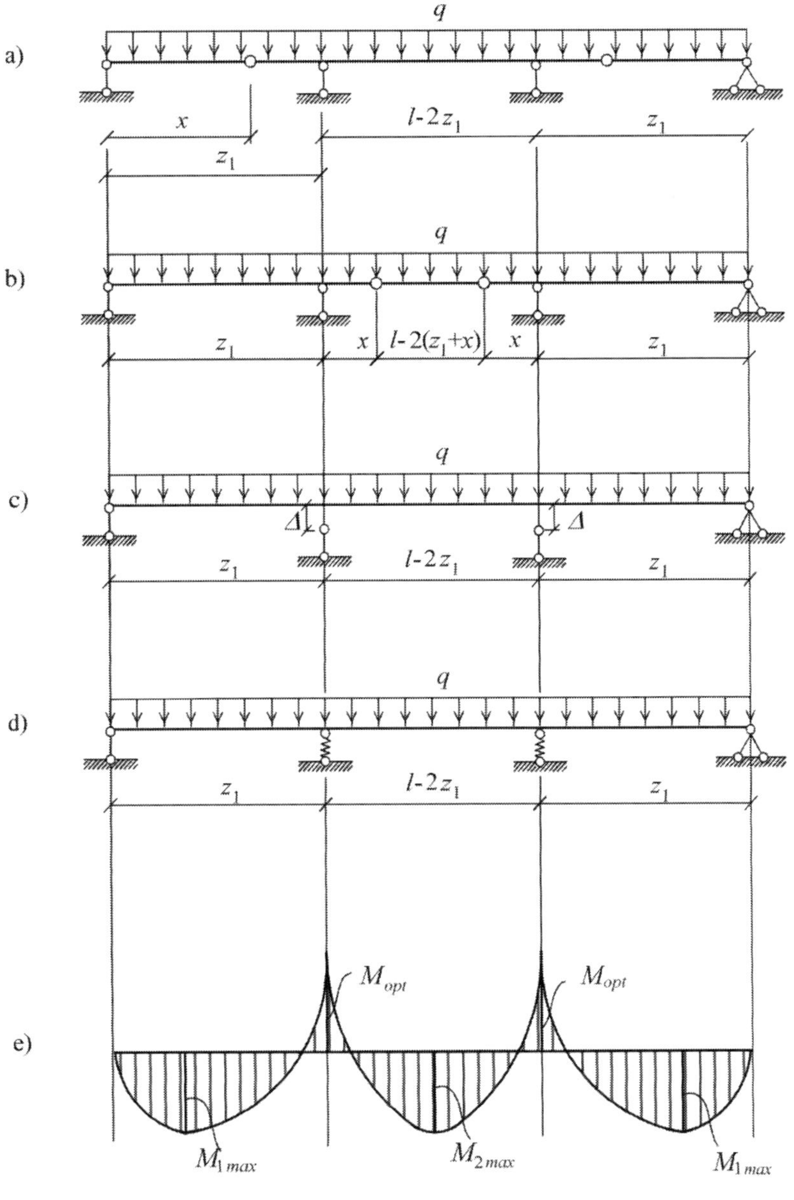

Fig. 1. Design options of the optimum structure and the optimum force distribution.

Table 1. The results of optimization of the bridge.

System type		z_1, m	x, m	k or cEJ, m³	M_{1max}, kNm	$\lvert M_{opt} \rvert$, kNm	M_{2max}, kNm
Statically determinate three-span girders	hinges in the end spans	11,351	9,404		11,054	11,050	11,054
	hinges in the middle span	11,352	1,947		11,058	11,049	11,049
Continuous girders	with unilateral support constraints	11,369		k 0,0271	10,994	10,996	10,989
	with resilient constraints	11,351		cEJ 40,151	11,053	11,052	11,052

$$M_{2\max} = \left[\frac{2\left(l^3 - 2lz_1^2 + z_1^3\right)k + l^3 - 4l^2z_1 + 4lz_1^2 - 2z_1^3}{8(3l - 4z_1)}\right]q,$$

where coefficient k determines part of the distributed load kq, at which the girder adjoins the supports, i.e. becomes statically determinate.

Parameters z_1 and k can be numerically determined by exhaustive search with 0,001 and 0,0001 increment, respectively, until the conditions (1), (2) are met.

3.6 Continuous girder on resilient supports

Let us write the equations for the structure shown in Fig. 1, d:

$$M_{1\max} = \frac{xX_1}{z_1} + \frac{qx(z_1 - x)}{2} \text{ at } x = \left(\frac{X_1}{z_1} + \frac{qz_1}{2}\right)/q, \quad (10)$$

$$M_{opt} = X_1;\ M_{2\max} = X_1 + \frac{q(l - 2z_1)^2}{8}, \quad (11)$$

where

$$X_1 = -\frac{(l - 2z_1)^3 + z_1^3 - 12cEJ(l - z_1)/z_1}{3l - 4z_1 + 6cEJ/z_1^2/(l - 2z_1)^2}\frac{q}{4}. \quad (12)$$

Parameters z_1, cEJ can be numerically determined by exhaustive search with 0,001 and 0,0001, respectively, until the conditions (1), (2) are met.

The results of optimization of all types of girders are summarized in Table 1. The calculations confirm the assertion that under constant load the optimal force distributions are practically equal for all four structures' types (Fig. 1, e).

4 Conclusions

As a result, the authors have proved that there are optimal solutions in the set of statically indeterminate systems both with unilateral and with elastic constraints, and under constant load they completely coincide with the solutions in the set of statically determinate systems.

References

1. A. Vinogradov, Problems of optimal design in structural mechanics (High school, 1973).
2. R. Bellman, Applied problem of dynamic programming (Science, 1965).
3. M. Reitman, Methods of optimal design of deformed bodies (Science, 1976)
4. N. Abovskiy, Proceeding international congress ICSS-98, 307–313 (1998).
5. N. Banichuk, Forms optimization of elastic solids (Science, 1980).
6. V. Gemintern, Methods of optimal design (Science, 1980)
7. M. Zhou, G. Rozvany, Comput Methods Appl Mech Eng, **89(1–3)**, 309–336, (1991).
8. M. Bendsoe, N. Kikuchi, Comput Methods Appl Mech Eng, **71(2)**, 197–224, (1988).
9. Y. Xie, G. Steven, Comput Struct, **49(5)**, 885–896, (1993).
10. K. Choi, N. Kim, Structural sensitivity analysis and optimization I-linear systems (Springer, 2005)
11. Z. Luo, N. Zhang, Y. Wang, W. Gao, Int J Numer Methods Eng **93(4)**, 443–464 (2013)
12. G. Vasilkov, Evolution theory of life cycle of mechanical systems (LKI, 2008).
13. Y. Kitov, G. Vatulia, M. Verevicheva, Collected scientific works of Ukrainian State University of Railway Transport, **143**, 124 – 131 (2014).
14. Y. Kitov, G. Vatulia, Collected scientific works of Ukrainian State University of Railway Transport, **125**, 24 – 33 (2011).
15. V. Shmukler, Collected scientific works of Ukrainian State University of Railway Transport, **167**, 54 – 69 (2017).
16. E. Chikhladze, G. Vatulia, Y. Kitov, Basis for calculation and design of composite and steel-concrete structures (Transport of Ukraine, 2006).
17. M. Gogol, Bridges and Tunnels: theory and practice, **3**, 33-38 (2012).

MATEC Web of Conferences 133, 03002 (2017) DOI: 10.1051/matecconf/201713303002
BulTrans-2017

Determination of the primary technical parameters of the test bench for controlling the temperature of rails and rail bars of continuous welded rail

Vitalii Tertychnyi[1], *Glib* Vatulia[1,*], *Oleksandr* Belorusov[1], and *Maksym* Kovalov[1]

[1]Ukrainian State University of Railway Transport, Structural Mechanics and Hydraulics Department, 7 Feuerbach Sq., 61050 Kharkiv, Ukraine

Abstract. One of the most important factors determining the reliability and safety of operation of the continuous welded rail track is the temperature of the rails. All conditions of laying, repair and maintenance of the track during the entire cycle of rails operation depend on it. The existing rail temperature monitoring method does not take into account a few significant factors and that leads to significant errors in determining the actual temperature of the rail bar. The authors propose a new rail temperature measurement method on a special test bench which maximally simulates possible environmental operating conditions for the rail bars, the profile of the roadway (embankment or cut), as well as the actual spatial latitude and meridian location of the rail slabs. At the bench, the studied rails are fastened with the help of rail braces on half ties, in the ballast bed, with sizes in accordance with the existing reference documents. At the first stage, the most common KB-65 intermediate fastener for the continuous welded rail in Ukraine will be used, while in subsequent stages KPP-5 fastening will be used.

1 Introduction

With the increasing speed of the railway transport everywhere in the world and in Ukraine, in particular, significant upgrade of the rail infrastructure is required, especially through the arrangement of the continuous welded rail track [1, 2]. However, during the operation, temperature variations cause considerable longitudinal forces and tensions in the rails. At present, there are numerous temperature monitoring and determination systems using a variety of non-destructive testing methods, each having both advantages and significant disadvantages [3-8].

According to [9], in Ukraine at all sections of the continuous welded track, the rail temperature should be continuously monitored directly at special temperature metering posts (or special benches located in latitudinal and meridional directions) of the track sections determined by the geophysical station of the railway. In most cases, these are guarded crossings, as well as benches of the meteorological stations of the railways. According to the requirements [10], the temperature post is half rail-track of the corresponding rail and tie type, the type of ballast bed used at the section, and consists of a short rail fixed on two half ties. All available means of rail temperature measurement provide for an alcohol thermometer hole drilled at the end of the rail head or on the working surface. In this case, the thermometer is covered with a special metal sheet casing to prevent sunlight exposure.

2 Setting objectives

It should be noted that most common benches used on the railways in Ukraine and in the CIS countries have significant disadvantages, such as:

- the whole range of the latitudinal and meridional positions of the rails on the lines cannot be taken into account;
- the specific configuration of the ballast bed (embankment or cut) cannot be taken into account;
- the natural heat transfer conditions existing in the rail bars both in the longitudinal direction and in the cross section cannot be taken into account;
- the conditions of heat transfer from the rails to the ballast bed are not taken into account;
- the condition of intermediate fasteners with regard to the tension of fasteners cannot be taken into account;
- the short rail length is not substantiated;
- the conditions of the heat transfer to the rails from the reflected solar radiation are not taken into account;
- heat transfer to the rails from the ballast bed ground cannot be taken into account;

In addition to the above given drawbacks, the principle of measurement in a single cross-section is not perfect, the thermal technical parameters of which fail to correspond to the reliable state of the railing bar [1, 3-5]. This leads to an error in determining the rail bar temperature of 12÷15°C, depending on the traction current type and operating mode (movement speed and intensity, axial loads, mode of movement or braking).

* Corresponding author: glebvatulya@gmail.com

The test benches of the track sections are basically designed to collect the data on rail temperature changes in order to predict its value for scheduling track works and determine their working values based on the air temperature, adding the constant value $\Delta t = 20°C$ to it (if the air temperature is above zero). This does not comply with the existing requirements to an improved track works practice, because it does not allow to determine the reliable values of the rail bar temperature in real operating conditions. For the routine maintenance or repair of the welded track, we need to know the temperature of the pre-weighed rail bars at the time of their laying and fixing for permanent operation or the carrying out of urgent, emergency works to eliminate track defects of geometric or physical origin.

3 Determination of the factors influencing rail temperature

Direct solar radiation, which is 90÷92 % of the total heat flow, is known to be the main heat source for rail heating, while scattered solar radiation is about 8÷10%. Scattered radiation reaches the entire free rail surface, while the direct one only the projection area of the rail normally oriented to the sun beams. In view of this fact and the size of modern ties, we can consider that part of the heat, which the rail receives from scattered radiation, to be up to 20÷25 %. Thus, the main factor influencing rail heating is latitudinal and meridional orientation of the short rails at the test bench, which is determined in four exposure options: Northern (N), Eastern (E), North-Eastern (N-E), South-Eastern (S-E). Scattered radiation is formed by the reflection of the sun beams from the grass, from the ballast bed surface, from the cut or embankment slopes, and only slightly from the trees or surrounding buildings and constructions (artificial constructions, platforms, overpass bridges, bridges, storehouses etc.).

Heat loss of the short rails at the test bench should correspond to the actual heat loss conditions of the existing track. For this purpose, the ends of short rails should be insulated by sticking foam polystyrene end insulation caps to them to prevent the rails from cooling on the ends and help to maintain a constant stable temperature of the entire short rail. Its actual (mean volume) temperature will be calculated based on the measurements of the prolonged short rail. Its length can be calculated using the formula (1) obtained from the known expression for determining free thermal distortion under exposure to heat

$$l_{rc} = \frac{\Delta \varepsilon_l}{\Delta \varepsilon_t} \frac{n}{\alpha}, \qquad (1)$$

where $\alpha = 11{,}8 \times 10^{-6}$ 1/°C is the mean estimated value of the temperature coefficient of the linear heat expansion of rail steel [1, 9]; $\Delta\varepsilon_l$ and $\Delta\varepsilon_t$ are errors in measurement of longitudinal and temperature prolongations; n – number of devices for controlling temperature prolongations.

Calculation using the formula (1) determined that the length of the test rail should be not less than 1,5 m.

According to the requirements of the regulatory document [9], the torque force on the terminal bolts of the intermediate rail fastening KB-65 should be at least 100 Nm, while on the insert bolts – at least 70 Nm.

We will accept the number of half-ties, type and thickness of the crashed stone ballast to be in accordance to the requirements [10]. It should be noted that the requirements [10] do not provide for any sand bed, but for the complete reproduction of the heat transfer conditions we arrange it as the rail pad with the standard thickness of 200 mm.

To reproduce the solar radiation flow, we provide for creation of loam embankment (cut) slopes subsequently seeded with the grass.

To reproduce the intensity of the solar radiation reflected from the embankment or cut slopes, we assume for the test bench a similar steepness of 1:1,5.

To reproduce the heat transfer process from the rail to the air and ballast bed, we assume that the ballast section shoulder is 350 mm.

The obtained principle technical parameters of the bench are summarized in Table 1. The embankment height and the cut depth at the test bench remain unspecified, since these parameters should be determined taking into account the requirements of the similarity theory [11] and the study of air flow modelling processes [12-14].

4 Determination of the embankment height of the cut depth at the test bench

The rail track is situated in ground atmospheric air, where wind velocity changes with height by logarithmic dependence, while the monthly mean velocities in the summer are 2÷4 m/s [15].

Depending on the ballast bed configuration [2], slope steepness and orientation have significant impact on heat delivery to the earth's surface which affect the thermodynamic condition and the small scale climate of the entire slope area (cut and embankment slopes), and, respectively, affect the rail temperature [9]. Thus, the available data on snow-wind transfer should be analyzed, which were did in detail while designing the embankments and cuts in the most dangerous areas, with various models and aerodynamic benches [12-14].

In [15] it was demonstrated that the air flows around a separate 100 mm high hill with a base length of 1500 mm, upper surface length of 500 mm and a span of 2000 mm, enabled the detection of a very slight flow disturbance (turbulence). Similar results were obtained with slope surface angles of up to 15-20°. This shows that since the wind velocity above the rail track in the summer is much lower (2÷4 m/s) as compared to that used with the model in the air tunnel (20 m/s), even with steeper slopes the flow will have minor disturbances.

The railway designing guidelines set forth a minimum embankment height of 1,0 m. The studies [12, 14] proved that with a cut slope of 30°, the wind velocity is 1,36% lower for each 100 mm of its

Table 1. Principal technical parameters of the test bench for temperature measurement.

No	Parameter	Parameter value under the natural conditions	Parameter value at the model	Determination method
1	Rail type	R65	R65	Analogue is accepted
2	Rail length	250÷800 m	1,5 m according to the method [10]	Calculation depends on the accuracy of the determining thermal distortion of the short rail
3	Intermediate rail fasteners	KB-65	KB-65	Analogue is accepted [9]
4	Tie	Sh 1-1 concrete tie	1,0 m long half-tie, three half-ties	Analogue for support of one rail is accepted [10]
5	Ballast	Crushed stone 25÷60 mm, layer thickness 350 mm	Crushed stone 25÷60 mm, layer thickness 350 mm	Analogue is accepted [9]
6	Sand bed	Thickness 200 mm	Thickness 200 mm	Analogue is accepted [9]
7	Ballast bed ground	Loam with humus	Loam with humus	Analogue is accepted [9]
8	Ground surface	Grass seeding	Grass seeding	Analogue is accepted
9	Slope steepness	1:1,5	1:1,5	Analogue is accepted [9]
10	Embankment height	0,9-30 m	To be studied	Simulation suing the similarity methods
11	Cut depth	0,9-30 m	To be studied	Simulation suing the similarity methods
12	Ballast section shoulder	0,35 m	0,35 m	Analogue is accepted [9]
13	Heat loss via the adjoining rail cross-sections	Rail bar	Foam polystyrene end insulation caps	Thickness at least 50 mm
14	Heat loss via he fixing elements, tie, ballast	Tension of each intermediate rail fastening bolt – at least 100 Nm (terminal bolts) and 70 Nm (insert bolts)	Tension of each intermediate rail fastening bolt at least 100 Nm (terminal bolts) and 70 Nm (insert bolts)	Analogue is accepted [1]
15	Cloud amount	Natural	Natural	Analogue
16	Atmosphere transparency	Natural	Natural	Analogue
17	Wind	Natural	Natural	Analogue

depth, i.e. for the depth up to 1,0 m it is about 14% less. It was also proved that there is a wind shadow behind the rail, with transverse air flow direction to the axis. Thus, at a certain distance from the rail, in the shadow, the wind velocity gradually drops to 0. For a R65 rail, the above distance changes from 720 mm to 900 mm. This fact makes it possible to plan placing a freely-supported short rail on the concrete half-tie, designed for measuring the temperature of the previously unloaded rail bars for further scheduled laying in the continuous welded track.

In [16], before the beginning of the experimental studies, the simulation conditions were analyzed to obtain the similarity criterion. For this purpose, it was assumed that the flow of fluid in all points of the natural and simulated flows is described with the differential equation of motion of incompressible viscous fluid (Navier-Stokes equation). Then we will find the similarity criteria as four groups of non-dimensional values: homochronicity number (H), Froude number (Fr), Euler number (E) and Reynolds number (Re).

It was demonstrated that no whirls occur in the cuts which are at least 2,0 m deep, or they can be considered whirls in which the velocity of air current only slows down. This is the factor influencing the rail temperature which in the cuts is higher than on the embankments.

The studies [16] proved that the thermal energy is carried mostly by the beams with a wave length of 0,4÷0,8 μm and IR beams with a wave length of 0,8÷3,0 μm. However, the earth surface radiates thermal energy with a wave length of 9÷12 μm. The numerical depth of the ground layer is very small for long-wave radiation (fraction of a millimetre) and too big (up to a few centimetres) for short-wave radiation. In [16, 17], the nature of the diurnal change of the radiation balance of the ground surface was determined. It was determined with the periodic function that changes the sign twice a day – in the morning and in the evening, while a zero balance is observed not during sunrise or sunset, but when it is 10÷15° above the horizon. This fact is critical for determining the time to begin measuring the thermal distortion of short rails, while it is the time when the rail temperature is equal to the air temperature. Since 1° above the horizon corresponds to one Sun disc, and the time for which the Sun moves by 1° is 4 min, thermal distortion of short rails shall be recorded at the test bench not later than 40÷60 min after the sunset.

Thus, four 1,0 m high embankments and four 1,0 m deep cuts, one for each of the chosen exposure direction, should be arranged at the test bench.

5 Conclusions

On the basis of the performed studies, the design of a test bench for thermometric studies of welded track was elaborated on. The findings of field tests are to be verified by comparing them with the results of calculations made using software complex based on the finite element method [18].

References

1. V. Albrekht, A. Kogan, Continuous welded rails, *Transport*, (2000)
2. E. Danilenko, Railway track: design, installation and operation, *Inpress*, (2010)
3. V. Rubkin, M. Arbuzov, Science and progress of transport, **9**, 74-77 (2005)
4. V. Pavlenko, Transport of the Russian Federation, **5(36)**, 30 – 31 (2011)
5. A. Wegner, *Proceedings of 10th International Heavy Haul Association Conference*, New Delhi, India, 160 – 168, (2013)
6. J. Zakeri, H. Xia, J. Fan, Latin American Journal of Solids and Structures, **6**, 89 – 104, (2009)
7. J. Zakeri, H. Xia, J. Fan, Journal of Northern Jiaotong University, **24(1)**, 50-55, (2000)
8. D. Bowness, A. Lock, W. Powrie, J. Priest, D. Richards, *Proceedings of the IMechE*, Part F: J. Rail and Rapid Transport, **221(F1)**, 13-22 (2007)
9. Technical requirements of the operation, repair and installation of CW rails in Ukraine, TSP0266, Kiev, (2012).
10. Technical recommendations of determination of CW rails temperature, Transport, (1980)
11. M. Kirpichev, *Theory of similarity*, Science, (1953)
12. L. Andreeva, *Protection of railway track and structures from natural factors, Moscow*, (2016)
13. A. Petrov, Construction and Architecture, **3**, 582-584 (2010)
14. A. Komarov, M. Istrapilovich, I. Veronskaya, Scientific Works of Novosibirsk Railway Institute, **130**, 168-176 (1971)
15. P. Tverskoy, *Meteorology*, Gidrometeoizdat, (1951)
16. D. Laihtman, A. Chudnoskij, *Physics of surface layer*, Gostechizdat, (1949)
17. A. Petrov, G. Petrova, I. Panchishkina, *Proc.12th Int. Conf. Atm. Electricity*. Versailles, France, 5, 2003.
18. H. Lee. *Finite Element Simulations with ANSYS Workbench 16*, SDC Publication (2015)

The impact of the type of operation on the parameters of a shunting diesel locomotive with hybrid power plant

Anatoliy Falendysh[1], *Mykyta* Volodarets[1], *Olha* Kletska[1,*], and *Viktoriia* Hatchenko[2]

[1]Ukrainian State University of Railway Transport, Department of Rolling Stock Operation and Maintenance, 7 Feuerbach square, 61050 Kharkiv, Ukraine
[2]State University infrastructure and technologies, Traction rolling stock of railways Department, Kyiv, Ukraine

Abstract. In designing a traction vehicle it is necessary to take into account the type of field operation it is meant for. It was found out that during shunting operation a ChME3 diesel locomotive is in idling mode for almost half of its general running time. That is why, the introduction of a less powerful engine along with an energy storage device for shunting operations at a shunting diesel locomotive is appropriate. The calculations were made using a modernized shunting diesel locomotive ChME3 with hybrid transmission of power as an example. The dependences of the general running time of a diesel locomotive on the position of the engine driver controller under performing shunting, transportation and hump operation were given in percentages. For every operation mode there was calculated the optimum power of a diesel-generator plant and the optimum energy capacity of an energy storage system taking into account the above-mentioned dependences. It was found out that due to the introduction of a hybrid drive for the section discussed above and the corresponding mass of a train the aggregate fuel consumption will be reduced by 25% and the efficiency of travel will increase by one third.

1 Introduction

When designing a traction vehicle, it is necessary to take into account the type of operation for which it is intended. From this point of view, it would be expedient to consider this issue taking as an example the shunting diesel locomotive, because it performs a wide range of jobs. Determination of the parameters of the hybrid diesel locomotive is considered in [1-9], but they have a number of shortcomings, namely: cost indicators, or mass-dimensional, or indicators of operation and repair, or a set of these indicators are not taken into account.

Consider, for example, a modernized ChME3 locomotive with hybrid power transmission of determining the parameters of a hybrid vehicle, depending on the type of work performed: maneuvering, on the hill and on the soles.

2 Research and discussion of the results

It is necessary to analyze the work of the diesel generator set (DGS) during the entire working cycle, taking into account both the permanent and the transient modes during acceleration and braking. In this case, the importance of the ratio of constant time and transients throughout the work cycle is of great importance. It was suggested in [10] that under the operating cycle there was a certain set of stable and transitional modes of operation of a diesel engine (DG) by a thermal engine during a relatively short (several minutes) segment of time, which was repeated throughout the period a particular technological operation.

In an operational cycle not only the ratio of the separate modes, but also their sequence, duration and frequency of changes are reflected. The conditional operational cycle is also the subject of research in the estimated way. In the case of compilation of the conditional operational cycle lasting for about 11 minutes long-term, research of the DGS shunting locomotive (SL) operational modes conducted by the Ural unit of the All-Russian Scientific Research Institute of Railway Transport, was considered. The number of changes of the modes (switching of the position of the locomotive controller (PLC)) and the relative time of stay of DGS shunting locomotive on each PLC corresponds to an average value of these variables in maintenance. For this cycle, the lifetime of the diesel engines of the ChME3 of position of the locomotive controller was developed for various types of work (shunting operation, work on the hill and export work), which can be expressed by the appropriate massifs $\tau_{1PLC}(\tau_{1PLCi})$, $\tau_{2PLC}(\tau_{2PLCi})$, $\tau_{3PLC}(\tau_{3PLCi})$, $i = 0...8$. In this case, $\Sigma(\tau_{1PLCi}) = \Sigma(\tau_{2PLCi}) = \Sigma(\tau_{3PLCi}) = 1$. In the case of the execution of shunting operation, we have a massif,

$$\tau_{1PLC} = (45{,}5;\ 4{,}9;\ 17{,}5;\ 17{,}5;\ 8{,}8;\ 4{,}1;\ 1{,}2;\ 0{,}3;\ 0{,}2),$$

for the corresponding positions of the controller of the driver,

* Corresponding author: gurao@ukr.net

 DOI: 10.1051/matecconf/201713303003

$\tau_{PLC} = (0; 1; 2; 3; 4; 5; 6; 7; 8)$.

The analysis has shown that nearly half of the overall operating time the locomotive works in idling mode. *Ne*$_{arly}$ 50% of time work from a position, the first on the fourth. Thus, most of the time the locomotive works in the uneconomical mode. That is introduction on shunting locomotive of the engine of smaller power together with accumulator of energy (AE) for shunting work expediently.

Dependence of power consumption AE with the corresponding restrictions and cost indexes of modernization from DGS power for performance of shunting work (Fig. 1) it is made taking into account a technique, which is presented in [11].

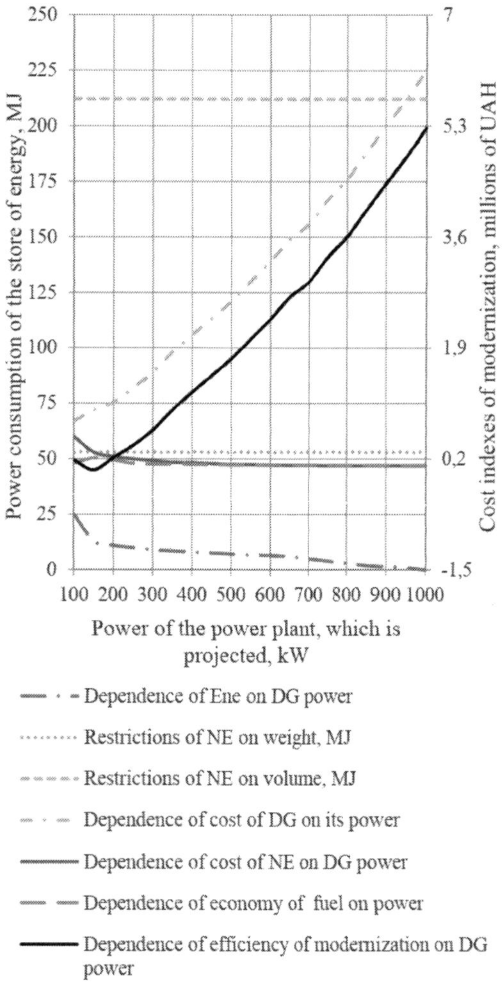

Fig. 1. Dependence of power consumption of the store with the corresponding restrictions and cost indexes of modernization from diesel generator power for performance of shunting work

From this drawing, it is obvious that the optimum power of DGS for the considered operating mode of a locomotive is about 150 kW and power consumption NE - about 20 MJ. The period of operation of the diesel on PLC for operational cycle shunting locomotive of the ChME3 series which works at a hump yard within one hour can be expressed through the massif,

$\tau_{2PLC} = (54,8; 2,3; 9,8; 13,4; 13,0; 4,5; 0,9; 0,6; 0,6)$.

This cycle has been received in [10] by processing and synthesis of data about the operation of shunting locomotive of this type on various railroads of Ukraine. In this cycle of work more than half of the total time the locomotive is in idling mode and more than 40% of time works with the first on the fifth PLC, that is most of the time the locomotive works in the uneconomical mode.

Similarly, with shunting work, there is dependence of power consumption AE with the corresponding restrictions and cost indexes of modernization from DGS power for performance of work on hill (Fig. 2).

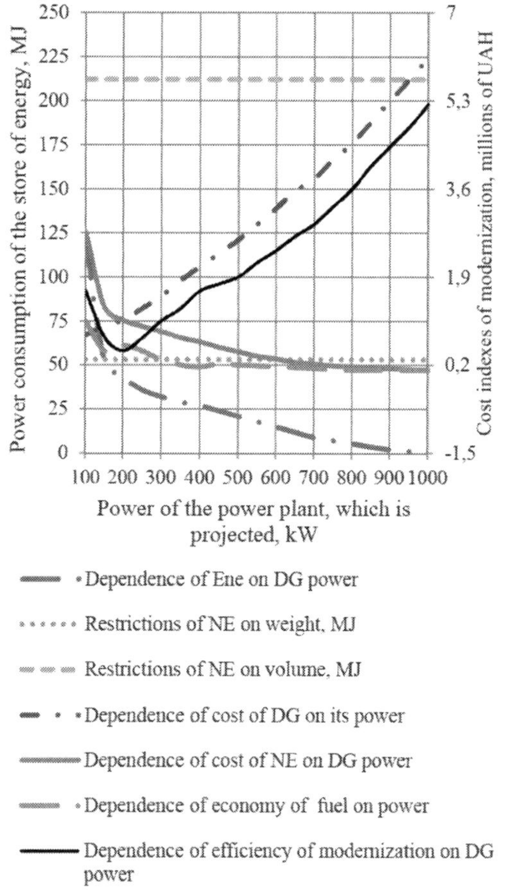

Fig. 2. Dependence of power consumption of the store with the corresponding restrictions and cost indexes of modernization from diesel generator power for performance of work on the hill.

It follows from the calculations that the optimum power of DGS for the considered operating mode of a locomotive is about 200 kW, and the necessary power consumption not is about 50 MJ. These parameters correspond to the smallest general expenses connected with modernization.

 DOI: 10.1051/matecconf/201713303003

The period of operation of the diesel on PLC for operational cycle of shunting locomotive ChME3 which performs export work can be expressed through the massif,

τ_{3PLC} = (44,1; 1,7; 5,4; 6,6; 9,5; 19,3; 10,4; 2,3; 0,8).

When performing export work for about 44% of the time the locomotive is in the idling mode, and about 40% of the time works on 4-6 PLC, i.e. a considerable part of time the locomotive works in an uneconomical mode. For this cycle we have carried out calculations of parameters of a hybrid locomotive and the results are given in the Fig. 3.

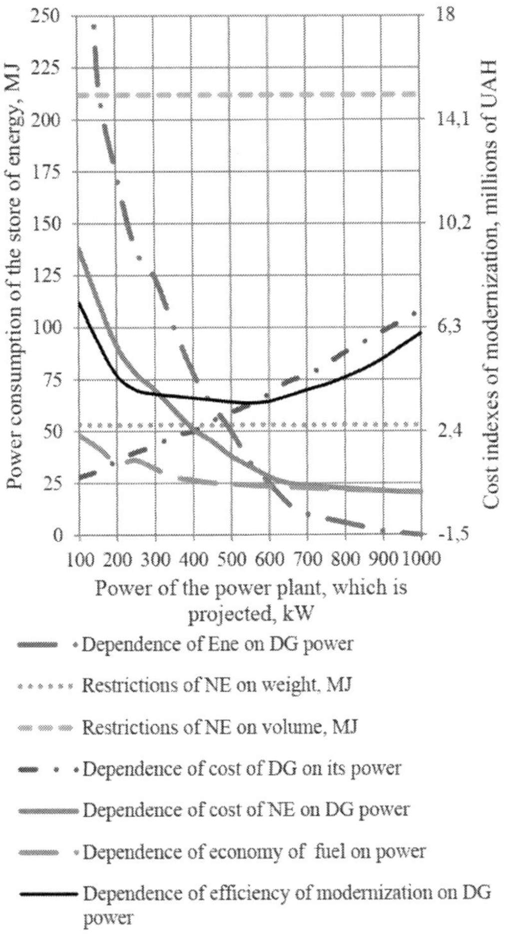

Fig. 3. Dependence of power consumption of the store with the corresponding restrictions and cost indexes of modernization from diesel generator power for performance of export work.

As a result of the approximation of the calculated values of power consumption AE *Ene* and power of DGS Neng it is revealed that generally with the size of reliability of approximation bigger 0,99 it is characterized by the following dependence, MJ,

$$E_{ne} = D\left(\frac{1}{N_{eng}}\right)^2 - E\left(\frac{1}{N_{eng}}\right)^2 + F\left(\frac{1}{N_{eng}}\right) - G, \quad (1)$$

where *D, E, F, G* - the parameters characterizing a look carried out works and service conditions of the locomotive.

For each type of works the following dependences are obtained:

- for shunting work

$$E_{ne} = 243163\left(\frac{1}{N_{eng}}\right)^2 - 51028\left(\frac{1}{N_{eng}}\right)^2 + +3465,4\left(\frac{1}{N_{eng}}\right) - 4,4354, \quad (2)$$

- for work on the hill

$$E_{ne} = 10^8\left(\frac{1}{N_{eng}}\right)^2 - 2\cdot10^6\left(\frac{1}{N_{eng}}\right)^2 + +21818\left(\frac{1}{N_{eng}}\right) - 17,489, \quad (3)$$

- for export work

$$E_{ne} = 10^8\left(\frac{1}{N_{eng}}\right)^2 - 4\cdot10^6\left(\frac{1}{N_{eng}}\right)^2 + +70590\left(\frac{1}{N_{eng}}\right) - 76,296. \quad (4)$$

The obtained dependences are fair for an interval of change of power of the power plant $N_{eng} \in [45... 993]$ where the lower restriction characterizes minimum necessary power of DGS for ensuring its own needs and start-off from the place of the locomotive in case refuses Ne and also considers extreme values of the DGS and AE mass-dimensional parameters. The top restriction is characterized by the most necessary power of DGS for performance of a certain type of works (undertakes the equal power of a locomotive of the ChME3 series which is taken as a basis). For the reviewed examples taking into account [11] and dependences (1-4) for dependence of the general expenses connected with modernization, *Uzag*, MJ, we have:

$$U_{zag_j} N_{eng} = A N_{eng}^2 + B N_{eng_j} + C + +u_2\left[D\left(\frac{1}{N_{eng}}\right)^2 - E\left(\frac{1}{N_{eng}}\right)^2 + F\left(\frac{1}{N_{eng}}\right) - G\right] - U_0 - -\left[\sum_{i=1}^{n}\left(Nf_i\, ge_0 \frac{\Delta\tau}{3600}\right) - \sum_{i=1}^{n} G_{i,j}\right]ct + +C\sigma\left(H N_{eng_j} + K - 1\right) \longrightarrow \min \quad (5)$$

where *A, B, C, D, E, F, G, H, K* - the coefficients characterizing the type of chosen DGS, type of work, service conditions and system of service and repair of a locomotive;

U_0 – the liquidating cost of the basic diesel generator, UAH;
Nf_i – the power of the power plant, kW;
ge_0 – specific fuel consumption the diesel of the basic locomotive, kg/kW h;
$\Delta\tau$ – a power differentiation interval during time, with;
$G_{i,j}$ – fuel consumption of a hybrid locomotive, kg;
ct – cost of fuel, UAH/kg;
$C\sigma$ – expenses for maintenance and repair of a basic locomotive, UAH.

Function (5) is a compound function, which includes functions continuous over a certain interval and discrete function. Therefore, in an analytical form it is not possible to receive optimum value of power N_{eng}. It is rational also to approximate the determined function of dependence of $G\ (N_{eng})$ as it does not allow to receive a simple universal formula for further calculations, therefore, fuel consumption considerably depends on the type of work performed, areas of operation, type and parameters of the AE hybrid locomotive. To obtain the needed level of reliability of approximation it is necessary to introduce a large number of additional parameters and too complicates function.

Therefore, the method of complete search as simple and such was applied to the solution of an objective of non-linear programming that yields quite exact results of computation. Its main shortcoming is that it requires a search of a large number of combinations, and it takes a lot of time for obtaining the decision. Therefore, target function of $U_{zag}\ (N_{eng})$ was most simplified and given to one independent variable, taking into account the restrictions superimposed on it allowed to reduce considerably quantity of combinations of search and duration of computation.

As a result of the calculations we have chosen the corresponding parameters of hybrid N_{engopt} locomotives – the optimum power of DGS and E_{ne} – optimum power consumption AE depending on the type of work performed, namely

- for shunting: N_{engopt} = 150 kW; E_{ne} = 20 MJ;
- for work on the hill: N_{engopt} = 200 kW; E_{ne} = 50 MJ;
- for export work: N_{engopt} = 530 kW; E_{ne} = 50 MJ.

Similar calculations have been carried out also for another type of hybrid drive of a locomotive with which the main source of energy is AE, and DGS is necessary only to resupply energy in AE (the advantage of this type of the drive is the fact that low-power DGS works most of the time in an economical operating mode. Depending on the type of work performed the following E_{ne} and N_{engopt} values are obtained:

- for shunting: E_{ne} = 370 MJ; N_{engopt} = 50 kW;
- for work on the hill: E_{ne} = 560 MJ; N_{engopt} = 80 kW;
- for export work: E_{ne} = 1300 MJ; N_{engopt} = 100 kW.

Also the cost and mass-dimensional indexes of DGS and AE for the hybrid shunting locomotive working in various operating modes have been calculated. It was found out that the second type of hybrid drive is less effective on cost indexes, and in the case of export work it does not satisfy also the mass-dimensional restrictions of shunting locomotive.

By means of traction calculations the comparative analysis of two engines of basic ChME3 and hybrid on its base was made. The calculations were carried out using the program Mathcad complex with train weighing 500 t for export work on a profile Stakhanov-Popasna [11]. What follows from the calculations is that for an almost identical time of movement fuel consumption hybrid of shunting locomotive was made by 126 kg that is 25% less, than the basic engine of the ChME3 series. It was found out that the trip efficiency of the modernized locomotive increased from 14,6% to 19,2%, which confirms the efficiency of introducing a hybrid drive on a shunting locomotive.

3 Conclusions

1. Dependences of percent of the general operating time of a locomotive on PLC when performing shunting work are given, export and works in the hill.

2. Taking into account these correspondences for each operating mode N_{engopt} – the optimum power of DGS, by kW, and E_{ne} – optimum power consumption AE, MJ has been calculated. It confirms the fact that for easy shunting work such powerful power plant which is installed on a ChME3 series locomotive is really necessary.

3. Traction calculations have been carried out using the program Mathcad complex for a profile Stakhanov-Popasna with train weighing 500 t for export work with parameters of the locomotive chosen through results of calculations.

References

1. P. Wolfs, AUPEC, September 25, 123-123 (2005).
2. C. Akli, Int. Journal of Applied Electromagnetics and Mechanics, **30**, 151-162, 2009
3. A. Lohner, *IEEE 35th Power Electronics Specialist Conference,* 672-676 (2004)
4. H. Yap, *IEEE Power Electronics, Machines and Drives Conference*, 61-66 (2004)
5. R. Cousineau, IEEE Instrumentation & Measurement Magazine, February, 25-29 (2006)
6. L. Liudvinavičius, Transport Problems: an International Scientific Journal, **6**, 135-142 (2011)
7. A. Falendysh, TEKA. Commission of motorization and energrtics in agriculture, **12**, 58-63 (2012)
8. A. Varakin, Transport science and technology, **12**, 34–40 (2007)
9. E. Kossov, Locomotive inform, **3**, 44-45 (2008)
10. M. Sergienko, *Selection and scientific substantiation of technical and economic indicators of diesel diesel engines for the locomotive park of Ukraine* (Kharkiv, 2000)
11. A. Falendysh, P. Kharlamov, O. Kletska, N. Volodarets, *Transportation Research Procedia*, **14**, 665-671 (2016)

MATEC Web of Conferences **133**, 03004 (2017) DOI: 10.1051/matecconf/201713303004
BulTrans-2017

Comparative analysis of the results from static strength calculations and strength tests of an Y25Ls-K bogie frame

Svetoslav Slavchev[1,*], *Vladislav* Maznichki[1], *Valeri* Stoilov[1], and *Sanel* Purgic[1]

[1]Technical University – Sofia, "Railway engineering" Department, 8 Kliment Ohridski Blvd., 1172 Sofia, Bulgaria

Abstract. The comparative analysis is based on the results of the static strength analysis of an Y25Ls-K bogie frame and the tests that have been carried out. Strength calculations have been made using the Finite Elements Method at the Department of Railway Engineering at the Technical University – Sofia and the tests of the bogie frame have been conducted in VUZ Testing laboratory in the Czech Republic. It was found out that the stress results are very similar, especially in the areas with maximal values. This proves that a suitable calculation model with a relatively small number of finite elements can be developed. This allows us to solve a wide range of problems concerning the improvement of the bogie frame with similar construction.

1 Introduction

This paper presents a comparative analysis of the static strength analysis and tests conducted on supporting steel structure of an Y25Ls-K bogie frame.

Strength calculations have been made using the Finite Elements Method [1, 2] at the Department of Railway Engineering of the Technical University – Sofia. The tests of the Y25Ls-K bogie frame have been carried out in the Výzkumný Ústav Železniční a.s.- VUZ Testing laboratory in the Czech Republic [3].

Theoretical and experimental studies have been done in full compliance with European standard EN 13749:2011 [4], EN 12663-2:2010 [5], TSI "Rolling stock" [6], and Leaflets 510-3 and 615-4 [7, 8] of the International Railway Union (UIC). According to those documents 13 load cases for study of stresses and displacements in the bogie frame construction are provided. Due to the limited size of the paper it presents only the results of five load cases marked in Table 1 with S1-S5. It should be noted that the number of points where the comparison is carried out is also limited. Only those that have relatively high stress values or are located in zones with section changes were studied.

2 Loading methodology

In accordance with the international requirements, each newly-built bogie is subjected to theoretical and native studies to demonstrate the strength of its structure. Types and conditions of loading are regulated by the European Standard EN 13749:2011 [4], the TSI [6] and the Leaflets of International Railways Union (UIC) 510-3 and 615-4 [7, 8]. According to the documents cited, the size and direction of the forces acting on the bogie depend on the load case (Fig. 1).

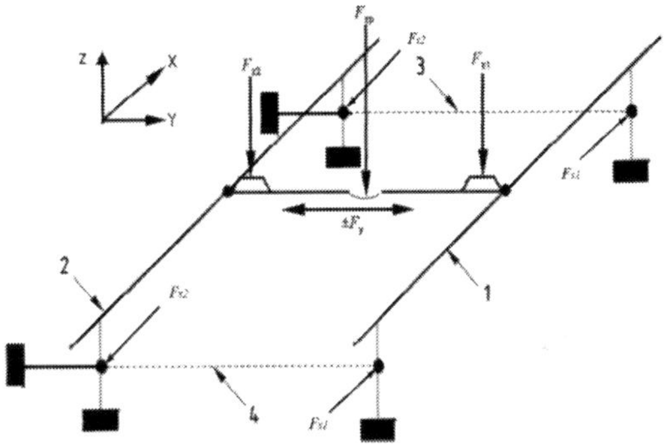

Fig. 1. Forces acting on bogie frame according to [4].

The forces acting on the bogie frame are defined in equations given in [4, 10], as follows:

• *Vertical forces*

$$F_z = 4Q_0 - m^+g. \quad (1)$$

In the case when the force is applied only to the pivot:

$$F_{zc} = 2F_z. \quad (2)$$

For cases when the force is applied to both the pivot and one sidebearer:

$$F_{zc1} = 1{,}5F_z(1 - \alpha); \quad (3)$$

$$F_{zc2} = 1{,}5F_z\alpha. \quad (4)$$

where: F_z – total vertical load supported by the bogie; Q_0 – static vertical force at the level of the wheel for a loaded wagon, for axle load p_0=22,5 t: Q_0 = 0,5p_0g = 110 kN; m^+ – bogie mass (4128 kg); g – acceleration due to gravity; F_{zc} – force applied only to the pivot;

* Corresponding author: slavchev_s_s@tu-sofia.bg

DOI: 10.1051/matecconf/201713303004

F_{z1} and F_{z2} – forces applied to the sidebearers. For UIC service with a distance between the sidebearers of 1,7 m, α is taken as 0,3. F_{bz} – vertical brake force defined in accordance to [10].

- *Transverse force*

$$F_y = 20 + (F_z + m^+g)/3. \qquad (5)$$

- *Longitudinal lozenging forces*

$$F_x = 0{,}1(F_z + m^+g). \qquad (6)$$

F_{bx} – longitudinal brake force defined in accordance with [10]. Brake force depends on the type of frame structure (axle box guidance, type of brake rigging etc. [10]).

- *Twist loading* g^+ – the bogie frame should withstand the loads resulting from a track twist of 10 ‰.

Summing up the results of the equations given above, the following load cases combining the calculated forces (in kN) are obtained (Table 1).

Table 1. Applied load cases. All forces in [kN].

	F_{z2}	F_{zc}	F_{z1}	F_y	$g+$	F_x	F_{bx}	F_{bz}
S1		801,9						
S2		421,0	180,4		+10 ‰			
S3		421,0	180,4		–10 ‰			
S4	180,4	421,0			+10 ‰			
S5	180,4	421,0			–10 ‰			
S6		421,0	180,4	167,2				
S7	180,4	421,0		-167,2				
S8		601,4				44,1		
S9		601,4				-44,1		
S10		400,9				202,6		
S11		400,9				-202,6		
S12		481,1					98,8	24,7
S13		481,1					-98,8	-24,7

3 Calculation models

From the analysis of the design documentation we have determined the bearing elements of the structure, the materials from which it is built and the technology of manufacturing (type of welds, additional machining, etc.). On the basis of this, the calculation models were designed [9] as shown in Fig. 2.

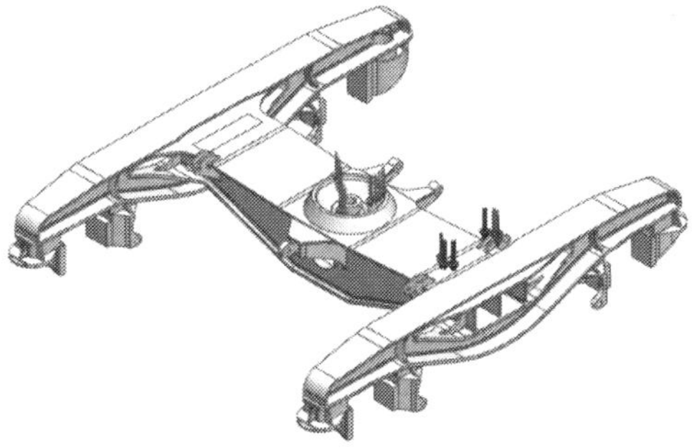

Fig. 2. Calculation model.

In this model the finite elements mesh is compressed (1 374 520 nodes and 843 616 elements), the maximum size of finite elements is 15 mm (Fig. 3), which shows a very good density of the analysed quantities.

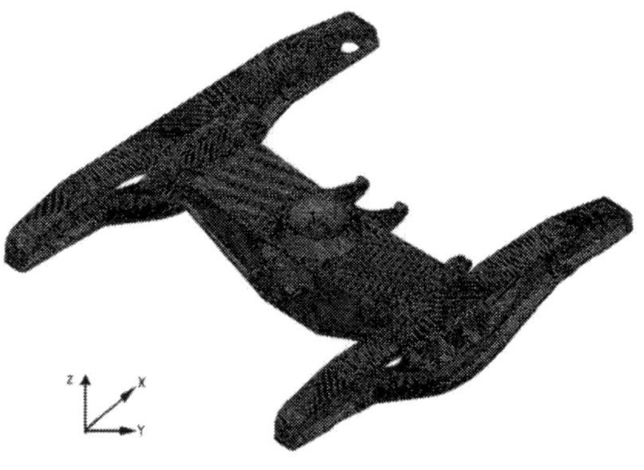

Fig. 3. Finite elements mesh of calculation model.

4 Results analysis

During the tests, stresses in terms of value and distribution, similar to those obtained theoretically, have been registered. A comparative analysis of the results is presented in Table 2 where the following indications are used: the strain gauge number/number of its corresponding element in the calculation model (Fig. 4), the test results in MPa are presented in the same row; σ_x, σ_y, and σ_z are the components of the stress tensor (7) on the main inertia axes x, y, z in MPa:

$$\sigma_{ij} = \begin{bmatrix} \sigma_x & \tau_{yx} & \tau_{zx} \\ \tau_{xy} & \sigma_y & \tau_{zy} \\ \tau_{xz} & \tau_{yz} & \sigma_z \end{bmatrix}; \qquad (7)$$

von Mises stress results from FEM-calculation, von Mises stresses calculated with equation (8);

$$\sigma_{vM} = \sqrt{\frac{(\sigma_x - \sigma_y)^2 + (\sigma_y - \sigma_z)^2 + (\sigma_z - \sigma_x)^2 + 6(\tau_{xy}^2 + \tau_{yz}^2 + \tau_{zx}^2)}{2}} \qquad (8)$$

S1- S5 are codes of analyzed load cases.

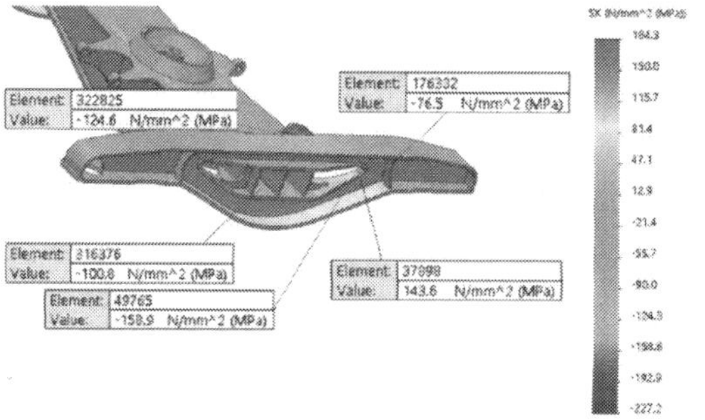

Fig. 4. Stress values obtained in elements corresponding to strain gauges

Table 2. Compared stress values.

DMS/Element	S1	S2	S3	S4	S5	max	min
T111/316376	-85	-47	-51	-70	-75	**-47**	**-85**
σ_x	-100,8	-61,4	-66,5	-87	-92,1	**-61**	**-101**
σ_y	-0,1	0,1	0,1	0	0,1	**0**	**0**
σ_z	0	0	0	-0,1	-0,1	**0**	**0**
von Mises	100,8	92,6	66,5	87	92,1	**101**	**67**
T112/322825	-119	-68	-65	-126	-119	**-65**	**-126**
σ_x	-124,6	-63,5	-57,5	-121,2	-115,2	**-58**	**-125**
σ_y	0	-0,8	-0,7	-1,4	-1,3	**0**	**-1**
σ_z	0	-0,1	-0,1	-0,1	-0,1	**0**	**0**
von Mises	124,6	113	57,1	120,4	114,5	**125**	**57**
T312/314726	-112	-123	-131	-64	-66	**-64**	**-131**
σ_x	-124,2	-121,2	-128,2	-59,6	-66,6	**-60**	**-128**
σ_y	0,1	-1,7	-1,7	-0,9	-0,9	**0**	**-2**
σ_z	0	-0,2	-0,2	-0,1	-0,1	**0**	**0**
von Mises	124,3	65,9	127,4	59,1	66,2	**127**	**59**
T213/176332	-76	-39	-24	-70	-58	**-24**	**-76**
σ_x	-76,5	-57,9	-36,9	-76,4	-55,4	**-37**	**-77**
σ_y	-0,8	-0,7	-0,3	-1	-0,6	**0**	**-1**
σ_z	0	-0,1	0	0	0,1	**0**	**0**
von Mises	76,1	73,7	36,7	76	55,1	**76**	**37**
T214/157611	-85	-76	-56	-91	-76	**-56**	**-91**
σ_x	-59,4	-42	-14,5	-74,6	-47	**-15**	**-75**
σ_y	-0,5	-0,7	-0,3	-1	-0,7	**0**	**-1**
σ_z	0	0,1	0	0,1	0	**0**	**0**
von Mises	59,1	65,4	14,4	74,1	46,7	**74**	**14**
T315/305644	-84	-77	-56	-37	-48	**-37**	**-84**
σ_x	-74,2	-61,9	-69,6	-45	-52,8	**-45**	**-74**
σ_y	-2,4	-2	-2,1	-1,7	-1,8	**-2**	**-2**
σ_z	0,2	0,5	1	0,1	0,6	**1**	**0**
von Mises	76,4	43,5	72,2	46,5	54,7	**76**	**44**
T316/309615	70	59	89	33	45	**89**	**33**
σ_x	73,1	65,4	81,4	30,4	46,4	**81**	**30**
σ_y	0	0	-0,1	0	-0,1	**0**	**0**
σ_z	17,5	16,1	22,1	6,8	12,8	**22**	**7**
von Mises	88,8	49,4	102	36,6	58,4	**102**	**37**
T119/246268	87	49	51	82	84	**87**	**49**
σ_x	119,4	64,4	64,3	108,7	108,7	**119**	**64**
σ_y	84,1	44,3	44,6	73,7	74,1	**84**	**44**
σ_z	-1	-1	-1	-1,5	-1,5	**-1**	**-2**
von Mises	107,2	97,1	58	97,6	97,6	**107**	**58**
T319/405348	92	87	88	52	52	**92**	**52**
σ_x	119,4	109,8	109,2	64,4	63,9	**119**	**64**
σ_y	83	72	70,7	42,2	41	**83**	**41**
σ_z	-0,1	0,1	0	0,1	0	**0**	**0**
von Mises	106,1	55,9	95,9	56,7	56,1	**106**	**56**
T322/238708	58	47	33	34	41	**58**	**33**
σ_x	27,9	47,6	42	0,7	-5	**48**	**-5**
σ_y	0,5	0,5	0,7	0,5	0,6	**1**	**1**
σ_z	1,1	2,2	1,8	-0,1	-0,4	**2**	**0**
von Mises	28,9	17,2	43,5	1,8	5,7	**44**	**2**
T32/467786	-117	-102	-163	-53	-83	**-53**	**-163**
σ_x	-130,5	-111,7	-147,2	-48,7	-84,2	**-49**	**-147**
σ_y	-0,6	-0,7	-1	-0,3	-0,6	**0**	**-1**
σ_z	0,2	0	-0,1	0	-0,1	**0**	**0**
von Mises	130,7	73	147,2	48,6	84,2	**147**	**49**
T325/263874	84	81	76	32	55	**84**	**32**
σ_x	65,7	54,9	59,4	36,4	40,9	**66**	**36**
σ_y	1,2	1,1	1,2	0,7	0,8	**1**	**1**
σ_z	12,9	11,4	12,8	7,3	8,8	**13**	**7**
von Mises	79	50,7	72,8	43,9	50,1	**79**	**44**
T229/198779	-89	-48	-57	-57	-69	**-48**	**-89**
σ_x	-23,2	-11,6	-14,4	-12,4	-15,2	**-12**	**-23**
σ_y	-53,9	-29	-31,5	-30,5	-33	**-29**	**-54**
σ_z	0,1	0	0	0,1	0	**0**	**0**
von Mises	75,5	46,3	44,8	42	47	**76**	**42**
T330/379240	82	52	41	43	47	**82**	**41**
σ_x	-1,5	-1,3	-1,5	-1,3	-1,5	**-1**	**-2**
σ_y	61,5	37,2	37,9	35,9	36,5	**62**	**36**
σ_z	0,7	0,3	0,4	0,3	0,4	**1**	**0**
von Mises	64,4	38,3	39,7	37,4	38,3	**64**	**37**
T430/377102	82	41	51	45	42	**82**	**41**
σ_x	-2	-1,7	-1,6	-1,7	-1,6	**-2**	**-2**
σ_y	58	37,5	36,6	36,5	35,6	**58**	**36**
σ_z	0,8	0,5	0,4	0,4	0,4	**1**	**0**
von Mises	61,1	38,7	38,4	38,4	37,3	**61**	**37**
T331/188444	108	21	218	18	123	**218**	**18**
σ_x	12,8	8,2	14,3	4,3	10,3	**14**	**4**
σ_y	129,7	91,5	141,3	47	96,8	**141**	**47**
σ_z	12,7	8	13,6	4,5	10,1	**14**	**5**
von Mises	137,7	104	150,4	49,7	104,1	**150**	**50**
T431/297427	128	206	2	112	5	**206**	**2**
σ_x	13,6	16,3	9,4	11,4	4,6	**16**	**5**
σ_y	132,6	146,1	95,9	97,9	47,7	**146**	**48**
σ_z	13,9	15,5	9,5	11	4,9	**16**	**5**
von Mises	141	49,1	100,3	105	50,1	**141**	**49**
T132/84072	-136	-67	-87	-118	-150	**-67**	**-150**
σ_x	-143,3	-77	-78,4	-131	-132,4	**-77**	**-143**
σ_y	3,9	-0,2	2,1	-1,4	1	**4**	**-1**
σ_z	-69,9	-39,1	-40,3	-65,8	-67	**-39**	**-70**
von Mises	195	179,7	106,8	175,5	179,9	**195**	**107**
T232/49765	-172	-111	-77	-174	-137	**-77**	**-174**
σ_x	-158,9	-81,2	-82,2	-136,3	-137,4	**-81**	**-159**
σ_y	5,6	3,5	-1,3	6,1	1,3	**6**	**-1**
σ_z	-70,8	-35,2	-33	-58,8	-56,7	**-33**	**-71**
von Mises	206,2	148,7	100,5	178,4	172,4	**206**	**101**
T237/37898	155	95	68	161	124	**161**	**68**
σ_x	143,6	77,8	80,4	136,6	139,2	**144**	**78**
σ_y	-12,3	-2,8	-5,3	-5,8	-8,4	**-3**	**-12**
σ_z	69,3	37,5	40,5	66,7	69,7	**70**	**38**
von Mises	204,6	162,9	114,7	191,1	197,6	**205**	**115**

The analysis of the data in Table 2 shows that there is a very good match of the results. Differences in stress values can be explained by the simplifications adopted in the modeling: the way of gluing and positioning of the strain gauges; measuring errors of equipment; inaccuracies in the manufacturing of the bearing structure; technological deviations of the material; the presence of serious stress notches etc.

Table 3 presents data for the forces applied in five load cases considered in the paper (Figures 5 and 6). In vertical direction, very good matching can be observed and the test forces differ slightly from the calculated ones. This can be explained by the errors of the measuring equipment and the asymmetry of the structure.

Table 3. Applied vertical forces in kN.

	S1	S2	S3	S4	S5
F_{z1}	-399,0	-223,5	-223,9	-377,0	-376,8
F_{z2}	-399,9	-376,3	-376,8	-222,4	-224,1
sum F_z	-799,0	-599,8	-600,7	-599,4	-600,9
F_z FEM	801,9	601,4	601,4	601,4	601,4

Table 4 shows the reaction data R_{z1}-R_{z4} in bearings in the vertical direction (Figures 5 and 6) for the asymmetrical load cases S2 to S5. A very good match can be seen, excluding the values of the bearing reaction, where the track twist is imitated.

DOI: 10.1051/matecconf/201713303004

Fig. 5. Test rig at Výzkumný Ústav Železniční a.s. (VUZ) [3].

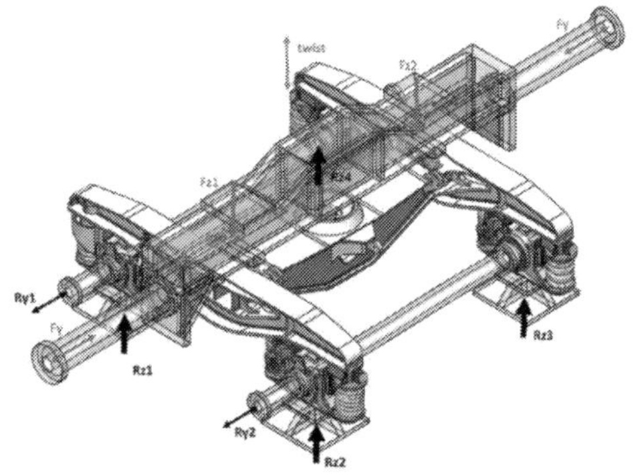

Fig. 6. Bearing reactions.

Table 4. Bearing reactions in [kN].

	R_{z1}	R_{z2}	R_{z3}	R_{z4}
S2	88,2	134,6	146,1	212,8
	101,5	123,2	172,0	224,3
S3	133,9	88,5	212,8	146,8
	123,2	101,8	204,7	151,7
S4	179,8	200,9	83,5	136,8
	177,4	199,2	96,2	148,6
S5	212,8	151,7	133,8	88,3
	199,0	177,8	128,9	76,0

This can be explained by a difference in the position where the frame displacement (as a consequence of the track twist) is applied. If the force is applied under the support of the measuring equipment, then the magnitude of the reaction would increase. Similarly, if the force is applied above the support, the magnitude of the reaction would decrease with the value of the force produced by the track twisting (Fig. 7).

5 Conclusions and recommendations

when summarizing the overall work done on this study the following conclusions and recommendations can be made:

1. Sophisticated calculation models for the strength analysis of the Y25Ls-K bogie frame have been developed, consisting of 843 616 finite elements that accurately describe the geometry of the object.

2. In the process of strength testing minor design changes of the Y25Ls-K bogie frame were made.

3. The analysis of the results in Tables 2 and 3 shows that the stress distribution obtained theoretically is very close to the stress distribution recorded in the test.

4. The relative error of stress results is relatively small, which allows for developed calculation models to be used in order to study other similar bogie frame designs as well as to optimize the tested bogie.

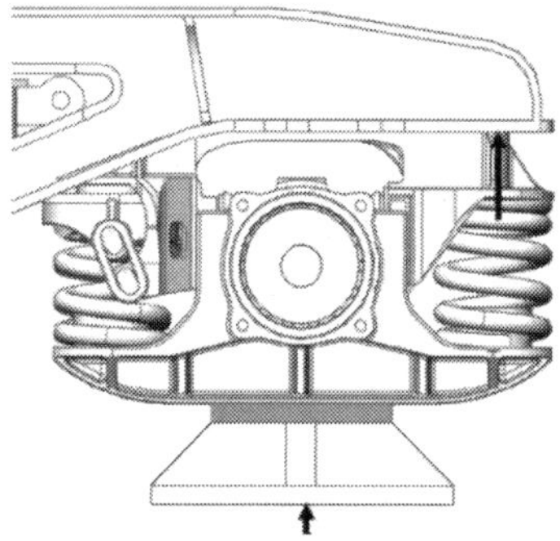

Fig. 7. Reaction under the support.

References

1. O. Zenkiewicz, *The Finite Element Method in Engeneering Science*, McGraw-Hill, London, (2002)
2. L. Segerlind, *Applied Finite Element Analysis*, 2nd edition, Wiley, New York, (1984)
3. VUZ Plc., Test report Strength tests of bogies and their components: Non-destructive testing – Magnetic particle test (MT) - Bogie frame Y25Ls-K, VUZ Plc., Czech Republic, (2017)
4. EN 13749:2011 Railway applications. Wheelsets and bogies. Method of specifying the structural requirements of bogie frames, (2011)
5. EN 12663-2:2010 Railway applications. Structural requirements of railway vehicle bodies. Freight wagons, (2010)
6. Technical specifications for Interoperability (TSI) - Rolling stock: Freight wagons. European Union, Agency for Railways, (2013)
7. UIC Leaflet 615-4, Motive Power Units - Bogies And Running Gear - Bogie Frame Structure Strength Tests, 2nd edition, UIC, (2003)
8. UIC Leaflet UIC 510-3, Wagons - Strength Testing Of 2 And 3-axle Bogies On Test Rig, UIC, (1989)
9. R. Tenchev, *User guide for COSMOS/M*, Technical University - Sofia, (2010)
10. K. Velkov, O. Krastev, *Technologies and systems for train control*, Technical University - Sofia, (2010)

 DOI: 10.1051/matecconf/201713303005

Methodology for profitability assessment in the case of old rolling stock replacement with new trains in Bulgarian State Railways

Valeri Stoilov[1,*], *Kiril* Velkov[1], *Oleg* Krastev[1], and *Svetoslav* Slavchev[1]

[1]Technical University – Sofia, Department of Railway Engineering, 8 Kliment Ohridski Blvd., 1000 Sofia, Bulgaria

Abstract: A methodology has been developed for assessing the profitability of passenger carriers in the event of a complete or partial replacement of the old rolling stock with new one. The evaluation has taken into account the energy costs, the cost of servicing the train stock, repair costs, the cost of infrastructure charges, the number of days for the movement of trains. A correction coefficient was introduced for bringing the analysed costs closer to the real ones reported in the accounting documents of BDZ-PP Ltd. It was found out that when the rolling stock is completely changed, company's expenses are expected to decrease by BGN 82 million per year. The most significant share is due to a decrease in energy resources – 51%, followed by repair – 39%, train staff – 17% and infrastructure fees – 16%.

1 Introduction

Railway rolling stock used for passenger services is of crucial importance for the profitability of railway operators.

Railway vehicles at present used in BDZ – Passenger Services Ltd. (BDZ-PS) fleet have been in service for more than thirty years, are too old and in bad technical condition, so they make worse economical indexes of the enterprise. In order to overcome this negative trend and make transport service more attractive and thus increase the number of passengers, the management of the national railway operator – BDZ-PS is planning to buy modern rolling stock. In accordance with this expert opinion from Technical University of Sofia, from Department of Railway Engineering in particular, was commissioned. The task was to develop methodology for profitability assessment in case of old rolling stock replacement.

2 Methodology for profitability assessment in case of old rolling stocks replacement with new trains

This methodology contains the main principles and algorithms for profitability estimation of a new trains purchase for BDZ-PS fleet. It is based on costs calculation using:

- railway vehicles in the fleet now;
- new trains which would replace the rolling stock in service now.

The principles for calculating the annual costs for the present rolling stock and the new trains are absolutely the same. For that purpose the costs for every train on the timetable are calculated. All trains costs and total cost for the BDZ-PS – R_{Σ} is the sum total of all trains costs. The costs for every train are calculated according to formula (1):

$$R_j = (R_{1j} + R_{2j} + R_{3j} + R_{4j})N_j k, \qquad (1)$$

where R_{1j} are energy (electric or diesel fuel) costs, lv.; R_{2j} –locomotive and train crew costs calculated for a particular train, lv.; R_{3j} –tractive rolling stock maintenance costs, lv.; R_{4j} – infrastructure charge for the train, lv.; N_j – number of the days for a given train movement, according to the timetable; k – correctional coefficient which adduces costs from the other parameters in (1) to the real ones accounted for in BDZ-PS reports. It is used because part of the real costs do not depend on features of service and it is not included in formula (1).

The energy costs R_{1j} are calculated according to the formula (2):

$$R_{1j} = W_j C, \qquad (2)$$

where W_j is the work done by the train, and it is defined by the multiplication of the gross weight of the train and transport distance, t.km; C – the value of the energy source, lv./t.km.

The paper uses average costs for the electric locomotives while the costs for the diesel locomotives are differentiated according to their series. This is official information from BDZ-PS and is shown in Table 1.

Locomotive and train crew costs R_{2j} are calculated according to formula (3):

$$R_{2j} = H_j \sum_{i=1}^{5} n_i S_i, \qquad (3)$$

* Corresponding author: vms123@tu-sofia.bg

where H_j is the work time for serving the train and it is defined as the sum of the train movement time, 1-hour time for preparation and 0,5-hour after train arrival time according to the timetable; n_i – the number of locomotive and train crew members such as a locomotive driver, train master, conductor and others; S_i – the hourly rate for the different crew members in BDZ-PS: locomotive driver – 11,28 lv./h, assistant locomotive driver – 8,14 lv./h, train master – 7,46 lv./h.

The tractive rolling stock maintenance costs R_{3j} are calculated according to formula (4):

$$R_{3j} = L_j S_r, \quad (4)$$

where L_j is the distance between the initial and final station of the train destination, km; S_r – the tractive rolling stock current maintenance costs, lv./km.

The infrastructure charge for each train on the timetable is defined by means of the National Railway Infrastructure Company methodology using formula (5):

$$R_{4j} = L_j S_v + M_j L_j S_b = L_j \left(S_v + M_j S_b \right), \quad (5)$$

where L_j is as in (4); S_v – the costs for a train per kilometer of the train movement, S_v = 0,7902 lv./train.km; M_j – the train gross weight, t; S_b – the costs rate per gross-ton-kilometer, S_b=0,0025, lv./gtkm.

The overall costs of BDZ-PS for transportation are a sum which includes the costs for all trains, (6):

$$R_\Sigma = \sum_{j=1}^{n} R_j, \quad (6)$$

where n is the number of the all trains.

Table 1. Value of the energy source.

Rolling stock	C, lv./t.km
Electric	0,01000
Diesel locomotive series 07	0,04649
Diesel multiple unit (DMU) series 10	0,01962
Diesel locomotive series 55	0,06587
Diesel locomotive series 75	0,05840
Diesel locomotive series 77	0,06056

3 Efficiency of replacement of the whole fleet

Efficiency of replacement of the whole fleet without available DMU series 10 and electric multiple unit (EMU) series 30 and 31 is defined for the different types of trains with the aim to:

- estimate the overall efficiency of tractive rolling stock (RS) replacement;
- suggest a strategy for priority replacement of the type of trains in order to fulfill the timetable requirements.

The methodology for profitability assessment described in part 2 was used to do the estimation. The data about the available rolling stock: the value of the energy source, the hourly rate for the different crew members etc. are shown in BDZ-PS official account report for 2016.

The information was used for the new trains as follows:

- the value of the energy source, lv./t.km – the values are averaged according to catalogue information from different leading producers. These values are on average by 20% less than the values for the electric locomotives series 44, 45, 46, 61, and EMU series 32 and diesel locomotives series 07, 55, 75, 77 and they are on average equal to the values for DMU series 10 and EMU series 30, 31. These locomotives, DMUs and EMUs are in service at present;
- hourly rate for crew members, lv./h – the values are the same as the values on the BDZ-PS report now. The number of crew members is changed because the new trains do not require an assistant locomotive driver;
- tractive rolling stock current maintenance costs, lv./km are the same as these for DMU series 10 and EMU series 30 and 31;
- the infrastructure charges are calculated according to (5).

The summarized results from the calculation are shown in Table 2.

On the basis of the data in Table 2 the following conclusions have been made:

- in case of all fleet replacement the costs of BDZ-PS would be reduced;
- the values of the costs reduction are as follows: energy source – 51%, rolling stock maintenance costs – 39%, train crew costs – 17% and infrastructure charges – 16%;
- transportation by fast trains with obligatory passenger registration is the biggest part in cost reduction – 52,5%, regional trains – 20%, suburban trains – 16%, and transportation on secondary railways – 11,5%. On the basis of this data the priority of replacement has been defined.

4 Profitability of rolling stock replacement until 2023

The profitability of RS replacement over the period of up to 2023 is calculated according to the methodology described in part 2 under the following conditions [1-5]:

- six new speed trains are bought;
- ten new EMUs with four wagons and with a high level of comfort are bought;
- twenty-five EMUs with two and three wagons are bought.

The new rolling stock makes it possible to serve 109 trains on the timetable [1, 2]. Their types and number are as follows:

- 12 speed trains with speed of operation 160 km/h;
- 26 fast trains with speed of operation 130÷140 km/h;
- 71 suburban, regional and trains on secondary railways.

Analysis of the results (Table 3) shows that the costs for all trains in the timetable have been reduced as follows:

- energy costs – by 20%;
- current maintenance costs – by 14,5%;
- infrastructure costs – by 6,9%;
- train crew costs – by 5,5%.

DOI: 10.1051/matecconf/201713303005

Table 2. Summarized results from the calculation.

Energy source value per day, lv.		Train crew costs per day, lv.		Current maintenance costs per day, lv.		Infrastructure costs per day, lv.		General costs per year, lv.		Costs reducing, lv.
Present	New	Present	New	Present	New	Present	New	Present	New	
Fast trains with obligatory passenger registration										
12 005	4 962	2 623	1 982	8 506	4 962	6 502	5 051	8 666 174	4 960 792	3 705 382
41%		24%		42%		22%		43%		
Fast trains										
68 028	27 466	20 312	16016	43 869	25 701	34 933	26 638	92 636 413	52 907 055	39 729 358
60%		21%		41%		24%		43%		
Suburban trains										
37 820	23 737	26 708	23 665	28 409	20 711	24 472	22 062	59 527 065	46 212 198	13 314 867
37%		11%		27%		10%		22%		
Regional trains										
28919	15 648	15355	12 228	25328	15 776	19154	16 538	49 981901	33 339 240	16 642 662
46%		20%		38%		14%		33%		
Trains on secondary rail ways										
27 638	13 358	12 826	10 655	8 621	3 397	7 640	7147	22 831040	13 572 488	9 258 552
52%		17%		61%		6%		41%		
General for BDZ - PS										
174 409	85197	77 824	64 547	115 032	70 546	92 701	77 436	233 642 594	150 991773	82 650 821
51%		17%		39%		16%		35%		

Table 3. Summerized results from the calculation in the case of rolling stock replacement until 2023 (stage 1).

Annual costs with available rolling stock, lv.	Annual costs with new rolling stock, lv.	Annual costs reduction, lv.	Costs reduction, %
233 642 594	199 481 068	34 161 526	14,6 %

5 Profitability of rolling stock replacement until 2028

The profitability of RS replacement in the second period from 2023 to 2028 is calculated according to the Methodology for profitability assessment described in part 2 of this paper under the following conditions [1, 2]:

- four new speed trains are purchased;
- ten new EMUs with four wagons and with a high level of comfort are purchased;
- thirty EMUs with two and three wagons are purchased.

The new rolling stock wil make it possible to serve 150 trains more than in stage 1 [1, 2]. The types and number of trains are:

- 10 speed trains with a speed of operation 160 km/h;
- 48 fast trains with a speed of operation 130÷140 km/h;
- 92 suburban, regional and trains on the secondary railways.

After stage 2 costs reduction would be about 35 million lv., as compared to stage 1, as a result of the replacement.

The calculations show that the costs for one destination of all trains in the timetable are reduced after stage 2 as follows:

- energy costs – by 18,9%;
- current maintenance costs – by 12,96%;
- infrastructure costs – by 6,9%;
- train crew costs – by 5,4%.

6 General profitability of the rolling stock replacement until 2028

The general profitability of RS replacement during the stage 1 and 2 is the sum total of the results calculated in part 4 and part 5 of this paper.

BDZ-PS costs reduction is about 70 million *lv.* annually and it means 30% less in comparison with 2016. The results of the calculation show that after stages 1 and 2 costs reduction in comparison with 2016 are as follows:

- energy costs – 39%;
- current maintenance costs – 27,5%;
- infrastructure costs – 12,4%;
- train crew costs – 10,9%.

7 Conclusions

1. Methodology for profitability assessment of the fleet replasment has been developed.

2. On the basis of this methodology the priority of the replacement is defined. It includes speed trains, regional, suburban and secondary railway trains.

3. BDZ-PS costs reduction levels are:

- up to 34 million lv. annually in comparison to 2016 year, after the end of the stage 1;
- additional up to 35 million lv. annully during stage 2 in comparison with stage 1;
- in general about 70 million annually in comparison with 2016 after stage 1 and 2.

4. Costs reduction due to the energy efficiency of the new trains, low level of current maintenance and infrastructure charges and crew costs.

The research shows that BDZ-PS has to replace RS in accordance with stage 1 and 2 [1-5]. As a result of this passenger transportation will be more comfortable and the return on investment will be within 6 to 8 years.

References

1. O. Krastev, K. Velkov, V. Stoilov, I. Petrov, S. Slavchev, Defining the number and the type of the BDZ – Passenger Services new trains, *BulTrans-2017*, (2017)
2. K. Velkov, O. Krastev, V. Stoilov, I. Petrov, S. Slavchev, Defining the main parameters of the BDZ – Passenger Services trains, *BulTrans-2017*, (2017)
3. K. Velkov, O. Krastev, V. Stoilov, B. Damyanov, Opportunities for improve locomotives maintenance, *BulTrans-2012*, 15-20, (2012)
4. O. Krastev, K. Velkov, V. Stoilov, An appropriate system for attendance trains with locomotive crews, *BulTrans-2012*, 21-28, (2012)
5. V. Stoilov, K. Velkov, O. Krastev, The main economical indexes analyses of the locomotive maintenance in Bulgarian state railways passenger transportation, *BulTrans-2012*, 29-37, (2012)

Research of oxyhydrogen gas mixture influence upon diesel engine performance

Evgeni Dimitrov[1,*], *Deyan* Deltchev[2], *Vladimir* Serbezov[3], and *Spas* Pantchev[1]

[1]Technical University – Sofia, Department of Combustion Engines, Automobile Engineering and Transport, 8 Kliment Ohridski Blvd., 1000 Sofia, Bulgaria
[2]New Energy Corporation Ltd., 2B Rozhen Blvd., 1000 Sofia, Bulgaria
[3]Technical University – Sofia, Department of Air Transport, 8 Kliment Ohridski Blvd., 1000 Sofia, Bulgaria

Abstract. The paper presents the results from testing a Volkswagen 1.9 D diesel engine on a test bench to work on gas-diesel cycle with oxyhydrogen gas mixture. Experimental research is done to show the impact of oxyhydrogen gas mixture on engine consumption and environmental indexes such as: fuel and specific fuel consumption; carbon monoxide; carbon dioxide; oxides of nitrogen; smoke emissions. The oxyhydrogen gas mixture delivered to the engine intake manifold with constant flow rate. The results are obtained under research contract № 6524-4/2016.

1 Introduction

Oxyhydrogen gas mixtures (OHGM) are also known as "Brown's Gas" (BG) or "Gas of Brown". According to data from internet sources – [1, 2] Ilia Valkov is considered to be Brown's gas discoverer. He emigrates in Australia at the end of the 1950s, where he lived and worked under the name of Yull Brown.

Oxyhydrogen gas fuel mixture is produced in generators, where electrolysis of sodium or potassium hydroxide water solution as well as some salts takes place [3, 4, 5]. As a result of the inside generator processes a gas mixture from hydrogen and oxygen is released in a volumetric ratio of two to one.

In burner combustion conditions Brown's gas flame temperature can reach very high values and for this reason it has found its application in a variety of technological processes like: cutting, melting, welding etc. – [3, 6].

In the past decade a number of companies have started manufacturing and offering oxyhydrogen fuel mixture generators, intended for mounting on transport vehicles with internal combustion engines. The generated OHGM is delivered to the common part of engine inlet manifold and serves as extra fuel [1, 5, 7]. A small gas volumetric fuel rate delivered from generator is the main problem of applying OHGM as extra fuel in internal combustion engines.

A Bulgarian company – "New Energy Corporation" has designed their own generator, with relatively big overall dimensions and with a maximum fuel flow rate of OHGM – 12 m^3/h (200 l/min). The generator with the abbreviation *VST-4C* is shown at Fig. 1.

Oxyhydrogen gas mixture production is carried out through the electrolysis of a five percent water solution of potassium hydroxide. Distilled water, purified in a special way, is used. Oxyhydrogen gas mixture is produced in the generator compartment, which consists of four working cells; gas separating section; main fluid circulation pump and a cooling system. The generator part of the system is powered with a single phase alternating voltage of 220 V. The maximum electrical power, consumed by the generator is 6 kW.

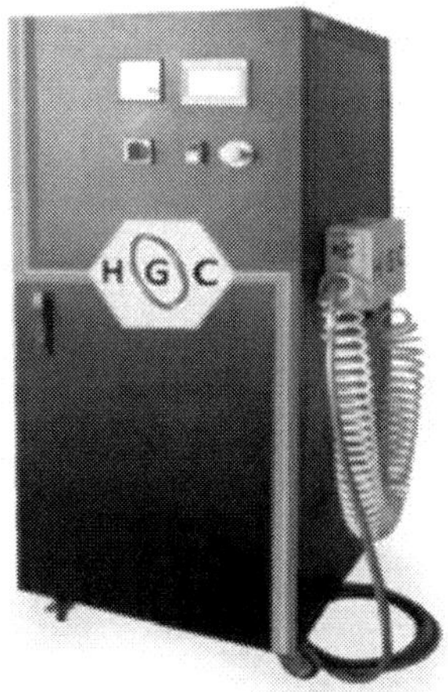

Fig. 1. Generator of oxyhydrogen gas mixture *VST-4C*, product of "*New Energy Corporation*".

The VST-4C generator is equipped with automatics that control and govern some crucial system parameters: system pressure; work processes temperature; flow rate of OHGM; pressure hysteresis of OHGM; electrolyte level; OHGM gas leakage; flowing electric currency magnitude; cooling system-turn on and off; functionality of electric driven elements etc.

* Corresponding author: etzd@tu-sofia.bg

2 Research point

The aim of the present research is to determine the impact of the oxyhydrogen gas mixture, delivered to common part of the inlet manifold, on diesel engine consumption and environmental indexes.

3 Object of research and methods

3.1. Object of research

Four cylinder four stroke diesel engine *Volkswagen 1,9 D* has been chosen for the experiment with parameters and indexes as follows: cubic capacity – $i.V_h = 1{,}896$ dm^3; compression ratio – $\varepsilon = 22{,}5$; brake power – $P_e = 47$ kW at speed $n = 4400$ min^{-1}. This is a swirl chamber engine equipped by the manufacturer with a distributor type fuel injection pump. Oxyhydrogen gas mixture is delivered to the inlet manifold entry through an additionally machined port mounted on the engine. A picture of the experimental engine is shown in Fig.2.

Fig. 2. Picture of the experimental engine.

3.2 Research method

To obtain four load characteristics at engine speed of rotation – $n = 2000$ min^{-1} is at the core of the research methodology. The first one is the so called basis – engine runs with standard diesel fuel and the other three are for comparison.

The data for two of the comparison load characteristics are taken with engine running on gas-diesel cycle (GDC) with oxyhydrogen gas mixture at two different flow rates of gaseous fuel.

In each of these characteristics OHGM volumetric flow is constant, despite the engine load. The third load characteristic data is from running the engine on standard diesel fuel, but after one hour operation on gas-diesel cycle with OHGM.

It is necessary to obtain data for the last characteristic so that the subsequent effect of OHGM on diesel engine effective indexes can be weighed up. It is stated [4] that OHG and water steam – the result of gas burning process, have the effect of cleaning soot and tar deposition on the piston, cylinder, engine valves and engine fuel system (injection nozzle holes). It is also known [8] that accumulation of carbon and tar deposit to above mentioned elements leads to worse engine technical condition.

From putting together the test results, the characteristics data are taken under the following conditions:

- same adjustment (according to manufacturer prescriptions) of valve-gear mechanism and engine fuel system;
- same engine heat condition (same temperature of cooling liquid and engine oil);
- same weather conditions i.e. obtain the characteristics data the same day.

4 Test results

The tests were carried out at 50 kW test laboratory of "Combustion Engines, Automobile Engineering and Transport" Department in TU-Sofia equipped with electric constant current dynamometer test bench type *SAK*-50. The test bench makes it possible to measure engine effective power indexes.

The calculation of fuel consumption and specific diesel fuel consumption is carried out by means of fuel flow meter that measures the time for the consumption of the exact volume of liquid fuel by the engine. The diesel fuel flow meter is engineered at the Department of Combustion Engines, Automobile Engineering and Transport in TU-Sofia, and the volume of consumed diesel fuel is read with an accuracy of 0,2 cm^3.

The oxyhydrogen gas volumetric flow rate is to be measured with gas flow meter type *SD 5000* of the *IFM* company (Germany).

Boston gas-analyzer of *Tecno control* (Italy) and smoke meter "*EFAW 68A*" of *Bosch* (Germany) are used for measuring the engine environmental indexes. The gas-analyzer reads the following components in the engine exhaust emissions: carbon monoxide – CO (accuracy 1 ppm); carbon dioxide – CO_2 (accuracy 0,1 %); nitrogen monoxide – NO (accuracy 1 ppm); oxides of nitrogen – NO_x (accuracy 1 ppm); oxygen with an accuracy of 0,2 points.

A type K thermocouple and electronic measuring instrument with an accuracy of 1 °C has been used to measure the exhaust gas temperature. The thermocouple is mounted on engine exhaust manifold, at a distance of 100 mm from the outlet collector flange.

Most of the results are shown in Fig. 3 to Fig. 12, where additional abbreviations are used:

- n – engine speed, min^{-1};
- p_e – mean effective pressure, MPa;
- B_{hdf} – engine fuel consumption (diesel fuel), kg/h;
- b_{edf} – specific fuel consumption (diesel fuel), g/kWh;
- T_{EG} – exhaust gas temperature, °C;
- Q_{HHO} – volumetric flow rate of oxyhydrogen gas mixture, l/min;
- R_b – smoke emissions (Bosch Smoke Number).

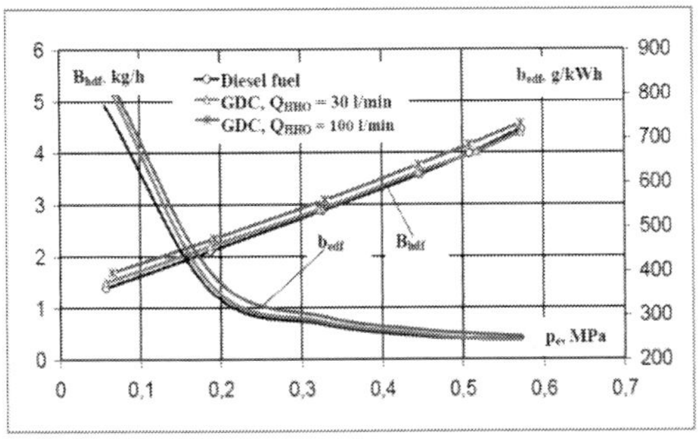

Fig. 3. Effect of OHGM on the fuel consumption of diesel engine *Volkswagen* – 1,9 D at speed n = 2000 min^{-1}.

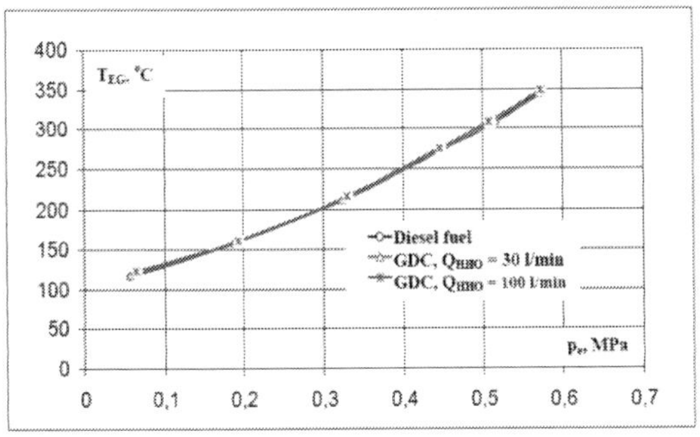

Fig. 4. Effect of OHGM on exhaust gas temperature of diesel engine *Volkswagen* – 1,9 D at speed n = 2000 min^{-1}.

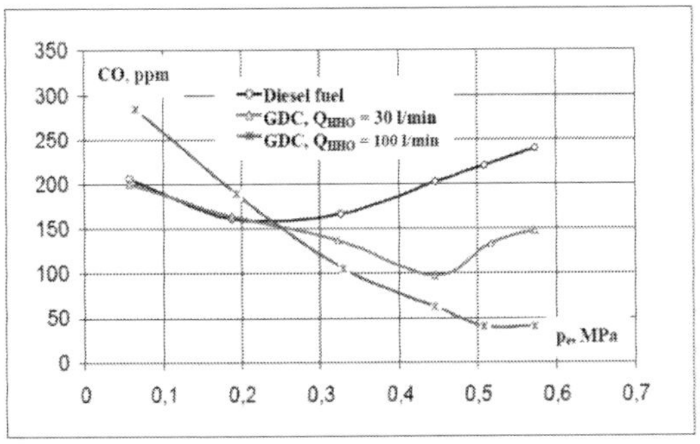

Fig. 5. Effect of OHGM on carbon monoxide emissions of diesel engine *Volkswagen* – 1,9 D at speed n = 2000 min^{-1}.

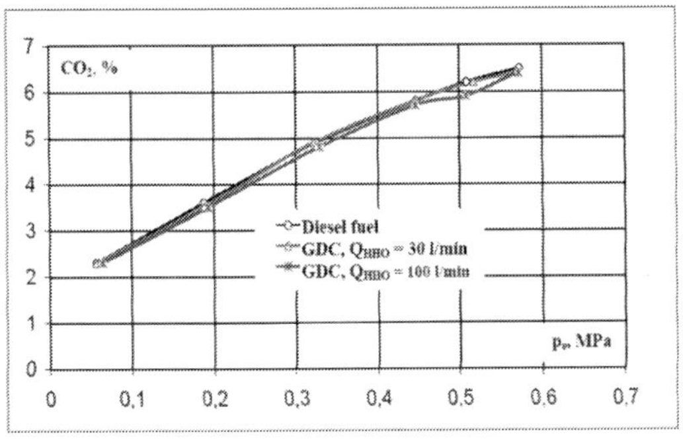

Fig. 6. Effect of OHGM on carbon dioxide emissions of diesel engine *Volkswagen* – 1,9 D at speed n = 2000 min^{-1}.

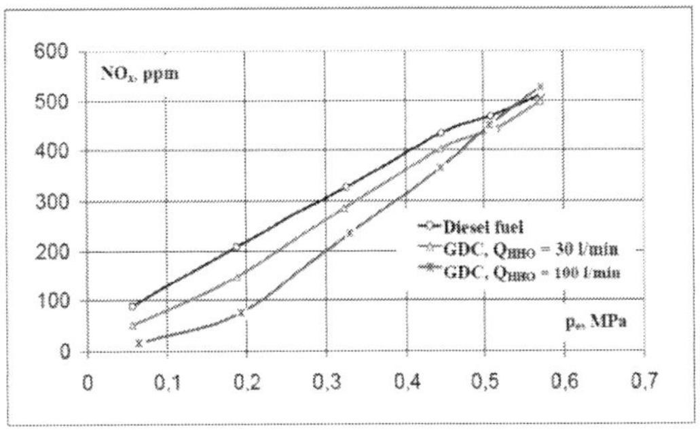

Fig.7. Effect of OHGM on oxides of nitrogen emissions of diesel engine *Volkswagen* – 1,9 D at speed n = 2000 min^{-1}.

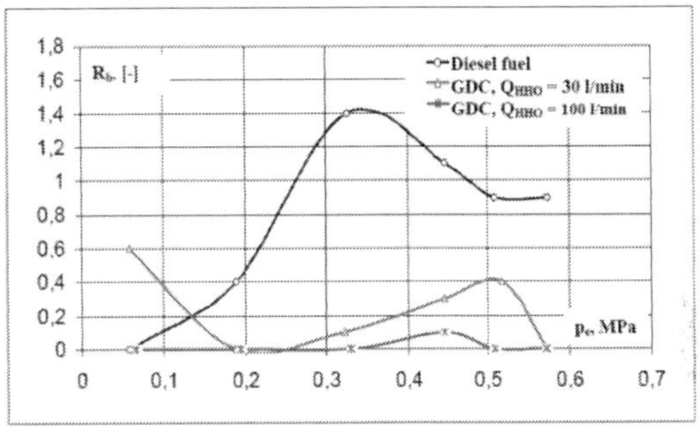

Fig. 8. Effect of OHG on smoke emissions of diesel engine *Volkswagen* – 1,9 D at speed n = 2000 min^{-1}.

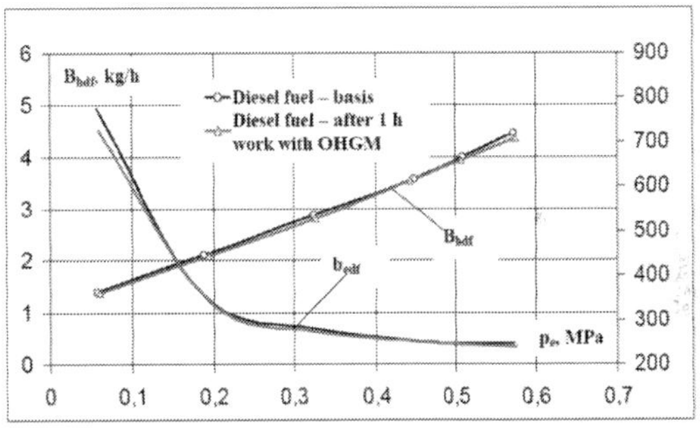

Fig.9. Comparison of fuel consumption of diesel engine *Volkswagen* – 1,9 D at speed n = 2000 min^{-1} after 1 hour running with OHGM.

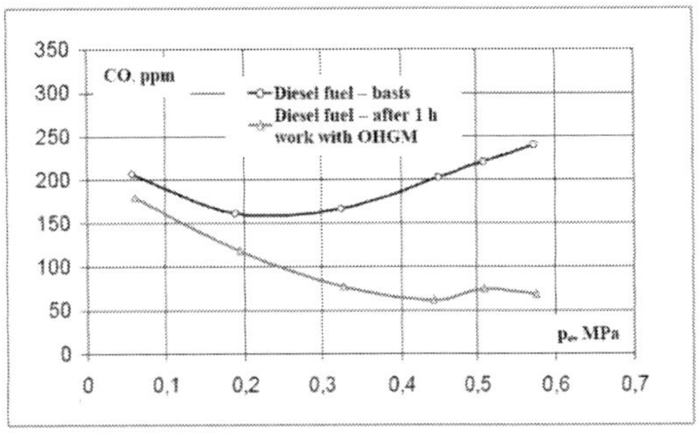

Fig.10. Comparison of carbon oxide emissions of diesel engine *Volkswagen* – 1,9 D at speed n = 2000 min^{-1} after 1 hour running with OHGM.

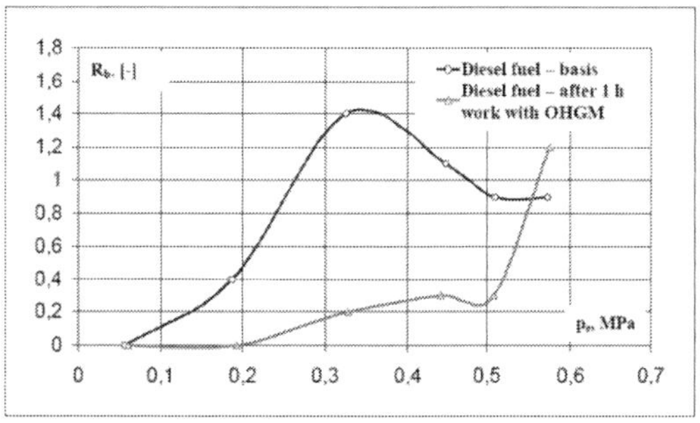

Fig.11. Comparison of exhaust gas smoke emissions of diesel engine *Volkswagen* – 1,9 D at speed $n = 2000$ min^{-1} after 1 hour running with OHGM.

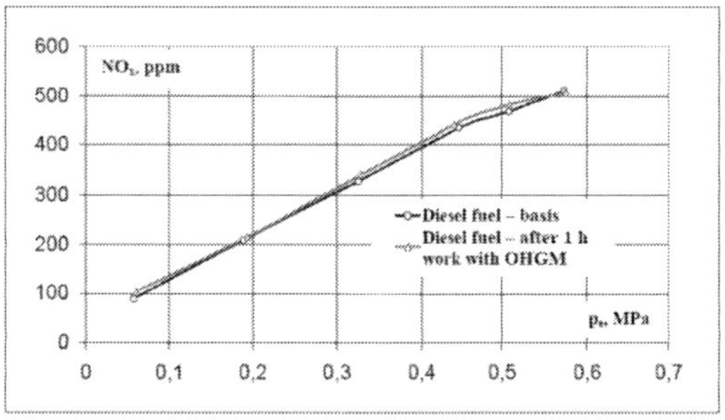

Fig. 12. Comparison of oxides of nitrogen emissions of diesel engine *Volkswagen* – 1,9 D at speed $n = 2000$ min^{-1} after 1 hour running with OHGM.

The impact of oxyhydrogen gas mixture on engine consumption indexes has been studied through a comparison of the corresponding diesel fuel consumption and specific diesel fuel consumption. This comparison method results from the fact that the composition and physical properties of the thus generated oxyhydrogen gas mixture are not defined in advance, so although the volumetric flow rate is known – the mass flow rate cannot be calculated. On the other hand, this allows for a comparison of the obtained data with other authors [9].

A watt-hour meter was used during the experimental research to measure the electric energy consumed by the oxyhydrogen gas mixture generator – E_{HHO}. With a volumetric flow rate of OHGM – $Q_{HHO} = 30$ l/min consumed electric energy is 0,82 kWh and with a volumetric flow rate of $Q_{HHO} = 100$ l/min – $E_{HHO} = 2{,}69$ kWh respectively.

The analysis of the experimental results of running the engine on a gas-diesel cycle with oxyhydrogen gas mixture at the thus examined engine rotation speed shows:

• at a volumetric flow rate of oxyhydrogen fuel gas mixture - $Q_{HHO} = 100$ l/min there is obvious reduction in fuel consumption – B_{hdf} and specific fuel consumption – b_{edf} of diesel fuel, respectively to 4 % and to 15,7 %, when running the engine on a gas-diesel cycle with OHG as compared to running with standard diesel fuel;

• at a volumetric flow rate of oxyhydrogen gas mixture – $Q_{HHO} = 30$ l/min there is no significant difference in engine consumption;

• change of carbon monoxide emissions – CO are not one-sided. At small engine loads carbon monoxide emissions are lowered by up to 29 % when running the engine with standard diesel fuel as compared to running the engine on a gas-diesel cycle with a volumetric flow rate of OHGM – $Q_{HHO} = 100$ l/min. Under the same load conditions the alterations of carbon monoxide emissions when running the engine with standard diesel fuel and running the engine on gas-diesel cycle with volumetric flow rate of OHGM – $Q_{HHO} = 30$ l/min are negligible. At middle and high engine loads considerable decrease of carbon monoxide emissions is observed – when running the engine at gas-diesel cycle with volumetric flow rate of OHGM – $Q_{HHO} = 100$ l/min as compared to running on standard diesel fuel and running on gas-diesel cycle with volumetric flow rate of OHGM – $Q_{HHO} = 30$ l/min – up to 6 times and up to 3,7 time respectively;

• oxides of nitrogen emissions NO_x – when running the engine on gas-diesel cycle with oxyhydrogen volumetric flow rate – $Q_{HHO} = 100$ l/min are significantly as compared to running with standard diesel fuel and running on gas-diesel cycle with OHGM volumetric flow rate – $Q_{HHO} = 30$ l/min – respectively: up to 2,7 times and up to 1,9 times;

• exhaust gas smoke emissions R_b – when the engine is running on gas-diesel cycle with volumetric flow rate of OHGM – $Q_{HHO} = 100$ l/min, as well as with volumetric flow rate of OHGM – $Q_{HHO} = 30$ l/min – they are significantly lower as compare to running the engine with standard diesel fuel – up to 1,4 times. When the engine runs on a gas-diesel cycle with volumetric flow rate of OHGM – $Q_{HHO} = 100$ l/min exhaust gas smoke – R_b is up to 30 % lower as compared to running the engine on gas-diesel cycle with volumetric flow rate of OHGM – $Q_{HHO} = 30$ l/min;

• emission variations of carbon dioxide – CO_2 and free oxygen O_2 in engine exhaust gas when running on standard diesel fuel and running on gas-diesel cycle with OHGM are defined by measurement accuracy;

• there are no variations in engine exhaust gas temperature when running with standard diesel fuel or gas-diesel cycle with oxyhydrogen gas mixture.

These experimental results of engine running with standard diesel fuel and speed of rotation $n = 2000$ min^{-1}, but after one hour running on gas-diesel cycle with OHGM, leads to the following conclusions:

• there is improvement in the engine consumption indexes of up to 3 %;

• there is essential decrease of emissions of carbon monoxide in engine exhaust gas – up to 70 %;

• many times decrease of engine exhaust gas smoke – up to 7 times;

• changes in the rest of the engine environmental indexes are negligible.

5 Conclusion

The analysis of the results of the experimental research leads to the conclusion that oxyhydrogen gas mixtures can be used in practice as extra fuel for diesel internal combustion engines, since the use of such gases leads to improvement of the engine environmental and, in the case of sufficient flow rate, consumption indexes. In the future there should be research of the methods and technical instruments for developing powerful oxyhydrogen gas mixture generators with acceptable dimensions so that they can be used on transport vehicles, set in motion by internal combustion engines.

References

1. I. Fudal, Using HHO Gas to Reduce Fuel Consumption and Emissions in Internal Combustion Engines, Thesis of M.Sc., University of Khartoum, Sudan, (2010)
2. https://littlebg.com/braunov-gas-koito-promeni-svetut.
3. R. Sterionov, Obtaining of Brown gas and possibilities for its application, Management and sustainable development **3-4**, 21, 228-230, University of Forester, Sofia, Bulgaria, (2008)
4. http://sintezgaz.org.ua/1_articles/127/chto-takoe-gaz-brauna.
5. E. Leelakrishnan, N. Lokesh, H. Suriyan, Performance and emission characteristics of Brown's gas enriched air in spark ignition engine, International Journal of Innovative Research in Science, Engineering and Technology, **2**, 2, 393-404, (2013)
6. K. Borse, J. Chavan, A. Ingale, K. Jejure, S. Bakal. Water powered hydrogen gas welding & cutting, *7th International Conference on Science, Technology and Management*, 214-218, India, (2017)
7. K. Ambarev, V. Nikolov, A. Ivanov. System for experimental study of diesel engine at work with Brownian gas, Machine Mechanics, **2**, 115, 20-23, Technical University Varna, Bulgaria, (2016)
8. I. Ivanov, Operating failures of internal combustion engine valves, *"ECO Varna"*, Bulgaria, (2017)
9. K. Ambarev, Experimental study of diesel engine at work with Brownian gas. Machine Mechanics, **2**, 115, 24-27, Technical University Varna, Bulgaria, (2016)

Pre-swirl mechanism in front of a centrifugal compressor: effects on surge line and on unsteady phenomena in surge area

Amélie Danlos[1,*], *Pierre* Podevin[1], and *Michel* Toussaint[2]

[1]Conservatoire National des Arts et Métiers, Laboratoire de Chimie Moléculaire, Génie des Procédés Chimiques et Energétique (CMGPCE – EA7341), 292 Rue Saint Martin, 75003 Paris, France
[2]Association Technique Turbine à Gaz (ATTAG), 15 rue Marat, 78210 Saint-Cyr-l'Ecole, France

Abstract. Using a pre-swirl mechanism upstream an impeller of a compressor allows to modify its characteristics curve, while weakly damaging its efficiency. Another consequence of the pre-swirl is to push back the surge line limit and to increase the operation zone towards the low flow rate limits. A centrifugal compressor has been modified in order to add a swirl generator device upstream the impeller. The incidence values of blades can vary from 0° (no pre-swirl) to ±90°. The variation of the stator blades incidence has several main consequences: to allow a flow rate adjustment with a good efficiency conservation, to increase the angular velocity with a constant shaft power, to produce a displacement of the surge line limit. In this paper, the results of experimental studies are presented to analyze the surge line and the intensity of unsteady phenomena when the compressor works in its surge area.

Nomenclature

- π compression ratio;
- q_m mass flow rate;
- N rotation speed;
- RPM rotation per minute;
- ω angular velocity;
- δ pre-swirl angle;
- P shaft power;
- U_1,U_2 respectively entrainment velocity at the inlet and the outlet of the compressor wheel;
- V_1,V_2 respectively absolute velocity at the inlet and the outlet of the compressor wheel;
- α_1 angle between U_i and V_i at the compressor wheel inlet;
- α_2 angle between U_i and V_i at the compressor wheel outlet;
- β_1 angle between U_i and W_i at the compressor wheel inlet.

1 Introduction

In order to reduce pollutant emissions and to decrease fuel consumption, a large part of the current automotive market is equipped with supercharged engines. Turbochargers used for these efficient systems have numerous advantages but their working range has two limits imposed by choke conditions for very high mass flow rates and surge phenomenon for low mass flow rates. This phenomenon is translated by instabilities in the turbocharger, with the appearance of reversed flow at the inlet. Deep surge is very dangerous for the system because of the mechanical damages caused by these instabilities. It is then necessary to find a way to push back this low limit of working range. Many devices have been studied [1, 2]. The pre-swirl, also called inlet guide vane, is the solution chosen for this study.

The pre-swirl, using stator blades with variable angle of attack at the upstream of a compressor wheel engine, is a technique mainly applied to improve the flow rate adjustment and the power of large impellers. This method allows to obtain an angular deviation of the absolute velocity of the fluid upstream the leading edge of the compressor wheel blades (Fig. 1).

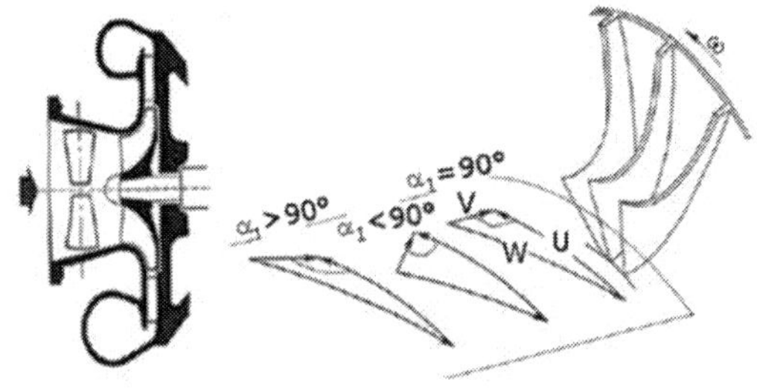

Fig. 1. Theoretical diagram of a pre-swirl system.

The consequence of this angular deviation is to modify, in accordance with Euler's equation, the compressor power input P with our experimental device (Equation 1).

$$P = q_m(U_2V_2\cos\alpha_2 - U_1V_1\cos\alpha_1). \quad (1)$$

Without pre-swirl, as α_1 is theoretically equal to 90°, the second term of Euler's equation is zero (at least at

* Corresponding author: amelie.danlos@lecnam.net

nominal point). If angle α_1 is no longer equal to 90° but takes the value imposed by the pre-swirl blades, it is immediately checked that the compressor power input increases or decreases according to the direction (angle δ) chosen by the user. In this paper, the positive pre-swirl represents values of δ which deflect the flow in the direction of angle β_1 of the blades at the compressor wheel inlet. Then, negative pre-swirl describes the deflection of the flow in the other direction. Experimental measurements presented in this paper have been carried out for three relatively low rotation speeds, in order to prevent compressor damage. Indeed, during measurements, when the compressor works in its surge area it is dangerous for the compressor itself but also for its associated circuit [3,4]. Thus, in this paper, we focus on compression ratio curves in order to visualize and analyze the surge line limit.

2 Compressor surge phenomenon

All compressors forced to operate at very low flow rate are submitted to a very unsteady and dangerous regime [3,4]. This aerodynamic instability has as its consequence, among others, strong fluctuations of pressure and flow rate. The inlet geometry of the circuit upstream the turbocharger has effects on the appearance of surge phenomenon [5]. It is then important to optimize this inlet.

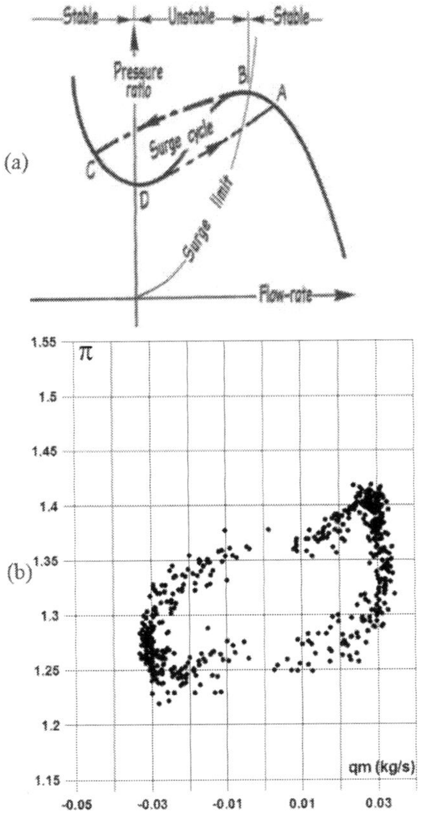

Fig. 2. Description of the surge loop: (a) Theoretical characteristic curve, (b) characteristic curve.

The unstable operating point characterizing deep surge describes a loop (Fig. 2) with an amplitude and a frequency dependent on the associated circuit [6-8].

3 Experimental device

The measurements have been conducted on a test bench composed of a pre-swirl located near the turbocharger inlet and a flow control valve placed downstream the turbocharger in order to control the flow rate introduced in the system. This device allows to modify the flow rate to use the turbocharger in its working range or even in the deep surge. Many sensors are used to measure different values at each component of the circuit: pressure sensors at the inlet and outlet of each component, temperature sensors and a flowmeter (placed upstream the pre-swirl). The pre-swirl device used has nine blades (Fig. 3). All blades are controlled by the same screw guidance to rotate all blades with the same angle to their own axes. The incidence angle of each blade can then be imposed by the user for different measurements. These blades can rotate at 360° around their axes.

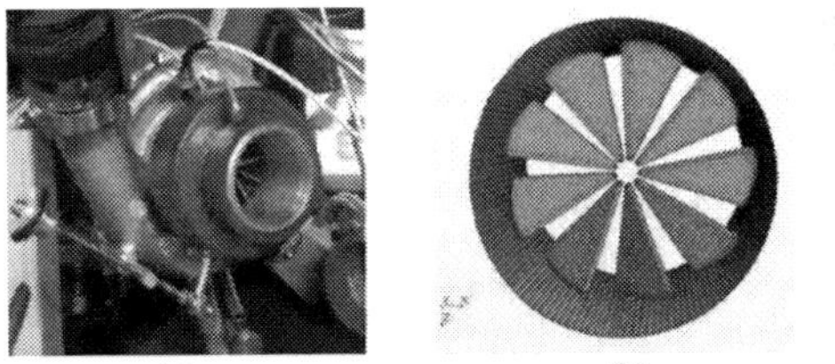

(a) (b)

Fig. 3. Pre-swirl device used at compressor inlet in the test bench: (a) picture of the pre-swirl inlet, (b) pre-swirl diagram in sketch.

4 Results

4.1 Effect of the pre-swirl on the surge line limit

The experimental results are presented in Fig. 4. These first measurements have been carried out for four arbitrarily chosen values of pre-swirl angle: 26°, 52°, –26° and –52°. A short analysis of these results confirms the effect of the pre-swirl of the flow upstream the compressor wheel on the appearance of the surge line limit. A pre-swirl in the flow direction at the compressor wheel inlet allows to push back this limit. This movement of the surge limit is accompanied by the light improvement of the compression ratio for a moderate pre-swirl angle.

In the case of a negative pre-swirl, the surge line limit is pushed back too for a considerable angle as for example with an angle of –52°. However, this angle of pre-swirl induces a large decrease of the compressor performances. On the other hand, a moderate negative pre-swirl has beneficial effects on the surge limit since it moves back in this case. The main challenge of this problem is to find the optimal angle to push back the appearance of surge phenomenon but also to improve compressor performances.

DOI: 10.1051/matecconf/201713304002

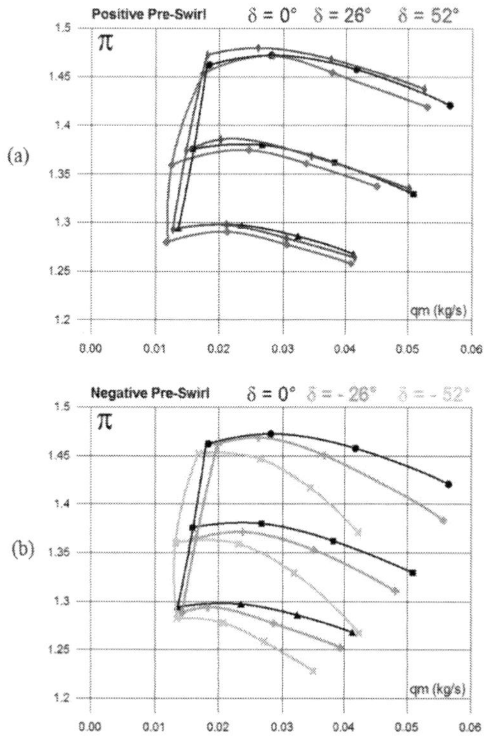

Fig. 4. Effect of different angles and direction of the pre-swirl on characteristic curves: (a) for a positive pre-swirl, (b) for a negative pre-swirl.

4.2 Effect of the pre-swirl on the intensity of instabilities in the surge area

The pressure and flow rate instabilities may be compared thanks to surge loops by analyzing the area. Indeed, the area of surge loops is sensitive to the intensity or to the dissipated energy during the compressor surge. Fig. 5 shows the evolution of pressures at the inlet and outlet of the compressor when the compressor operates in its surge area. The dissipated energy is then quantifiable with the loop describing the operating point.

Previous measurements, corresponding to five positions of the pre-swirl and three rotation speeds of the compressor, have been carried out by the surge loops recording (Figures 6 and 7). It progressively shuts the vane at the outlet of the compressor until a deep surge is reached simultaneously acting on the valve before the turbine to maintain a constant speed. Pressure measurements permit to calculate instantaneous compression ratio and air flow.

This zone corresponds to unstable operating points for which flow rate and compression ratio oscillate with a periodic signal described by a stable frequency.

In the case of a positive pre-swirl we can conclude that for any rotation regime, surge loop area decreases for a moderate pre-swirl at $\delta = +26°$ but seems to increase slightly for $\delta = +52°$. On the other hand, in the case of a negative pre-swirl, the surge loop area increases for $\delta = -26°$ and falls rapidly for $\delta = -52°$.

The measurement of the unsteady flow rate seems to be biased for some values of the pre-swirl angle. The location of the inclined blades of the pre-swirl between the flowmeter and the compressor disturbs the flow when it becomes reversed during the surge instabilities. The differential sensor associated to the flow rate measurement is not able to evaluate clearly the pressure variations due to the pressure reducing orifice.

5 Conclusions

The measurements obtained on our surge test bench show that the pre-swirl does not only affect the appearance of the surge phenomenon, as it was shown in the literature for other different systems, but also on the intensity of the instabilities in the deep surge range. This information is a new important result that implies further studies of the optimal incidence angle of pre-swirl blades are needed. In order to improve our study, new experiments are in progress. In order to analyze more precisely the flow upstream blades of the compressor wheel, a pressure measurement can be insufficient. An optical access allowing PIV (Particle Image Velocimetry) or LDV (Laser Doppler Velocimetry) measurements should be considered to have more precise results which will be completed with CFD (Computational Fluid Dynamics) simulations.

Our first analysis shows that a positive pre-swirl is beneficial for pushing back the surge line limit and for increasing the compression ratio in the case of a low incidence angle of pre-swirl blades (+26° for example). It also appears that for this angle, the adiabatic efficiency is slightly increased.

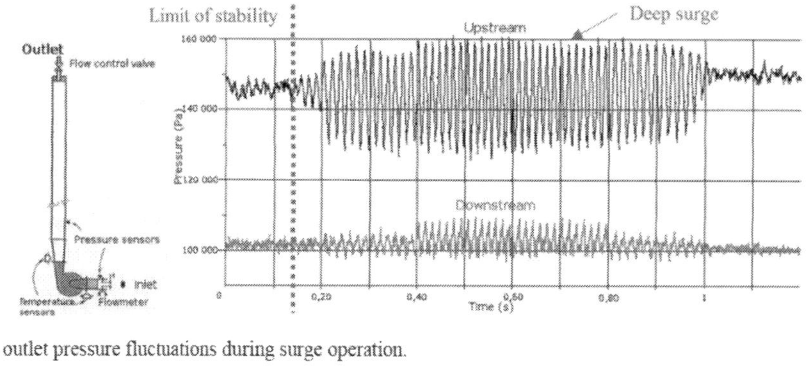

Fig. 5. Inlet and outlet pressure fluctuations during surge operation.

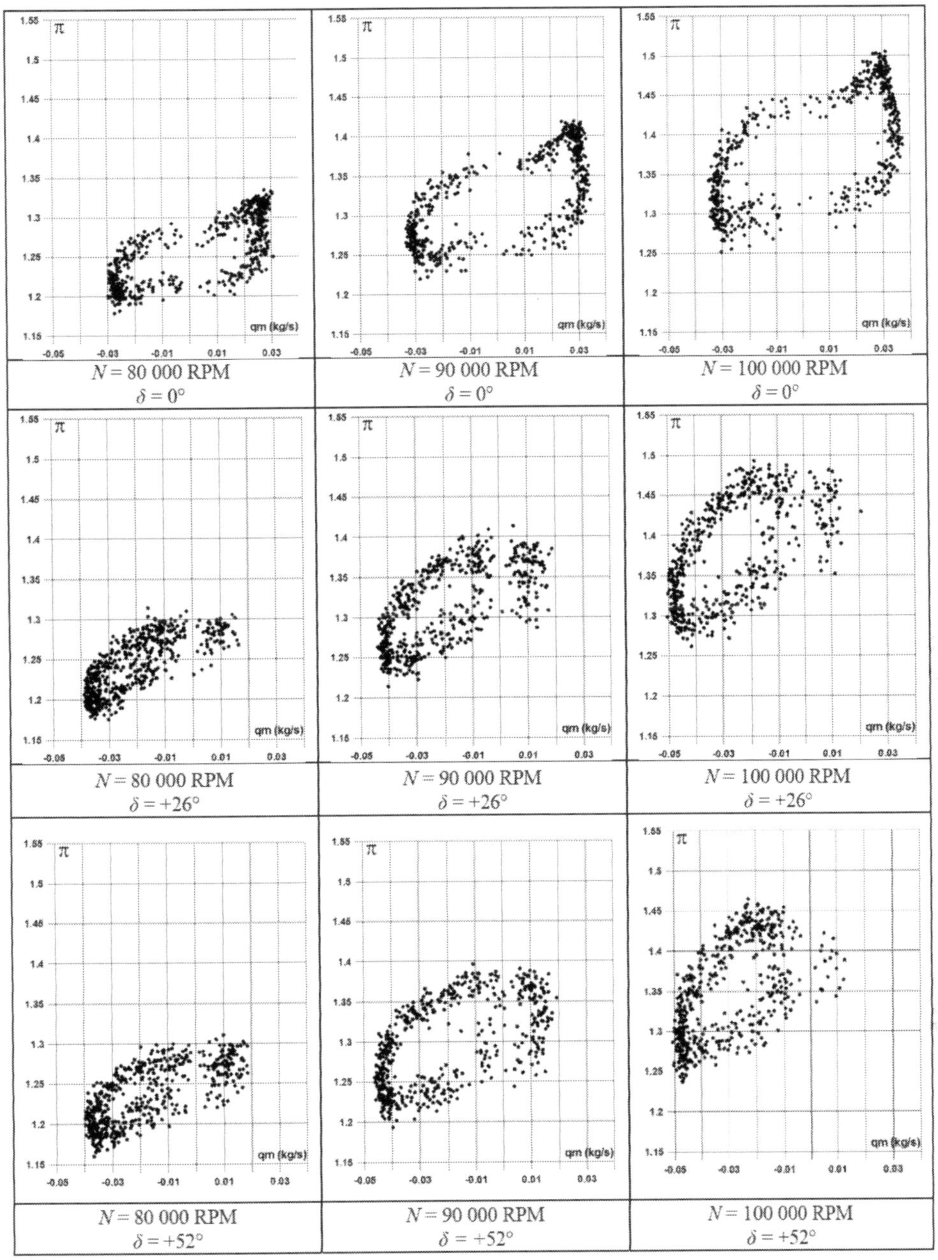

Fig. 6. Characteristic curves obtained in surge operation for a positive pre-swirl.

The authors would like to thank the Association Technique pour les Turbines A Gaz for its interest and for its logistic support.

References

1. R. Bidard, *Compte-rendu de l'Académie des Sciences*, Paris (1946)
2. J. Galindo, J. Serrano, X. Margot, A. Tiseira, N. Schorn, H. Kindl, Int. J. Heat Fuid Flow, **28** (2007)
3. J. Galindo, A. Tiseira, R. Navarro, D. Tari, C. Meano, App. Therm. Eng., **110** (2017)
4. J-M. Pugnet, Techniques de l'Ingénieur (2000)
5. M. Toussaint, M. Boudfar, *CIEM2007*, Bucarest, Romania (2007)
6. M. Toussaint, C. El Kamel, P. Aumoite, *ISROMAC13*, Honolulu, Hawaii (2010)
7. M. Toussaint, A. Lefebvre, G. Descombes, A. Marchal, *COFRET12*, Sozopol, Bulgaria (2012)
8. W. Yuchuan, T. Lei, Z. Baoshan, C. Shuliang, W. Binbin, JMES, **229**, 18 (2015)

DOI: 10.1051/matecconf/201713304002

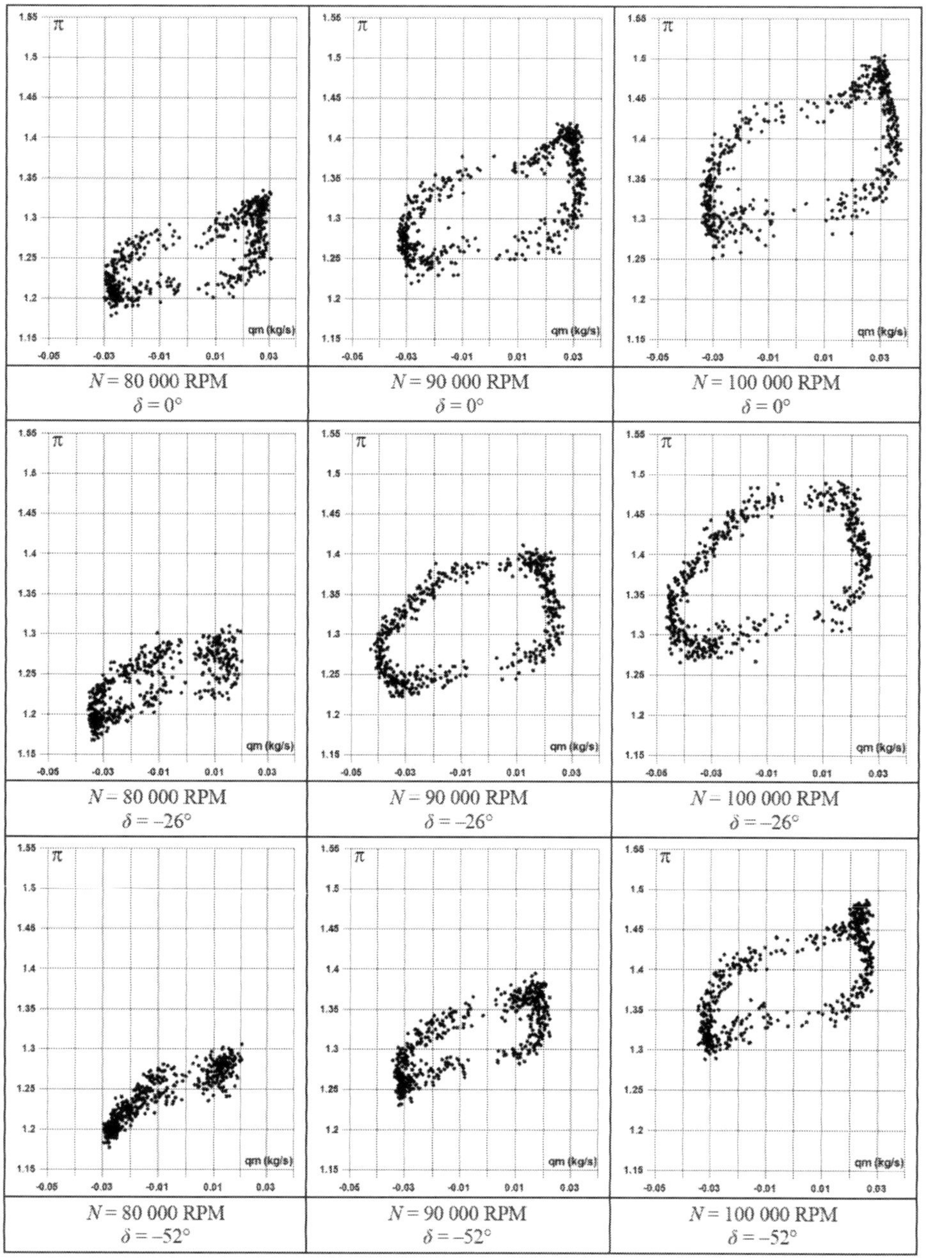

Fig. 7. Characteristic curves obtained in surge operation for a negative pre-swirl.

MATEC Web of Conferences **133**, 04003 (2017) DOI: 10.1051/matecconf/201713304003
BulTrans-2017

Experimental study on Rankine cycle evaporator efficiency intended for exhaust waste heat recovery of a diesel engine

Nikolay Milkov[1,*], *Quentin* Danel[2], *Plamen* Punov[1], *Christelle* Perilhon[2], *Pierre* Podevin[2], and *Teodossi* Evtimov[1]

[1] Technical University – Sofia, Department of Internal Combustion Engines, Automobiles and Transport, 1000 Sofia, Bulgaria
[2] Conservatoire National des Arts et Métiers, CMGPCE, 292 rue Saint Martin, 75003 Paris, France

Abstract. The paper pressents an experimental study of Rankine cycle evaporator efficiency. Water was chosen as the working fluid in the system. The experimental test was conducted on a test bench equipped with a burner charged by compressed fresh air. Generated exhaust gases parameters were previously determined over the diesel engine operating range (28 engine operating points were studied). For each test point the working fluid parameters (flow rate and evaporating pressure) were varied. Thus, the enthalpy flow through the heat exchanger was determined. Heat exchanger was designed as 23 helical tubes are inserted. On the basis of the results, it was found out that efficiency varies from 25 % to 51,9 %. The optimal working fluid pressure is 20 bar at most of the operating points while the optimum fluid mass flow rate varies from 2 g/s to 10 g/s.

1 Introduction

Waste heat recovery by means of Rankine cycle is a promising approach for achieving significant reduction in fuel consumption and CO_2 emissions of the vehicles. The efficiency of Rankine cycle strongly depends on the heat transfer in heat exchanger from exhaust gases to the working fluid.

There are many types of heat exchangers depends on their design. As a function of flow direction they can be separated as: co-current and counter-current. Usually the opposite direction of the fluids (counter-current) leads to higher efficiency. According to the design, heat exchangers can be classified as: shell and tube, spiral tube, finned plate and finned tube heat exchangers [1, 2]. In order to select suitable heat exchanger many parameters have to be taken into account such as: heat exchange surface, operating pressure, operating temperature, mass flow rates, hydraulic losses, size, mass, price and etc. Usually, the higher heat exchange surface and the minimum heat exchange resistance are preferable.

Glavatskaya, in [3] presented a comparative study of the technologies for the Rankine cycle evaporator. In this study three heat exchangers were recommended: plate and fins exchanger, pipe bundle type exchanger and shell and tube exchanger. As a result of this study it has been found that for the same transferred energy, the tube bundle type technology is the most compact and it offers a moderate exhaust gas pressure drop. As a part of this study two types of condensers have been selected for use in a Rankine loop: a tube and fin exchanger for external air cooling, and a plate exchanger for using glycol water as a well cold. The comparative study of the condensers on these operating conditions showed that the water-glycol cooled plate condenser has the best ratio between size and performance. In the work of Danel [4] was presented experimental and numerical study of a heat exchanger constructed by 23 helical tubes situated along the evaporator which provides the heat exchange surface approximately 1,8 m^2. The results show that its efficiency varies from 30 % to 42 %.

On the base of the reviewed articles [5-13] the tube heat exchangers were most commonly used. The efficiency varied from 24 % to 76 % depending on the design and operating conditions.

2 Experimental set-up and methodology

2.1. Experimental set-up

In order to conduct experimental research a test bench of Rankine cycle waste heat recovery system was used. The cycle including the following components: burner; heat exchanger (evaporator); cooler (condenser); reservoir and pump (Fig. 1).

All of the components are connected by means of stainless steel tubes. The test equipment consists of two fluid paths: the first is the exhaust gases flow which comes from the burner and goes through the heat exchanger while the second is working fluid flow through the pump, heat exchanger, cooler and reservoir. Water was used as working fluid in this cycle. However, the test equipment can work with different fluids such as ethanol and etc. The working fluid was cooled down as cold water with variable mass flow rate circulates through the condenser. In this test bench an expansion machine was not used. In order to keep the working fluid pressure an electro vane was used.

* Corresponding author: nikolay.milkov@tu-sofia.bg

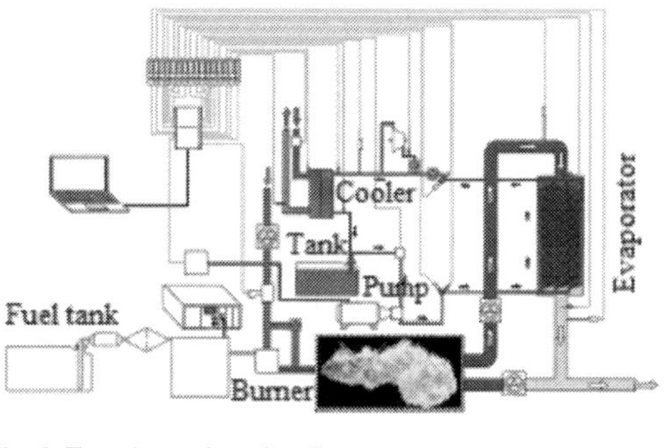

Fig. 1. Experimental test bench.

The purpose of the burner is to produce exhaust gases with the same parameters of that produced by a modern diesel engine. The burner operates with diesel fuel while the fresh air comes from a compressor. The enthalpy of exhaust gases can be varied by variation of the fuel and air mass flow in the burner. In order to simulate various combination of temperature and mass flow rate a part of fresh air passes outside the burner. Fresh air mass flow is setting by a vane, the main part of this flow goes to the burner and a second part is mix with burned gas to set the temperature. The fuel is injected in the burner by a pump and an internal by-pass is used to set the fuel mass flow.

The test bench is equipped with thermo-couples type K for measuring the temperature of the working fluid and exhaust gases, pressure sensors for acquiring the pressure drop on the exhaust gases flow and working fluid through the heat exchanger as well as the pressure in the condenser. Mass flow sensors are used for measuring the mass flow rate of working fluid, cooling water as well as of the air and fuel through the burner. In order to collect the experimental data a National Instruments DAQ model 6218 and a LabView based program were used.

For experimental test a counter-current flow heat exchanger was developed. The working fluid circulates inside the helical tubes situated along the evaporator. The number of tubes is 23 which provides the heat exchange surface approximately 1,8 m^2.

2.2 Methodology

The experimental study was carried out with the exhaust gas parameters (temperature and mass flow) previously determined at different operating points of the diesel engine. Thus, 28 engine operating points were selected (Fig. 2). For each point the exhaust gases temperature (T_g) and mass flow were measured on the engine test bench at the exhaust system 1,5 m downstream the turbocharger.

The burner injector can operate with four orifices with different size. Therefore, the burner is limited in terms of the fuel flow rate. For that reason, within the engine operating area with brake mean effective pressure (BMEP) of less than 6,6 bar, the fuel flow rate did not allow a combustion process with demanded exhaust gas parameters (temperature and flow). Thus, the engine operating field with BMEP below 6,6 bar was not studied.

The experimental test was carried out as follows:

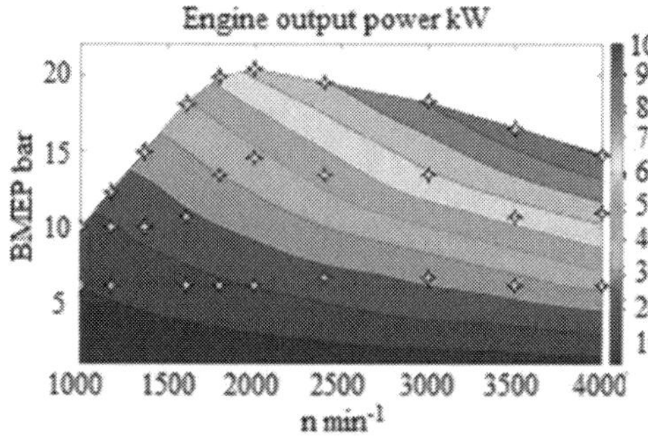

Fig. 2. Distribution of the studied points on the engine operating map.

- On the Rankin cycle based test bench, the exhaust gases with the same temperature and mas flow rate, determined for each of the selected engine operating point (Fig. 2) were simulated.
- The flow rate and pressure of the working fluid in the system were varied. The mass flow rate of the working fluid (m_{wf}) was set up to 2, 4, 6, 8 and 10 g/s, while for the pressure (p_{wf}) was 5, 10, 15 and 20 bar. These values were selected in order to meet the safety requirements of the system and to meet the requirements of the experimental test bench.
- After establishing the steady heat transfer in the Rankin cycle for each studied point, the temperature and pressure of the working fluid, the exhaust gases temperature as well as the mass flow rate of both fluids were measured.

The results which are close to the saturation curve or even on the left side of the curve are not presented, because in these cases the working fluid is not completely evaporated. For this reason, results are not presented for the engine rotation speed (n) below 1200 min^{-1} and for some points with low engine load.

3 Experimental results

Results present the variation of the heat exchanger outlet exhaust gases temperature and outlet working fluid temperature as a function of the working fluid pressure. Here, the results at maximum engine load and speed of 1200, 2000, 2800, 3500, and 4000 min^{-1} are presented.

Therefore, the variation of the exhaust gases temperature at the outlet of the heat exchanger (evaporator) as a function of the working fluid pressure are presented in the Fig. 3 to 7.

The variation of the exhaust gas temperature (Fig. 3) is within the range of 216 °C to 236 °C. The maximum value is at 20 bar working fluid pressure.

The exhaust gases temperature (Fig. 4) varies from 242 °C to 250 °C. The maximum value was obtained at 5 bar operating pressure.

In Fig. 5 the exhaust gases temperature at the outlet of the evaporator ranges from 311 °C to 322 °C. The maximum value was measured at 5 bar operating pressure.

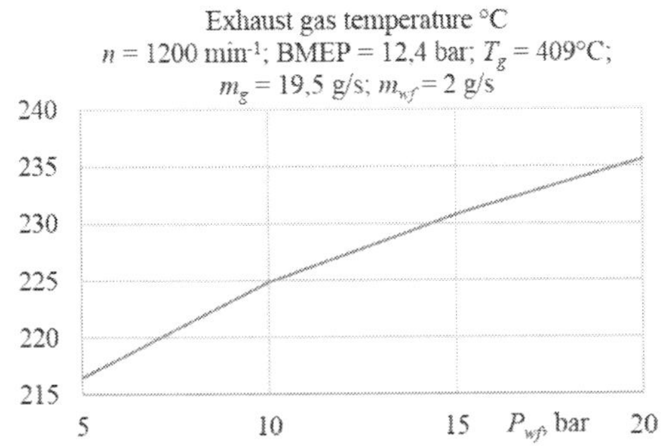

Fig. 3. Exhaust gas temperature after evaporator.

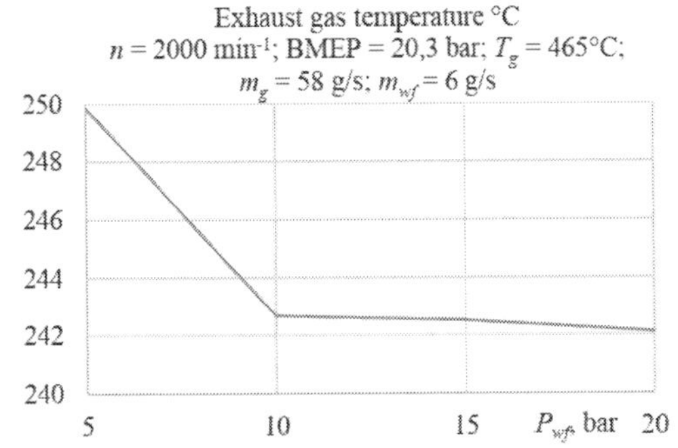

Fig. 4. Exhaust gas temperature after evaporator.

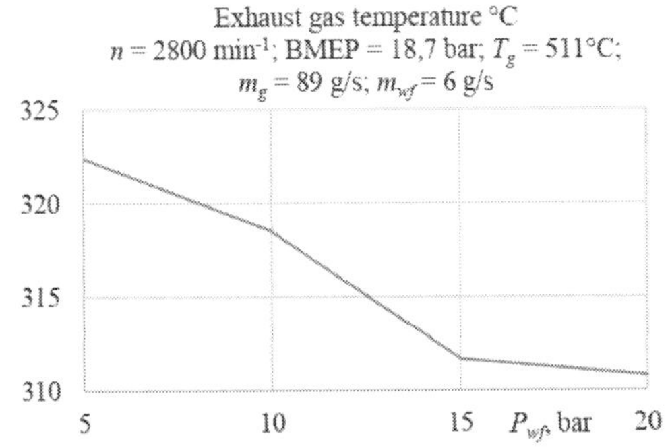

Fig. 5. Exhaust gas temperature after evaporator.

The temperature variation in Fig. 6 is within the range of 271 °C to 281 °C. The maximum value was measured at working fluid pressures 10 bar and 15 bar.

In Fig. 7 the exhaust gases temperature at the outlet of the evaporator ranged from 273 °C to 281 °C. The maximum value is at a working fluid pressure of 10 bar.

In Fig. 8 to 12, the variation of the working fluid temperature at the outlet of the evaporator depending on the operating pressure in the system is presented.

The variation of the working fluid temperature at the outlet of the evaporator shown in Fig. 8 is within the range of 155 °C to 215 °C. The values varied as a function of the pressure of the working fluid. For each pressure value the temperature is close to evaporation temperature, which means low degree of overheating. It can be explained by the low exhaust temperature at this operating point.

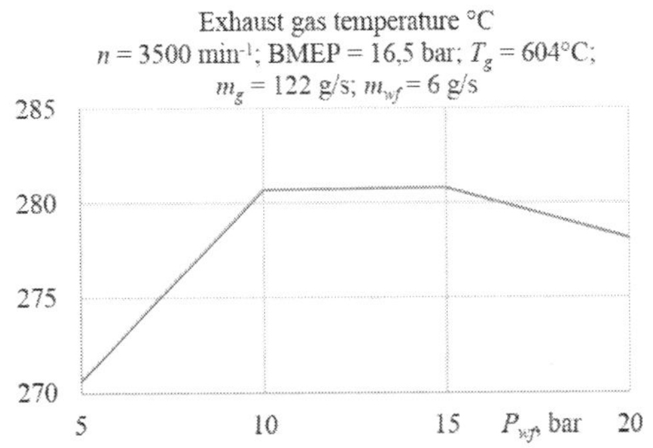

Fig. 6. Exhaust gas temperature after evaporator.

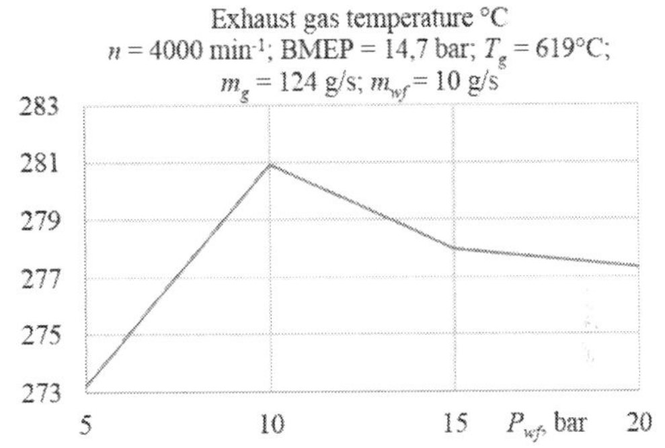

Fig. 7 Exhaust gas temperature after evaporator.

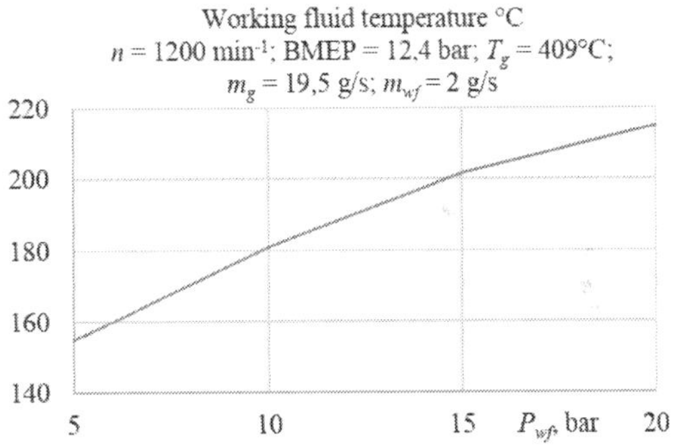

Fig. 8. Working fluid temperature after evaporator.

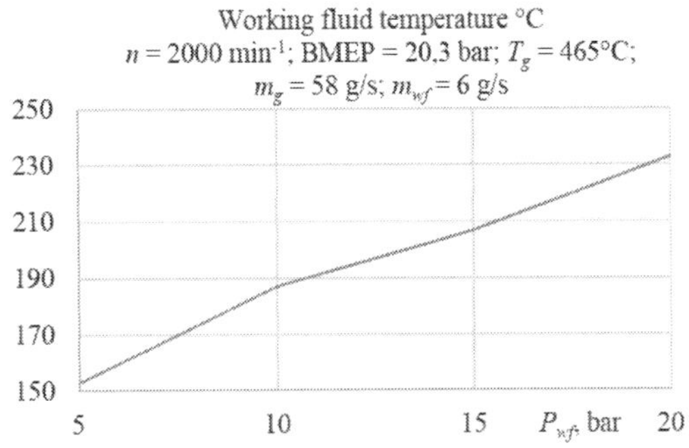

Fig. 9. Working fluid temperature after evaporator.

The temperature of the working fluid presented in Fig. 9 varies from 154 °C to 232 °C. In this engine operating

DOI: 10.1051/matecconf/201713304003

point the exhaust gases enthalpy is higher which lead to higher working fluid temperature.

The temperature presented in Fig. 10 is in the range of 161 °C to 300 °C. There is a significant increase in the working fluid temperature at this engine operating point. Here, it can be stated that the full evaporation of the fluid occurs.

The results in Fig. 11 present that the temperature varies from 159 °C to 240 °C.

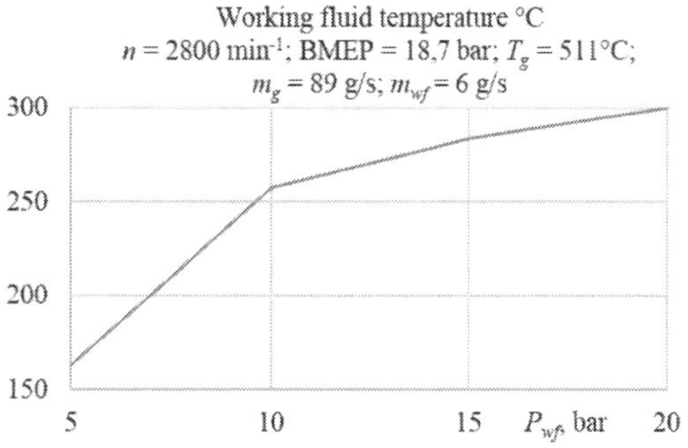

Fig. 10. Working fluid temperature after evaporator.

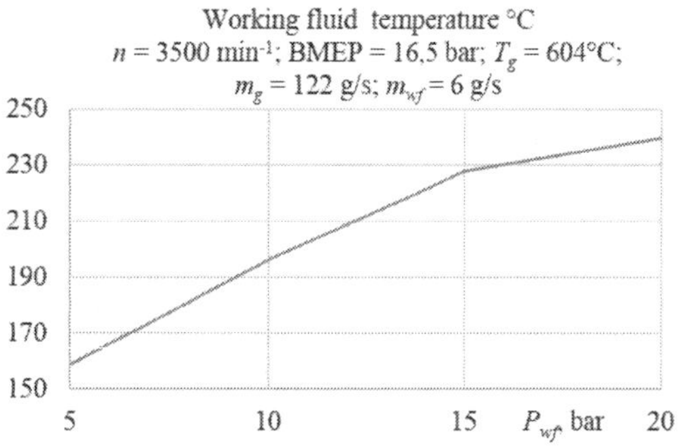

Fig. 11. Working fluid temperature after evaporator.

The results in Fig. 12 present that the outlet working fluid temperature varies from 156 °C to 234 °C.

The results obtained from the test were used in order to assess the heat transfer from the exhaust gases to the working fluid and then the efficiency of the heat exchanger was calculated.

The evaporator efficiency is calculated as follows:

$$\eta_{HE} = \frac{m_{wf}.(h_{fs}-h_{fe})}{m_g.c_{pg}(T_{ge}-T_0)}.100, \%, \quad (1)$$

where m_{wf} – mass flow rate of the working fluid, g/s; m_g –mass flow rate of the exhaust gases, g/s; h_{fs} – specific enthalpy of the working fluid at the outlet, kJ/kg; h_{fe} – specific enthalpy of the working fluid at the inlet, kJ/kg; c_{pg} – specific heat capacity of the exhaust gases, J/kg.K; T_{ge} – temperature of the exhaust gases at the inlet, K; T_o – ambient temperature, K.

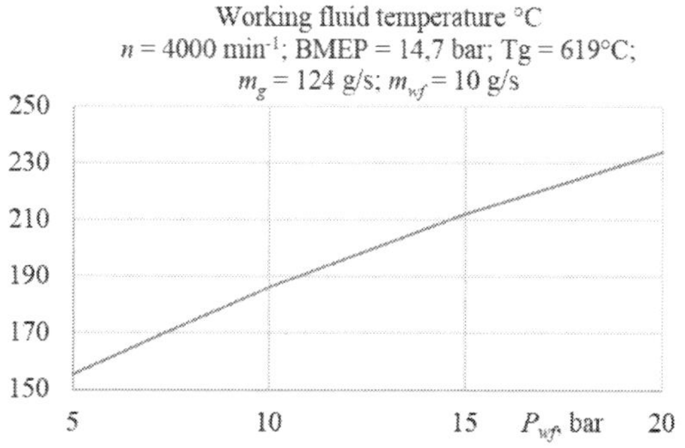

Fig. 12. Working fluid temperature after evaporator.

Fig. 13 to 17 show the efficiency of the evaporator in that operating points in which the working fluid temperature and exhaust gases temperature at the evaporator outlet are presented in the previous part. The efficiency is presented as a function of working fluid pressure.

In Fig. 13 the efficiency of the evaporator slightly decreases as increasing the working fluid pressure. Here, the efficiency ranged from 40 % to 44 %.

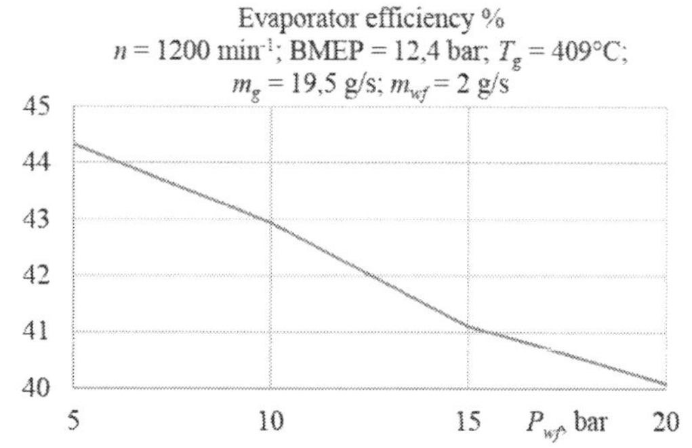

Fig. 13. Evaporator efficiency.

The efficiency of the evaporator at the engine operating at 2000 min^{-1} (Fig. 14) is in range from 49,6 % to 50,8 %. The maximum was observed at working fluid pressure of 10 bar.

In Fig. 15, the efficiency varied from 34,4 % and 36,8 %, the maximum value corresponds to a pressure of 20 bar.

The efficiency of the evaporator shown in Fig. 16 is within the range of 32,9 % to 35,3 %, with the maximum value at 5 bar operating pressure.

The efficiency of the evaporator at 4000 min^{-1} and full engine load (Fig. 17) varied from 34,4 % to 36,1 %, the maximum value was observed at a working fluid pressure of 5 bar.

At full engine load the evaporator efficiency ranged from 32 % to 50,8 %. The maximum value was observed at engine operating point corresponds to maximum engine at engine speed of 2000 min^{-1} and BMEP of 20,3 bar.

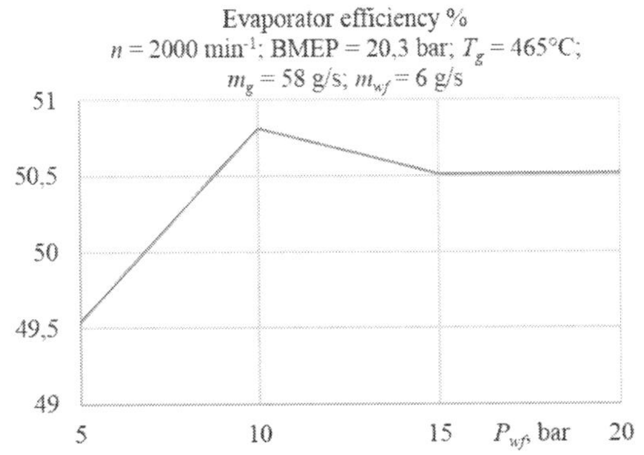

Fig. 14. Evaporator efficiency.

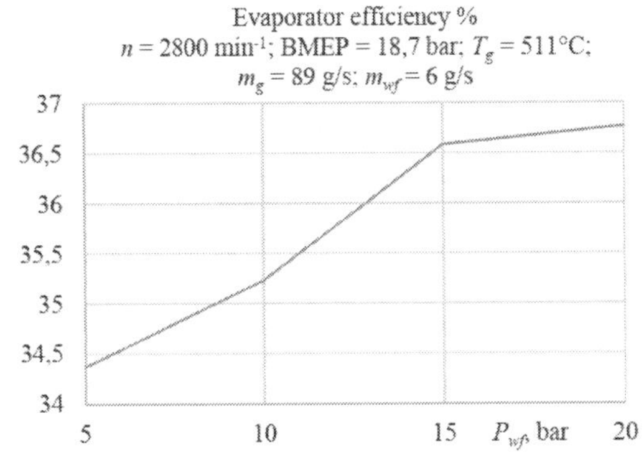

Fig. 15. Evaporator efficiency.

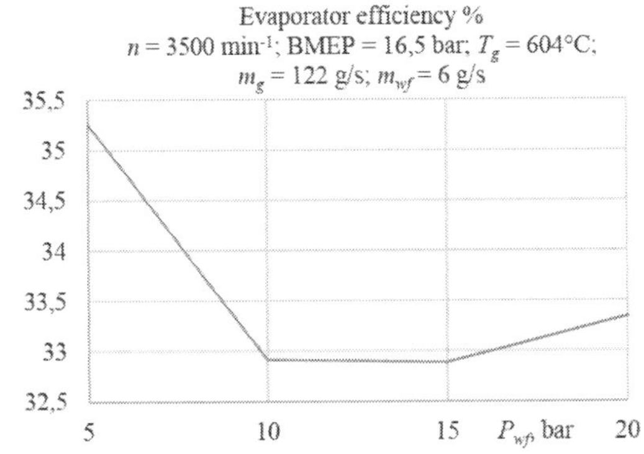

Fig. 16. Evaporator efficiency.

Fig. 18 shows the evaporator efficiency in a part of the engine operating map. Evaporator efficiency was estimated to be within the range from 24,8 % to 51,9 %, with the maximum value at 2000 min^{-1} and BMEP of 15,2 bar. The minimum value was achieved at 3500 min^{-1} and 6,6 bar BMEP. On the base of the results the evaporator efficiency is higher at the engine operating range with maximum efficiency of the engine.

Fig. 19 shows the variation of the thermal power of the working fluid after the evaporator, corresponding to the maximum values of the evaporator efficiency, presented in the engine operating map.

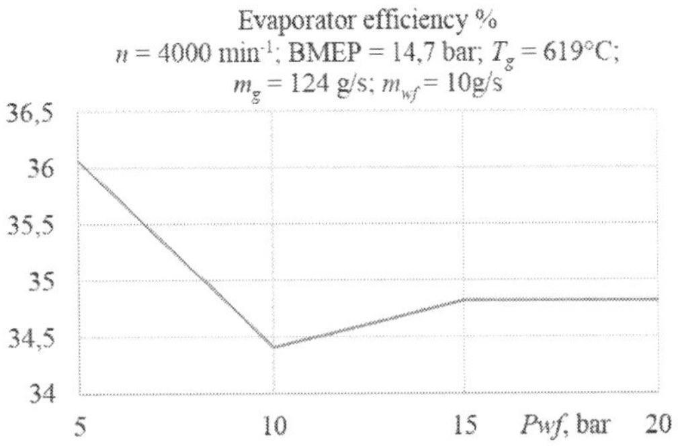

Fig. 17. Evaporator efficiency.

The thermal power of the working fluid at the outlet of the evaporator increases as increasing the engine speed and load. The maximum power of the working fluid was estimated to be 35 kW at the engine operating corresponds to maximum engine power (4000 min^{-1} and BMEP 14,7 bar).

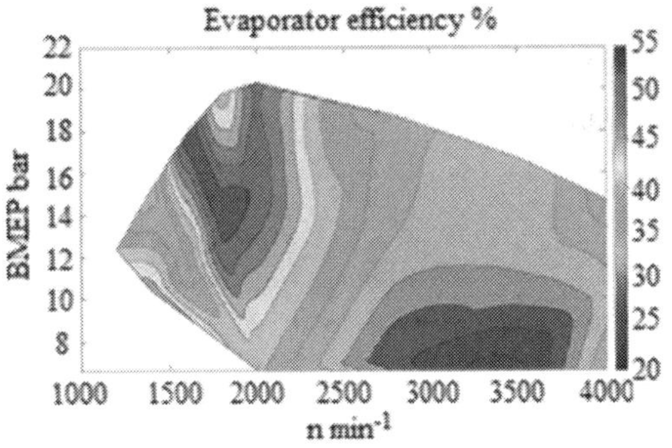

Fig. 18. Evaporator efficiency over the engine operating map.

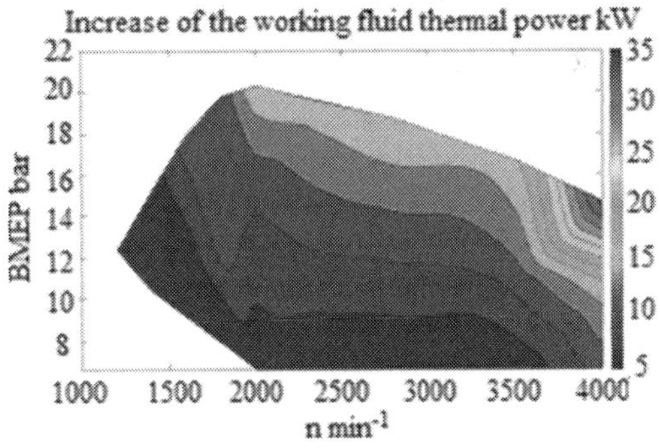

Fig. 19. Working fluid thermal power at evaporator outlet.

On the basis of this experimental study, the optimal value for the working fluid mass flow and pressure in the Rankine cycle were determined.

Fig. 20 presents the variation of the working fluid mass flow that corresponds to the maximum efficiency of the evaporator. The flow rate of the working fluid varies from 2 g/s to 10 g/s The mass flow rate increases as increasing of the load and engine speed. The maximum value was observed at the maximum power of the engine.

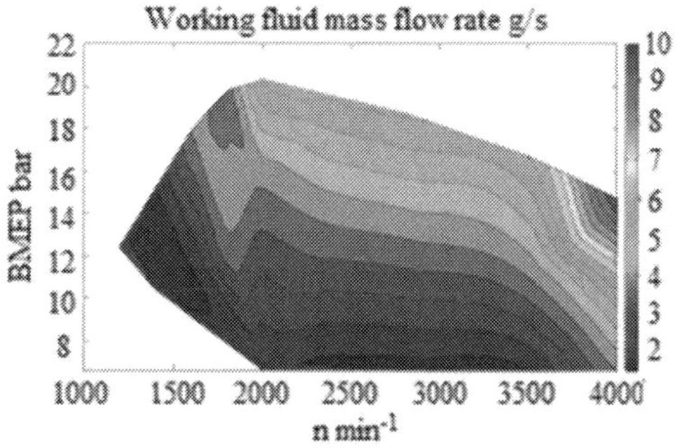

Fig. 20. Working fluid mass flow rate over the engine operating map.

The working fluid pressure variation is shown in Fig. 21. The optimal pressure varies from 5 bar to 20 bar. As it can be seen on the graph the working fluid pressure of 20 bar is optimal in large area on the engine map.

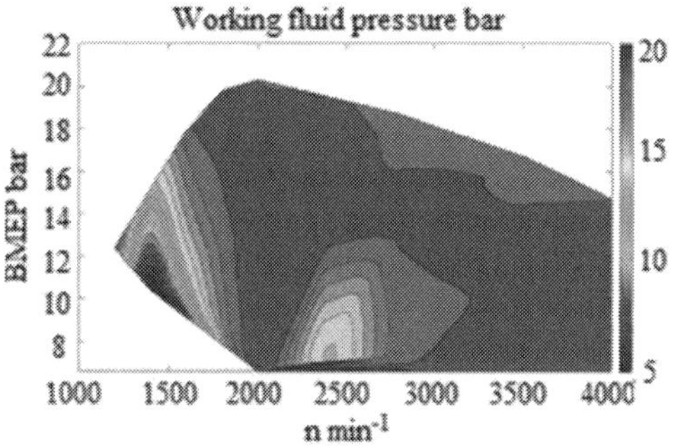

Fig. 21. Working fluid pressure over engine operating map.

4 Conclusions

Studied waste heat recovery system based on the Rankin cycle does not allow complete evaporation of the working fluid at all engine operating points. In the case of low energy of the exhaust gases, it is not possible to fully evaporate the working fluid.

The efficiency of the evaporator prototype varies from 25 % to 51,9 % in the studied engine operating range. The maximum efficiency was in the area with the maximum engine efficiency.

The variation of the working fluid thermal power after the evaporator depends on the exhaust gases energy at the evaporator inlet and increases with increasing engine speed and load.

On the basis of the experimental test, the optimal values for the mass flow rate and pressure of the working fluid in the system based on the Rankin cycle were determined.

The experimental study which is presented in this article was done at the Laboratoire du Chimie Moléculaire, Génie des Procédés, Chimiques et Energétiques (CMGPCE) – Cnam Paris.

References

1. M. Podhorsky and H. Krips, *Heat exchangers : a practical approach to mechanical construction, design and calculations,* NewYork, Wallingford, UK: Begell House Inc, (1998)
2. T. Kuppan, *Heat exchangers design handbook,* New York: Marcel Dekker, Inc., (2000)
3. Y. Glavatskaya, *Conversion de l'énergie thermique des gaz d'échappement en travail mécanique par un cycle de Rankine afin de réduire les émissions des gaz à effet de serre*, PhD thesis, (2012)
4. Q. Danel, *Étude numérique et expérimentale d'un cycle de Rankine-Hirn de fable puissance pour la récupération d'énergie*, PhD thesis, (2016)
5. V. Pandiyarajan, M.C. Pandian, E. Malan, R. Velraj, R.V. Seeniraj, *Experimental investigation on heat recovery from diesel engine exhaust using finned shell and tube heat exchanger and thermal storage system,* Applied Energy, **88**(1), 77-87, (2011)
6. S. Lee and C. Bae, *Design of a heat exchanger to reduce the exhaust temperature in a spark-ignition engine,* Int. Journal of Thermal Sciences, **47**(4), 468-478, (2008)
7. H. Zhang, E. Wang, B. Fan, *Heat transfer analysis of a finned-tube evaporator for engine exhaust heat recovery,* Energy Conversion and Management, **65**(0), 438-447, (2013)
8. M. Ghazikhani, M. Hatami, D. Ganji, M. Gorji-Bandpy, A. Behravan, G. Shahi, *Exergy recovery from the exhaust cooling in a DI diesel engine for BSFC reduction purposes,* Energy, **65**(0), 44-51, (2014)
9. T. Wang, Y. Zhang, J. Zhang, G. Shu, Z. Peng, *Analysis of recoverable exhaust energy from a light-duty gasoline engine,* Applied Thermal Engineering, **53**(2), 414-419, (2013)
10. S. Hossain, S. Bari, *Waste heat recovery from the exhaust of a diesel generator using Rankine Cycle,* Energy Conversion and Management, **75**(0), 141-151, (2013)
11. S. Bari, S. Hossain, *Waste heat recovery from a diesel engine using shell and tube heat exchanger,* Applied Thermal Engineering, **61**(2), 355-363, (2013)
12. S. Bai, H. Lu, T. Wu, X. Yin, X. Shi, L. Chen, *Numerical and experimental analysis for exhaust heat exchangers in automobile thermoelectric generators,* Case Studies in Thermal Engineering, **4**(0), 99-112, (2014)
13. J. Dong, X. Zhang, J. Wang, *Experimental investigation on heat transfer characteristics of plat heat exchanger applied in organic Rankine cycle (ORC),* Applied Thermal Engineering, **112,** 1137-1152, (2017)

 DOI: 10.1051/matecconf/201713305001

Stress analysis of a link for slat chain conveyor

Veselin Tsonev[1,*], *Georgi* Stoychev[1], *Emanuil* Chankov[2]

[1]Technical University – Sofia, "Strength of Materials" Department, 8 Kliment Ohridski Blvd., 1000 Sofia, Bulgaria
[2]Visteon Electronics Bulgaria

Abstract. This paper analyzes stresses in a link of a slat chain conveyor loaded with rock mass. Tensometric data and results from numerical simulation with Finite Element Method (FEM) are used. The tensometric measurements are performed in real life conditions. For numerical simulation an adequate model with FEM has been created. The adequacy of the model is confirmed by the comparison of the experimental and numerical results. The factor of safety used in the design of the conveyor links is determined.

1 Introduction

There are many investigations devoted to stress analysis of roller chains. The roller chains are widely used as driving members of escalators and conveyors.

The stress analysis of chain links using Finite Element Method (FEM) was examined by Antonescu [1]. The results were verified with photo-elastic data.

In [2] the FEA was applayed in order to replace mild stel roller conveyor links by different composite materials.

Noguchi et al. investigated static stresses and some weight saiving design of a link plate roller chain using FEM [3].

Jagtap et al. applied strain gages to analyze the behavior of roller conveyor chain strip [4].

Slat chain conveyors for transportation of rock material work in severe, abrasive environment. This in turn is associated with an increased factor of safety in their design [5]. The normal operation of the conveyor is of the highest priority due to the high cost of the down time and the interruption of the production process. Proper slat chain conveyor design is essential to ensure their long service life.

The reason for studying the conveyor in present paper is failure related to the destructions of a conveyor links.

The conveyor takes material from a bunker and transfers it to a belt conveyor for further processing. The leading wheel of the conveyor is at the top. Fig. 1 shows the model of the conveyor.

The main characteristics of the conveyor are:

- Number of links on the loaded part – 101 per chain;
- Weight per link – 35 kg;
- Number of slats on the loaded part – 101 per chain;
- Weight per slat – 300 kg;
- Length of the conveyor – 26 m;
- Length with material – 25 m;
- Declination – 24°;
- Material (width, height, mass density) – 2 m, 1,5 m, 1,8 t/m;
- Capacity – 1800 t/h.

Fig. 2 shows a link of the conveyor. The chain link is manufactured from steel with the following mechanical properties:

- Elasticity modulus – $E = 2{,}1.10^{11}$ Pa;
- Poisson's ratio – $\nu = 0{,}28$;
- Yield strength – $R_{p0,2} = 640$ MPa.

The task is to assess the factor of safety and to compare it with the one required in the design of slat chain conveyors.

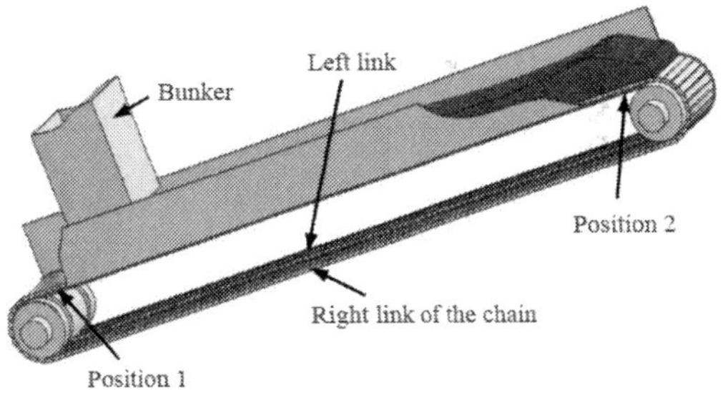

Fig. 1. Model of the conveyor.

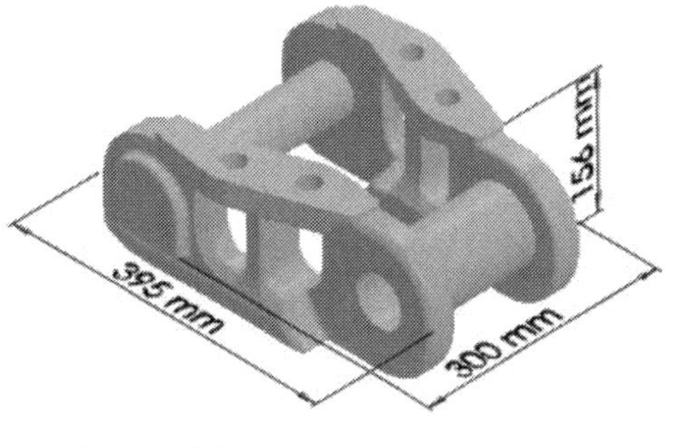

Fig. 2. Conveyor link.

* Corresponding author: gstojch@tu-sofia.bg

DOI: 10.1051/matecconf/201713305001

2 Experimental study

The method of calculating the stresses from the measured strains in free of loading plane is widely used in engineering practice [6]. In this case the principal stress in the perpendicular to this plane direction is equal to zero. The normal strains (ε_a, ε_b, ε_c) at a point on the free surface in directions *a*, *b* and *c* are determined using strain rosette. The two other principal stresses (σ' and σ'') are calculated. The principal stresses are arranged in descending order $\sigma_1 \geq \sigma_2 \geq \sigma_3$. Using the principal stresses von Mises stress (σ_{ekv}^{VM}) is calculated.

Fig. 3 shows pictures of broken conveyor links. The destruction is in the areas opposite to the holes of the conveyor plates at the locations of the geometric stress concentrators. These places are not suitable for mounting strain gauges. Fig. 4 shows the selected zone of the left and the right conveyor links in which the strain gauges are installed.

Fig. 5 shows schemata of strain gauge measurement. A rectangular strain gauge is mounted on the selected area. To measure strains in directions *a*, *b* and *c* Wheatstone quarter bridge is used. The quarter bridge consists of stressed strain gauge and unstressed strain gauge for temperature compensation. The strain gauge for temperature compensation is mounted on an unstressed steel plate. The quarter gauge bridges are connected to multi-channel strain gauge Data Acquisition system – 1, which is connected to PC – 2.

Table 1 shows designations of the mounted gauges.

Fig. 3. Pictures of broken conveyor links.

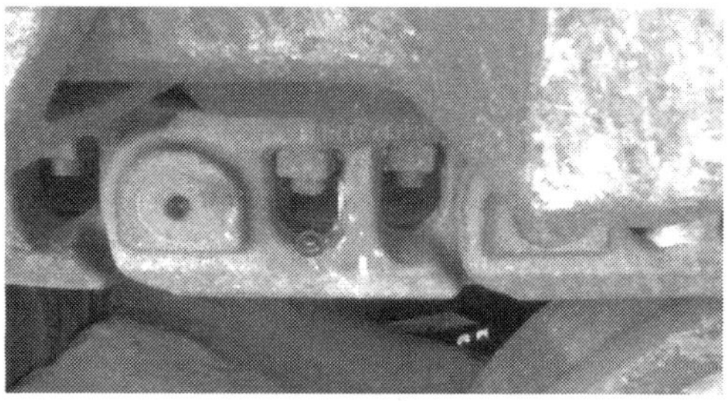

Fig. 4. Zone in which the strain gauges are installed.

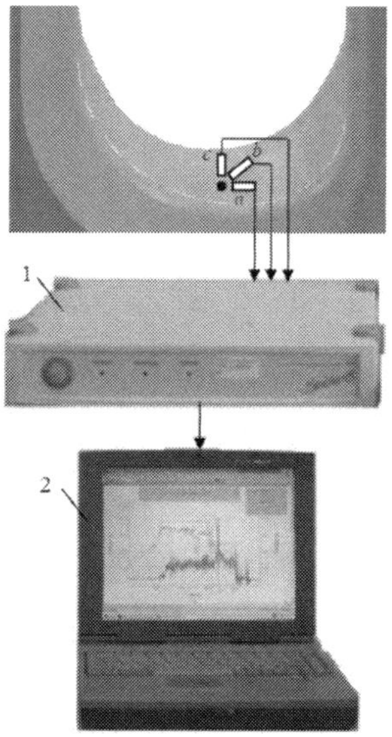

Fig. 5. Scheme of strain gauge measurement.

Table 1. Designations of the mounted gauges.

Link	Direction	Designation
Right	45° with respect to the axis of the chain (direction *b*)	CH1D45
	Aligned with the axis of the chain (direction *a*)	CH2D0
	90° with respect to the axis of the chain (direction *c*)	CH3D90
Left	45° with respect to the axis of the chain	CH4L45
	Aligned with the axis of the chain	CH5L0
	90° with respect to the axis of the chain	CH6L90

The following measurements were made:

Measurement 1 – The links on which the gauges are mounted are located at the bottom end of the conveyor before the bunker. The bunker is filled and the machine starts pouring material onto the conveyor. The measurement of the strains starts from position 1 (Fig. 1), in which the links are slightly after the lower shaft, to position 2 (Fig. 1), in which the links are slightly before the upper shaft;

Measurement 2 – The links on which the gauges are mounted are located at the bottom of the conveyor before the bunker. The machine runs with a full bunker and material loaded on the conveyor. The measurement of the strains starts from position 1, located slightly after the lower shaft, to position 2, located slightly before the upper shaft.

Figures 6 and 7 show the results from the measurements.

 DOI: 10.1051/matecconf/201713305001

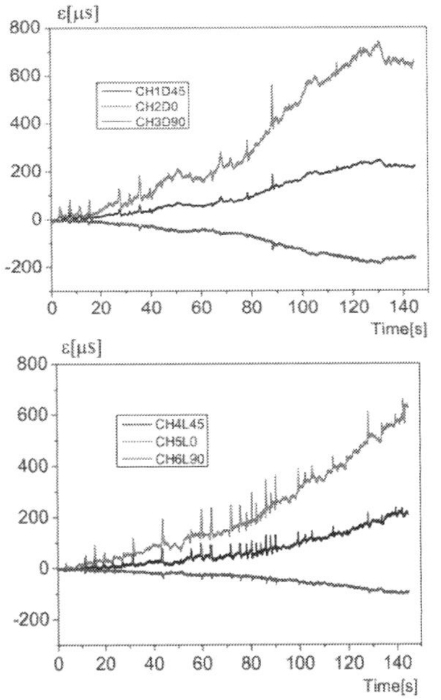

Fig. 6. Results from measurement 1.

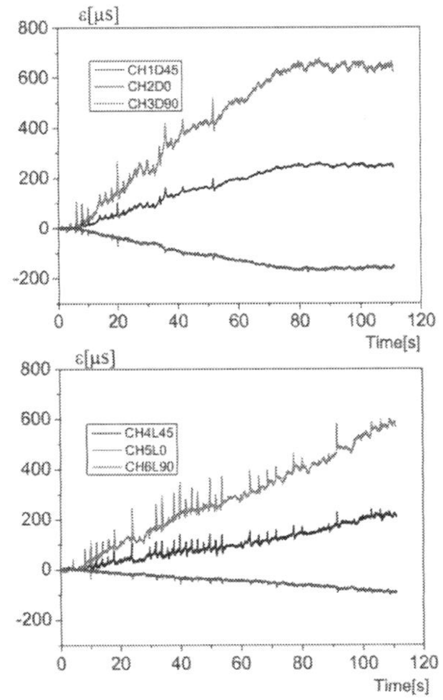

Fig. 7. Results from measurement 2.

The obtained results show that the dynamics of the process does not have any significant impact on the measured strains. The largest strains are at the end of the measurement (position 2).

Table 2 shows the results from the measured strains at the end of the data acquisition (position 2), the calculated principle stresses and the calculated von Mises stress in the vicinity of the test points.

Table 2. Results at the end of the data acquisition.

№	Right link	Left link
1	$\varepsilon_a = 660.10^{-6}$ $\varepsilon_b = 224.10^{-6}$ $\varepsilon_c = -161.10^{-6}$	$\varepsilon_a = 632.10^{-6}$ $\varepsilon_b = 214.10^{-6}$ $\varepsilon_c = -99.10^{-6}$
	$\sigma' = 140{,}1\ \text{MPa} = \sigma_1$ $\sigma'' = 5{,}4\ \text{MPa} = \sigma_2$	$\sigma' = 138{,}3\ \text{MPa} = \sigma_1$ $\sigma'' = 17{,}2\ \text{MPa} = \sigma_2$
	$\sigma_{ekv}^{VM} = 137{,}5\ \text{MPa}$	$\sigma_{ekv}^{VM} = 130{,}6\ \text{MPa}$
2	$\varepsilon_a = 655.10^{-6}$ $\varepsilon_b = 254.10^{-6}$ $\varepsilon_c = -162.10^{-6}$	$\varepsilon_a = 580.10^{-6}$ $\varepsilon_b = 222.10^{-6}$ $\varepsilon_c = -89.10^{-6}$
	$\sigma' = 138{,}9\ \text{MPa} = \sigma_1$ $\sigma'' = 4{,}9\ \text{MPa} = \sigma_2$	$\sigma' = 126{,}6\ \text{MPa} = \sigma_1$ $\sigma'' = 16{,}6\ \text{MPa} = \sigma_2$
	$\sigma_{ekv}^{VM} = 136{,}5\ \text{MPa}$	$\sigma_{ekv}^{VM} = 119{,}2\ \text{MPa}$

3 Numerical analysis

The numerical analysis was performed with FEM, using SOLIDWORKS software.

The most appropriate model for FEA is the model shown in Fig. 8, which includes a link from the chain and the associated slat, using symmetry in geometry and loading. The mesh of finite elements and the boundary conditions are shown. The mesh of elements is created with mesh control in areas with a high stress gradient.

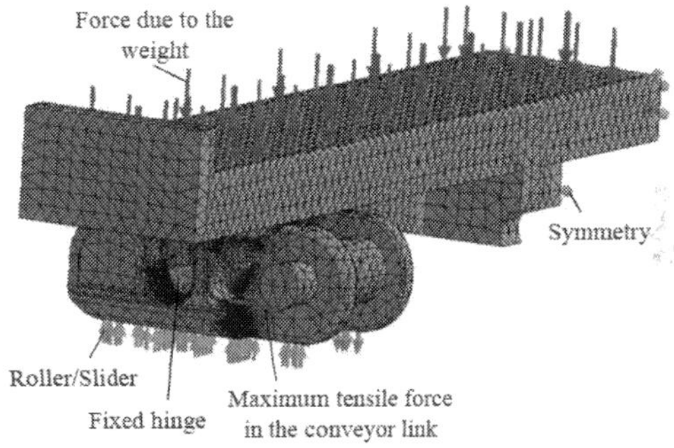

Fig. 8. FEA model.

The maximum tensile force in the conveyor link is calculated analytically [7]. It is calculated by taking into account the pull forces along the conveyor chain. The maximum of the force ($0{,}426.10^6$ N) is in the top shaft.

Fig. 9 shows the von Mises stress from the numerical analysis. Values of the stress in the range of 115 – 130 MPa are determined in the experimentally studied area. The areas with maximum stress correspond to the areas of destruction of the chain link. In these areas the highest stress values are determined in the range of 270 – 435 MPa.

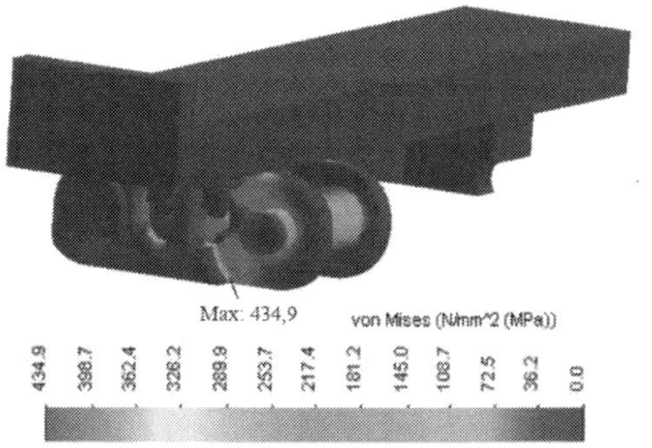

Fig. 9. Von Mises stress from the numerical analysis.

4 Conclusion

The comparison between the numerical analysis and experimental study shows a good match between the results. That gives us reason to assume that the results obtained with the numerical analysis are accurate. The numerical solution makes it possible to estimate stresses in all dangerous areas with geometric stress concentrators.

For a material yield strength of 640 MPa, the factor of safety turns out to be approximately 1,5. This value is very low. The recommended value of the factor of safety, especially when the conveyor operates in aggressive dust and other contaminants, is in the order of 8 [1].

It is interesting to investigate the chain link with respect to fatigue.

References

1. N. Antonescu, Stresses in roller chain link plate, *Rev. Roum Sci. Techn.-Mec. Appl.*, **20**(2) 311-322, (1975)
2. S. Kale, R. Navthar, Analysis and optimization of chain conveyor outer link, Int. J. of innovative research in science, engineering and technology, **4**(8), (2015)
3. S. Noguchi, K. Nagasaki, S. Nakayama, T. Kanda, T. Nishino, T. Ohtani, Static stress analysis of link plate of roller chain using finite element method and some design proposals for weight saving, Journal of advanced mechanical design, systems, and manufacturing, **3**(2), 159-170, (2009)
4. M. Jagtap, B. Gaikwad, P. Pawar, B. Ronge, Use of strain gages to analyze the behavior of roller conveyor chain strip, *International conference on computer science and mechanical engineering*, Jaipur, India, 35-38, (2014)
5. R. Jeffrey, *Conveyor chain designer guide*, http://www.renold.com/upload/renoldswitzerland/conveyor_chain_-_designer_guide.pdf.
6. L. Lazov, I. Slavov, *Strength of materials*, „Tehnika", Sofia, in Bulgarian, (1992)
7. S. Nachev, *Machines and equipment for continuous transport*, Sofia, in Bulgarian, (1981)

Experimental study of impact properties of aluminium honeycomb sandwich structure

Ondrej Flasar[1,*], *Vaclav* Triska[1], and *Milan* Junas[1]

[1]University of Defence, Department of Air Force and Aircraft Technology, Faculty of Military Technology, Kounicova 65, 662 10 Brno, Czech Republic

Abstract. Aircraft grade of structural sandwich plates has been studied to determine the influence rate of particular face sheets adjustments on impact energy absorption. The face sheets were fabricated by carbon fibre reinforced plastic (CFRP) laminas using unidirectional reinforcement with either cross-ply $[0/90]_n$ or angle-ply $[\pm 45]_n$ orientation of fibres. A core of the sandwich plates was then formed by aluminium honeycomb. The impact properties of small unnotched sandwich specimens were measured by instrumented Charpy impact test device. Assessment of force-displacement curve, loading process and failure mode was carried out for each specimen type to determine a general trend for increasing the impact properties of sandwich structures.

1 Introduction

Sandwich structures composed of strong skins and a low-density core provide very stiff structure units in many applications including aerospace, train industry, maritime etc. Except for their high strength property sandwiches excel at impacts and vibrations absorption. The damage response of sandwich panels depends on the mechanical properties of the core material and the fracture characteristics of the skin [1].

When observing energy absorbing properties, there are a few damage processes which can deal with impact loading such as core buckling, cells or foam folding, delamination, fibre fracture etc. [2-4]. Research can thus focus on either different core materials including foams, honeycomb, cork, balsa wood, coconut and other artificially constructed truss or sheet structures [4-11] or different face sheets as aluminium, carbon, glass, aramid reinforced composites etc. [3, 12]. Most of the studies use drop weight methods for the evaluation of sandwich structures in the range of low-medium velocity impacts [2, 3, 8, 12, 13]. There can be found also impact test apparatuses such as the pendulum impactor [14]. However, this study prefers the instrumented Charpy pendulum hammer due to the emphasis on deformation behaviour, total absorbing energy and test repeatability.

Since there does not exist general standard or advisory material for generic sandwich beam testing using a pendulum impactor, the only relative standard is ASTM D 6110 [15] concerning notched plastic material testing.

Due to lack of information about the differences in the dynamic behaviour of structural sandwiches with different stacking sequences of their face sheets, this paper focuses on determining the behaviour and trends affected by various adjustments of the tested sandwiches. For that purpose we used pure CFRP (carbon fibre reinforced plastics) materials and subsequently sandwich plates with face sheets made of CFRP with similar stacking sequences.

2 Material and method

2.1 Test specimen

Both pure CFRP and sandwich specimens have an unnotched beam shape derived from specimen described in the standard (standard). The CFRP material contains unidirectional reinforcement cured by epoxy resin. The fibre volume fraction of this material is 60%. Two stacking sequences with orientation either cross-ply [0/90] or angle-ply [±45] through the whole thickness were applied. The dimensions of CFRP specimens are as follows: length $l = 80$ mm, width $w = 10$ mm and thickness $t = 6$ mm.

The sandwich specimens are then laid up by CFRP face sheets and aluminium honeycomb core. The used CFRP sheets with fibre volume fraction $FV = 50$ % are cross-ply $[0/90]_2$, angle-ply $[\pm 45]_2$ and thicker angle-ply $[\pm 45]_4$. The aluminium honeycomb density is 32 kg/m^3 and the hexagonal cell has a size of 6,4 mm. The dimensions of the sandwich specimens are the same (illustrated in Fig. 1) as those of CFRP specimens except for width which is extended to $w = 20$ mm. All observed composite or sandwich versions had five tested specimens. The applied impact force is always oriented normal to the specimen plane.

* Corresponding author: ondrej.flasar@unob.cz

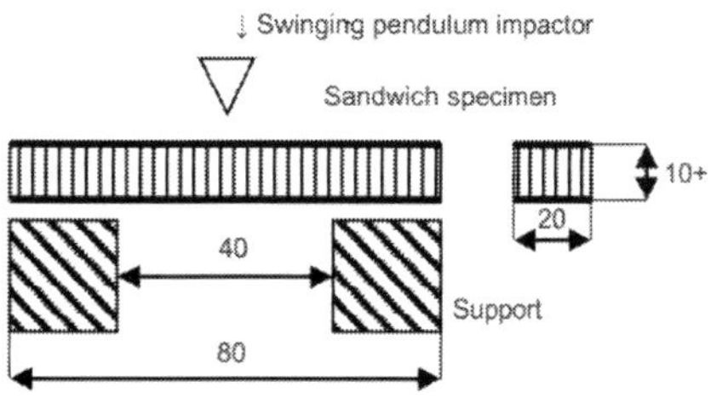

Fig. 1. Specimen dimensions and mutual position of the specimen and testing device.

2.2 Instrumented impact test

The device used for impact testing was instrumented Charpy pendulum Zwick-Roell 300. The pendulum hammer reaches a peripheral velocity of 5,234 ms^{-1} and a kinetic energy of 300 J at the impact point. The length of the support span is 40 mm. The instrumented pendulum records the whole force-displacement curve which is then used for results assessment. An integration of the curve gives a total value of absorbed energy W_t during specimen deformation:

$$W_t = \int_0^s F s\,ds, \tag{1}$$

where F is force and s is standard travel. The differences caused by different face sheet thickness are suppressed by the introduction of a value of absorbed energy per unit area a_k (cross-sectional):

$$a_k = \frac{W_t}{wt}. \tag{2}$$

In order to have more unbiased result assessment specific absorbed energy, which is dependent on both cross-section and weight of specimen, is then applied:

$$a_m = \frac{W_t}{wtm}, \tag{3}$$

where m is the weight of the specimen. All results, force-displacement curves and optical micrograms of specimens after failure were assessed during the post-processing.

3 Results and discussion

The results are studied in two phases. The first phase describes results gained during measurement of CFRP specimens. The other phase elaborates on the results of selected sandwich structures with respect to knowledge of pure CFRP behaviour.

Fig. 2 shows results of averaged absorbed impact energy a_k by comparing two CFRP stacking sequences. The angle-ply [±45] laminate exhibits a significantly higher absorbed energy (328,7 kJm^{-2}) than the cross-ply [0/90] laminate (221,8 kJm^{-2}). Nevertheless, the stiffness and strength of fibres oriented in 0° direction demonstrate its capability when comparing maximum reached load F_m (Fig. 3). In this case, the cross-ply [0/90] laminate shows more than two times higher value of averaged ultimate load (4557 N) while angle-ply [±45] laminate reached only 2187 N. Accordingly, it should be noted that total absorbed energy is not directly proportional to ultimate load.

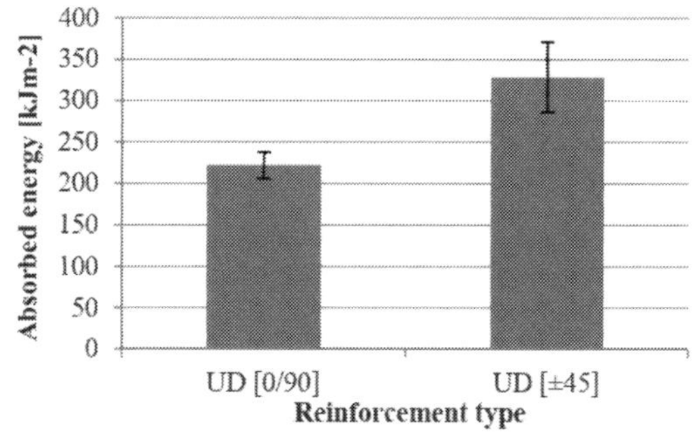

Fig. 2. Specific absorbed impact energy of CFRP specimens, error bars included.

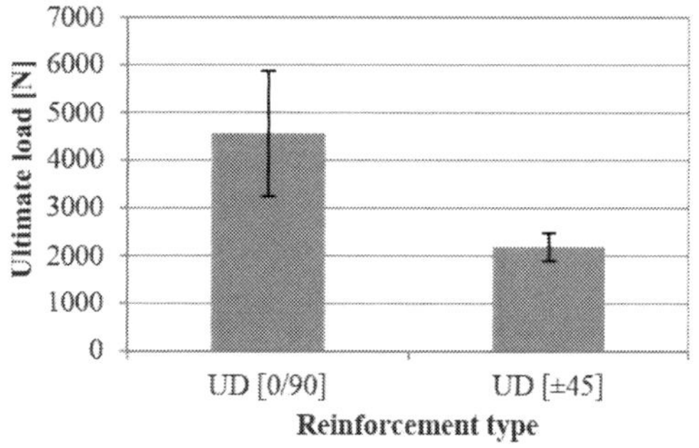

Fig. 3. Ultimate load of CFRP specimens, error bars included.

The fact of the low absorbed energy of cross-ply [0/90] laminate is obvious from Fig. 4. Even if this specimen reaches high load, the overall behaviour is quite vulnerable. The growth of the applied force in the beginning of loading exposes very stiff property. However, after reaching its maximum there is a steep fall to lower values with the very unstable process until the total failure. The sharp peak indicates brittle crack of the several plies on the top. Subsequently, the following unstable deformation characterizes numerous delaminations and matrix cracks at 90° plies. This process can be described as explosive fraying of the specimen edge.

The other angle-ply [±45] laminate exhibits rather plastic behaviour. The smooth force-displacement curve shows short lasting on maximum load value and a following gradual decrease to zero value. The transition between elastic deformation and constant phase of the force-displacement curve can be labelled as yield point, which identifies the moment where irreversible deformation occurs. This type of CFRP is characterized by extensive fibre pull out failure which causes its resilient post-failure integrity. This material behaviour can thus be called tough.

 DOI: 10.1051/matecconf/201713305002

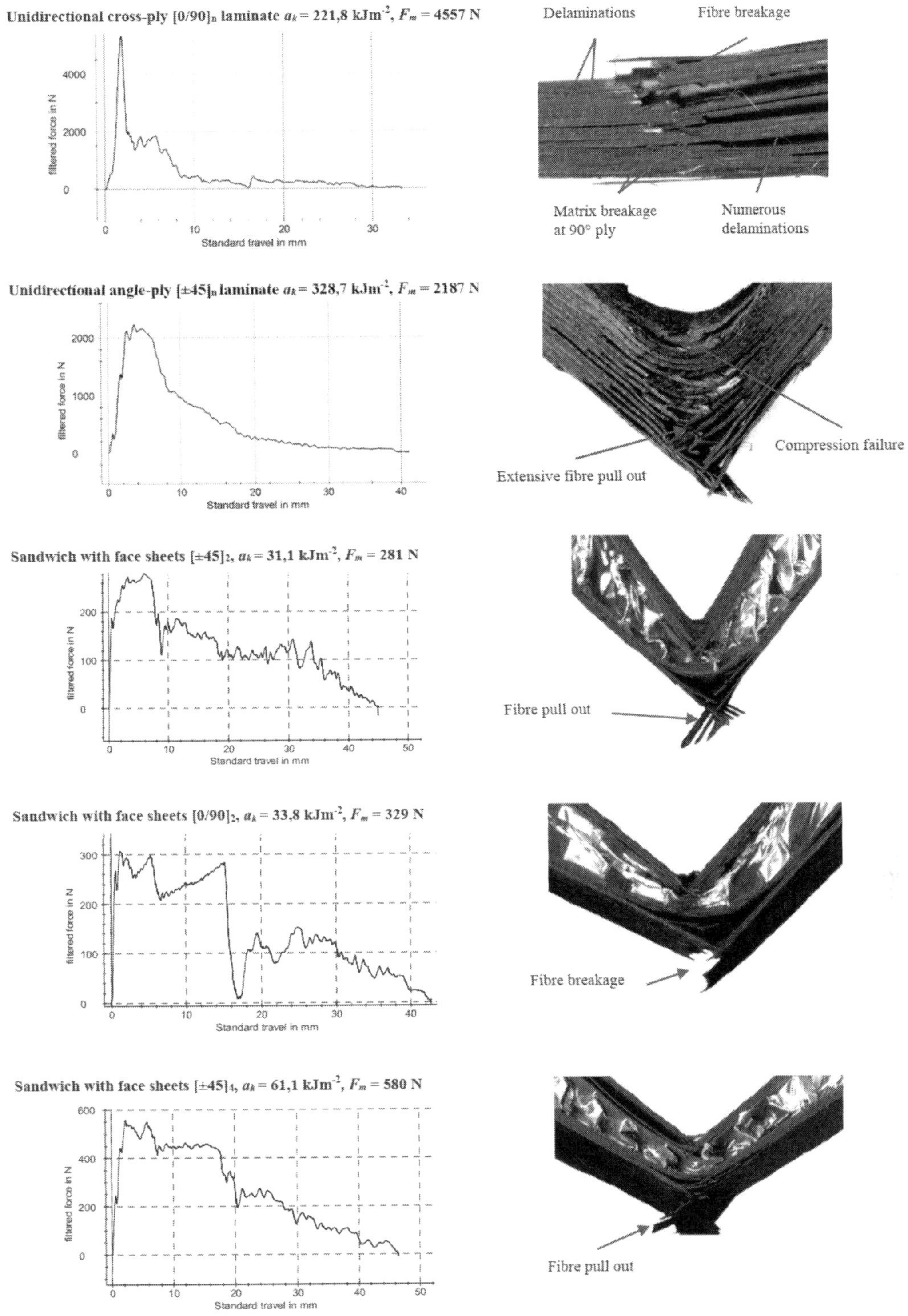

Fig. 4. Results of impact test.

 DOI: 10.1051/matecconf/201713305002

There was hereby established the behaviour of CFRPs with different stacking sequence. The research carries on testing of sandwich panels. Obviously, the sandwich with 8-ply face sheets gains the highest values of absorbed energy a_k (Fig. 5). When comparing sandwiches with four plies in the skin, the $[0/90]_2$ sandwich exhibits slightly better properties than the $[\pm 45]_2$ sandwich. Nevertheless, specific absorbed energy a_m (Fig. 6) related to the weight of each specimen provides an interesting comparison. This parameter demonstrates considerable increase of the weight of the thicker sandwich specimen, which means that the total difference is more than 80 % a_k (27 kJm^{-2}) as compared to 4-ply skin sandwich $[0/90]_2$, which is reduced to about 7 % a_m (468 kJ/kg^{-1}m^{-2}) when taking the weight into account.

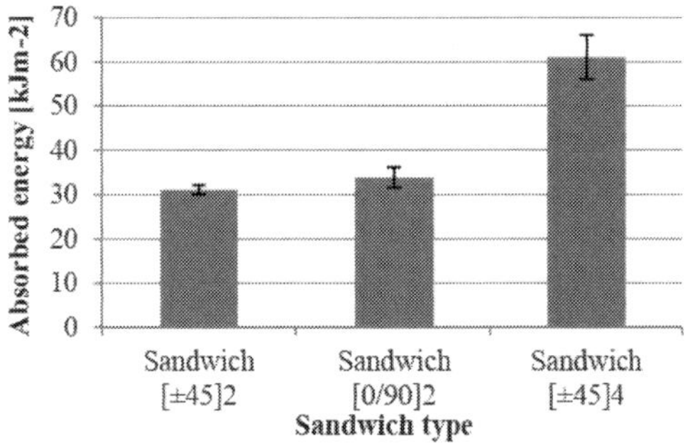

Fig. 5. Specific absorbed impact energy of sandwich specimens, error bars included.

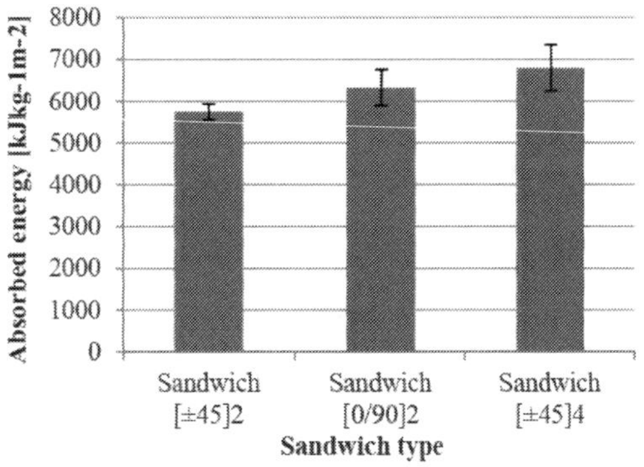

Fig. 6. Specific absorbed impact energy considering the weight of sandwich specimens, error bars included.

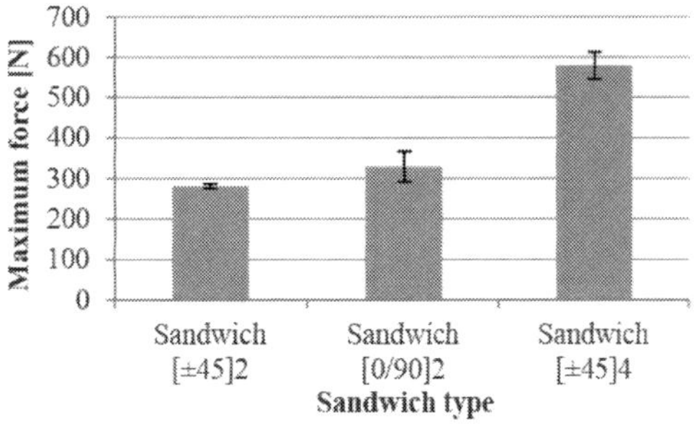

Fig. 7. Ultimate load of sandwich specimens, error bars included.

The $[\pm 45]_4$ sandwich still attains the highest results, but the increase of more than 60 % of the weight (Table 1) makes it less profitable. Next Fig. 7 shows the results of the maximum force reached F_m. There is an obvious similar trend in value increase as in Fig. 5. Force-displacement curves and optical micrograms give a more detailed view of failure behaviour of the tested sandwich specimens again (Fig. 4).

The first sandwich with face sheets $[\pm 45]_2$ exhibits similar behaviour with CFRP [±45] regarding fibres pulled out of the matrix. The force-displacement curve of the sandwich shows typical bench shape up to deformation 8 mm on the level of near maximum force. The beginning of bench means loss of stability of cells compressively loaded under impactor, whereas the end of bench characterizes the loss of stability of the whole beam and beginning of plastic bending. The rest of the curve produces a shear folding of honeycomb cells and a skin bending.

Furthermore, the sandwich specimen with face sheets $[0/90]_2$ demonstrates unstable and vulnerable failure behaviour. On the force-displacement curve it occurs as the first brittle crack of upper skin with compressive deformation of cells characterized by an unstable short bench near force maximum. The gradual compressive folding of cells and elastic bending of lower skin cause the subsequent increase of applied load. However, the brittle crack of the lower skin produces steep decrease to nearly zero values of force. The rest of the following unstable process induces shear folding of cells. When comparing sandwich with $[0/90]_2$ skin with pure [0/90] CFRP, there does not occur explosive delamination but only single delamination due to the low number of plies. The last thicker sandwich specimen $[\pm 45]_4$ exhibits behaviour similar to thinner sandwich $[\pm 45]_2$. Obviously, the whole loading process is moved to higher values of the applied force. The bench shape of the curve on values close to maximum force is extended to deformation 16 - 20 mm unlike the $[\pm 45]_2$ sandwich. There also occurs an extended region of compressive failure of honeycomb cells. This is caused by thicker face sheet which is able to spread impact force better to the core of the sandwich.

In this way the failure behaviour of different sandwich structures was described. All covered sandwiches achieve conformity in the case of core deformation where compressive deformation occurs first and then shear deformation. The force-displacement curves also reveal the difference between failure behaviour of the sandwiches with $[0/90]_2$ and $[\pm 45]_2$ face sheets. Even if the sandwich with $[0/90]_2$ face sheet reached higher values of absorbed energy a_k and maximum force F_m, due to its unstable and brittle behaviour it is less suitable as an energy absorber or a damper.

4 Conclusion

An overview of the comparison of the listed pure CFRPs and the sandwiches could state that the behaviour of specimens with a stacking sequence produced in a similar way is analogous. Both sandwiches with skin orientation in [±45] exhibit a plastic type of force displacement curve unlike the sandwich with skin orientation in [0/90], which is combined with brittle traces. Hence, the [±45] sandwiches and CFRPs are better for application as an energy absorber or a damper.

When comparing sandwiches only, the thickest sandwich with face sheets $[\pm45]_4$ has gained the highest average of absorbed energy a_k. However, the weight increase degrades its advantage. The research will thus continue with an analysis of sensitivity on various influences created by changing thickness skin and core thickness using constant materials.

References

1. S. Abrate, *Impact on composite structures*. New York: Cambridge University Press, (1998)
2. Y. Shen, F. Yang, W. Cantwell, S. Balawi, Y. Li, *Journal of Reinforced Plastics and Composites*, **33**(12), 1148-1157. DOI: 10.1177/0731684414524465, (2014)
3. A. Hazizan, W. Cantwell, *The low velocity impact response of foam-based sandwich structures*, DOI: 10.1016/S1359-8368(02)00009-4, (2002)
4. S. Pingle, N. Fleck, V. Deshpande, H. Wadley, Int. J. of Solids and Structures, **48**(25-26), 3417-3430, DOI: 10.1016/j.ijsolstr.2011.08.004, (2011)
5. S. Wang, S. Li, J. He, *Composite Structures*, **163**, 312-324, DOI: 10.1016/j.compstruct.2016.12.050, (2017)
6. T. George, V. Deshpande, K. Sharp, H. Wadley, *Composite Structures*, **108**, 696-710, DOI: 10.1016/j.compstruct.2013.10.002, (2014)
7. T. Liu, S. Hou, X. Nguyen, X. Han, *Composites Part B: Engineering*, **114**, 328-338, DOI: 10.1016/j.compositesb.2017.01.035, (2017)
8. H. Wang, K. Ramakrishnan, K. Shankar, *Materials and Design*, **99**, 68-82, DOI: doi.org/10.1016/j.matdes.2016.03.048, (2016)
9. G. Sun, S. Li, Q. Liu, G. Li, Q. Li, *Composite Structures*, **152**, 969-993. DOI: 10.1016/j.compstruct.2016.06.019, (2016)
10. G. Kooistra, H. Wadley, *Materials*, **28**(2), 507-514. DOI: 10.1016/j.matdes.2005.08.013. ISSN 02613069, (2007)
11. R. Mines, S. Tsopanos, Y. Shen, R. Hasan, S. Mckown, Int. J. of Impact Engineering, **60**, 120-132, DOI: 10.1016/j.ijimpeng.2013.04.007, (2013)
12. V. Birman, *Composites Part B: Engineering*, **35**(6-8), 665-672, DOI: 10.1016/j.compositesb.2003.08.012, (2004)
13. M. Meo, R. Vignjevic, G. Marengo, Int. J. of Mechanical Sciences. **47**(9), 1301-1325, DOI: 10.1016/j.ijmecsci.2005.05.006, (2005)
14. D. Zhang, Q. Fei, P. Zhang, *Composite Structures*, **168**, 633-645, DOI: 10.1016/j.compstruct.2017.02.053, (2017)
15. ASTM D6110-10, Standard Test Method for Determining the Charpy Impact Resistance of Notched Specimens of Plastics, ASTM International, West Conshohocken, PA, www.astm.org, (2010)

New solution for transport and industrial noise protection through reflective noise barriers

Ivan Kralov[1,*]

[1]Technical University – Sofia, 8 Kliment Ohridski Blvd., 1000 Sofia, Bulgaria

Abstract: A new solution for protection of transportation and industrial noise through reflective noise barriers is proposed and investigated in this study. The new solution combines the advantages of the known barriers and has its own advantages in addition. The preliminary results show a very good level of noise reduction for this type of barriers.

1 Introduction

Acoustic noise reduction is a very important process due to its direct impact on the environment. The main problem in this area is the noise created by the specific functional processes of the machines. In most of these cases the typical solution is to reduce its propagation by different types of barriers, absorbers or active noise reduction solutions. The topicality of the problem mentioned above is proven by the enormous number of recent research work and publications [10, 11, 12, 13]. In many cases, expensive and sophisticated design and technological solutions have to be created in order to meet the noise reduction standards. The methods for achieving noise reduction are mainly divided into passive, active and combined. Passive methods generally block the spreading of sound waves and decrease the noise levels by reflecting and absorbing their energy [1, 2, 3, 10, 11, 12, 13]. On the other hand, active methods input additional energy that is used to eliminate the acoustic energy of the source by creating significant noise background, propagation of sound waves in the opposite phase etc. [1, 2, 13]. Combined methods include both of the previously mentioned ones.

The subject of the present research is part of the passive methods of noise reduction because of their low cost, manufacturing technology, installation and usage, as well as the opportunity for high feasibility in industry. The main principle in the considered patent solutions is reflecting the sound waves, generated from the source, using profile barriers. In most of these cases there is also a considerable absorption of the barrier.

A classical example of a reflective barrier is described in Patent № UD 96/16230 – 30.05.1996 [4]. Convex arc-shaped reflecting profiles are fixed to the vertical concrete columns as shown on Fig. 1. Part of the incoming sound waves are reflected by the barrier and these waves interfere with the original ones close to the barrier surface performing noise reduction. Another part of the source sound waves is absorbed by the barrier while a third part passes through it. Some of the advantages of this solution are: a wide area of application, a stable construction consisting of one type of reflecting elements and relatively low cost of production. Different columns in case of different heights of the barrier, expensive transportation and assembly are the main disadvantages. From an acoustic point of view, the edges of the profiles joints are concentrators of source and reflected sound energy and in this capacity they could be treated as a secondary line source of sound. In this way the noise level near the barrier could be significant.

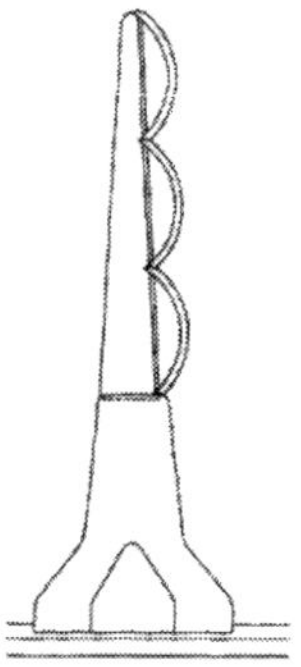

Fig. 1. Fixture of convex arc-shaped reflecting profiles to vertical concrete column.

Patents №№ US № 6305492 B1 / 23.10.2001 [5] and US 7789193 B2 / 07.09.2010 [6] are similar to the above described ones, but their reflecting surfaces have concave arc profiles – Fig. 2 or hexagonal concave profiles – Fig. 3. The main advantage in both cases is reduction through interference of the source and reflected waves near the barriers. The disadvantages are expensive production, transportation and high assembly costs especially for the second solution.

The noise barrier under Patent № AT 513615 A4 2014-06-15 [7] is built from elastic elements made from used car tires – Fig. 4. They are assembled to be able to make small vibrations. This construction has a few

* Corresponding author: kralov@tu-sofia.bg

advantages: use of materials normally considered waste (used car tires), higher degree of absorption due to internal damping in the rubber during vibrations and reduction through interference of the source and reflected waves near to barriers. Some problems of this solution are connected with how they look as well as their low climate resistance.

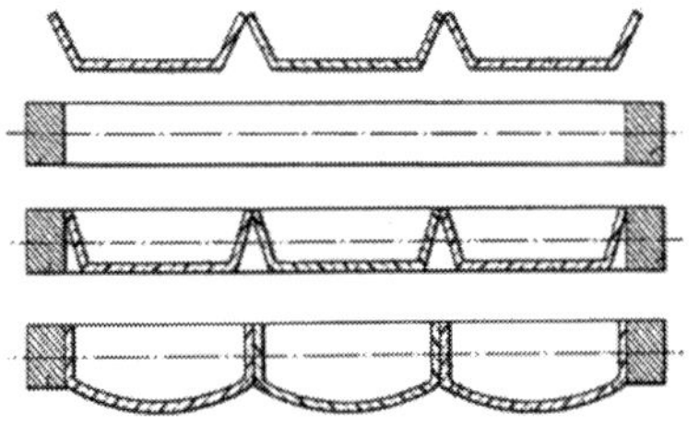

Fig. 2. Reflecting surfaces with concave and arc profiles.

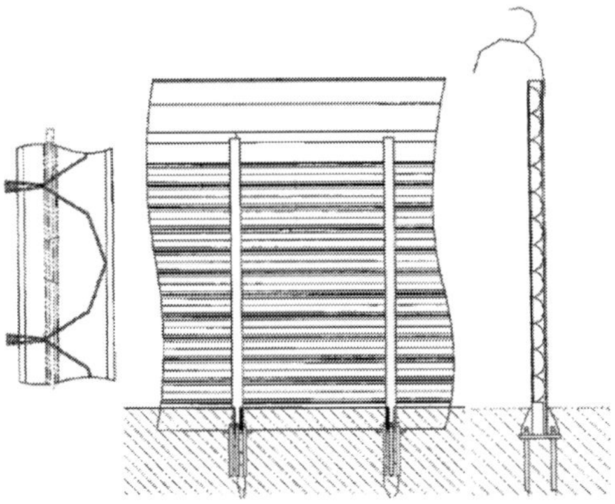

Fig. 3. Reflective barriers with hexagonal concave profiles.

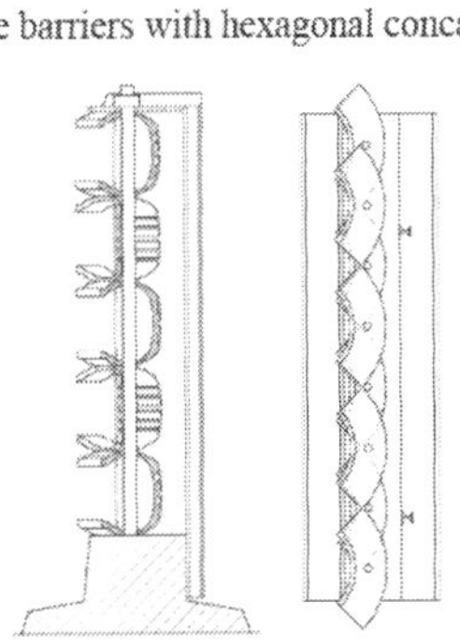

Fig. 4. Barriers with elastic elements made from used car tires.

Patent solutions №№ 89203035.4 / 29.11.1989 [8] (Fig. 5) and DE 42 20 547 A1 / 07.01.1993 [9] (Fig. 6) cover preliminarily automotive and railway transportation noise insulation. A specific characteristic of the first one is the possibility to change the angle of the reflecting surface. This is a simple construction, but very expensive to assemble and maintain. The second solution has a construction which allows to add heavy absorbing material layers to the barrier. This is also a simple, relatively cheap and noise-reduction effective solution. The main disadvantage is that it is not applicable in the open air due to the absence of climate resistance /rain, snow etc./ of the absorbing materials.

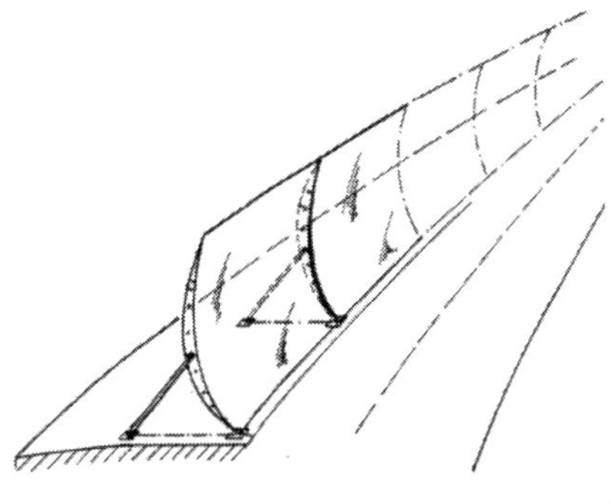

Fig. 5. Preliminarily automotive and railway transport noise insulation.

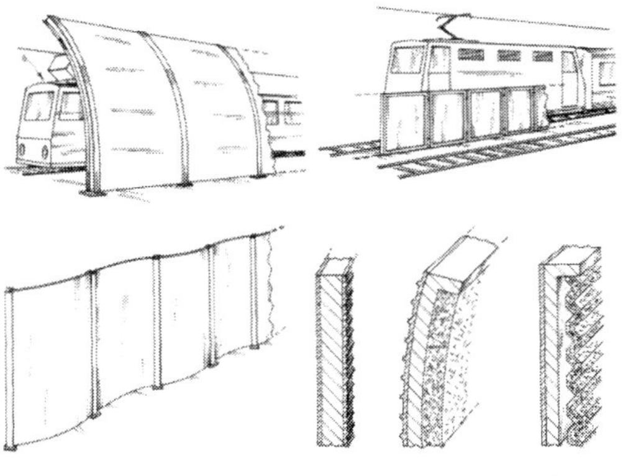

Fig. 6. Preliminarily automotive and railway transportation noise insulation.

Nowadays, barriers, similar to patent solution 10 2014 2017 767.7 / 10.03.2016. [10] (Fig. 7) are used widely. Barriers are made of long metal panels, lined one over the other and fixed between vertical metal columns. Some variants of the solution are perforated panels, or/and presence of absorbing material inside. This type of construction is effective, easy to transport and assemble, climate and fire resistant, but it has very high production cosst. This costs are due to the used material – aluminium, the expensive technology for mechanical perforation of the panel, as well as the assembly of the absorbing material.

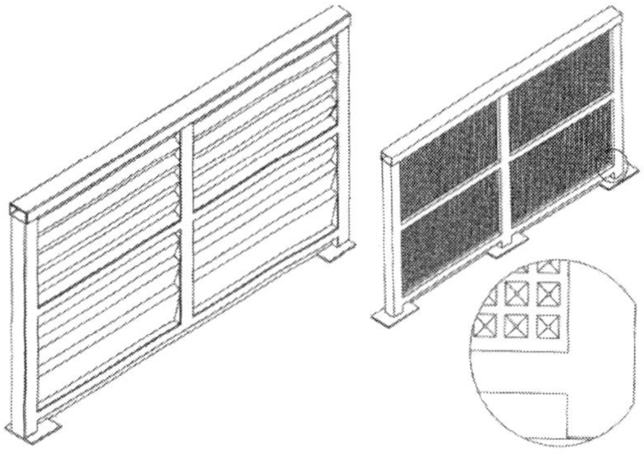

Fig. 7. Commonly used preliminarily automotive and railway transportation noise insulation.

On the base of the analysis of the above described main patent solutions for passive noise reduction barriers, the following conclusions could be made:

The great number of patent solutions for passive noise reduction barriers are based on the use of different profile reflecting surfaces.

Some noise barriers are made by assembly elastic elements in the construction. Their small vibrations excited by the sound energy increase the degree of absorption. In a few cases these elements are industrial waste and the ecological effect is a bonus.

The efficiency of the passive noise barriers is increased by adding additional layers of highly absorbing material.

A significant part of the proposed and used solutions for passive noise reduction have good level of insulation, but their production, transportation, assembly and maintenance costs are very high.

Taking into consideration the above conclusions, the aim of this study is defined as the synthesis of a new passive noise barrier construction, which has a minimum of the described disadvantages. To reach the final result three main problems need to be solved: define the design principles, design the construction and determine and verify its main acoustic parameters.

2 New construction of a noise insulating panel and barrier

As a result of the analysis above the following principles for the new barrier synthesis are used:

- passive noise barrier for transportation and industrial noise;
- level of noise reduction equal or better to that of the existing barriers;
- relatively low costs of production, transportation, assembly and maintenance;
- relatively high level of climate change resistance, wind-proof and fire-proof construction.

As a result of the synthesis the new noise insulating panel and barrier constructions are shown on Fig. 8. Noise insulating panels are made from non-metal materials. They are one of many horizontal channels arranged one over the other, and all their concave arc profiles have one and the same direction – to the sound source waves. The essential characteristic of the construction is that the concave profile is exactly semi-cylindrical and the border area between two neighbor channels is a sharp edge. These channels are arranged into H-shaped metal columns fixed to the ground. The channels are fixed vertically to the columns through non-metal semi-disks of external diameter, equal to the internal diameter of the channels. The semi-disks are fixed to the columns by bolt or stud connection.

The main expected effect of the new solution is that the reflection of the sound waves and the interference between the source and reflected ones is close to the barrier. In this way the noise reduction by converting sound energy into heat will not affect the noise sources. The degree of reflection depends mainly on the channel material, sound waves frequency and acoustic pressure. An important parameter of the barrier construction is the absence of reflecting surface, perpendicular to the source sound waves. The exact semi-cylindrical profile of the channels focuses reflected waves to its geometrical center. This reduces the area of interference and noise reduction.

Geometrical parameters of the barrier can be designed for specific cases to achive better performance. The width of the profile and the radius of the channel could by varied according to the sound power and frequency.

The modular construction has a lot of advantages. First, the non-metal channels are technologically easy to produce, for example by extrusion. The uniform elements allow for big series production. This reduces production costs. Second, the elements are easy to be transported, assembled and maintained. Assembly does not require neither special machines nor expensive equipment, and there is no need for the working team to posess any special skills. Third,the modular construction makes it possible to vary easily the height of the barriers.

By using appropriate materials some additional requirements like climate resistance and fire-proofing could be fulfilled. Other common requirements like construction stiffness, wind resistance etc. have to be designed using the common rules.

The wide area of application of the new barriers is an additional advantage. They could be used for noise insulation of automotive and railway roads, for industrial sites, for buildings, etc.

3 Numerical investigation procedure verification

Comsol multiphysics® acoustic module will be used for numerical investigation of the new barrier noise reduction. Before this step, an experimental setup as well as its numerical model were created to verify the numerical procedure used.

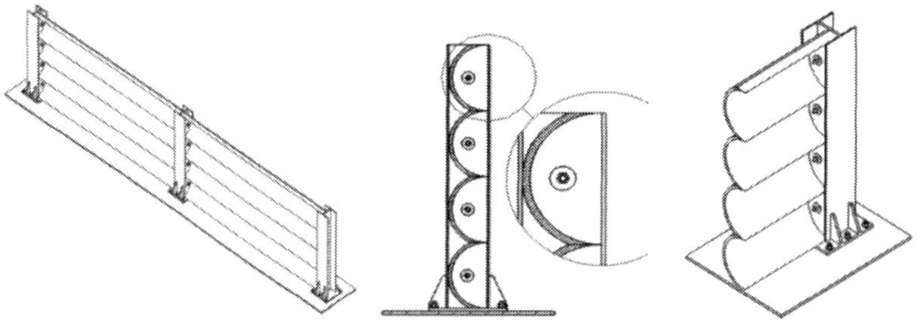

Fig. 8. New noise insulating panel and barrier.

This experimental setup and the testing procedure of the absorber are shown in Fig. 9. The noise, generated by the noise generator and powered by the amplifier, is emitted in the 30 dB acoustic chamber. The front side of the chamber is open. At a distance of 0,5 m the microphone registers the noise levels. In this way the noise levels of the source without a barrier are recorded. After the first experiment, the testing absorber is attached to the chamber front side (Fig. 9).

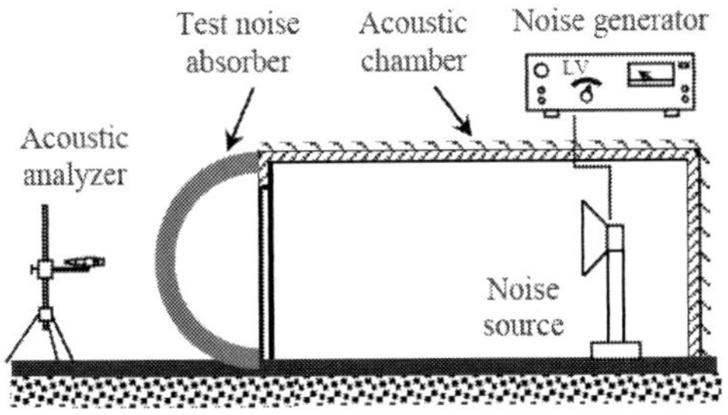

Fig. 9: Scheme of the noise absorber testing procedure.

The absorber is made of high density polyvinile (HDPV) material and is shaped as a semi-circular tube with a diameter of 500 mm and a width of 10 mm. All holes are insulated and covered by high-level absorbing materials. The same noise levels are generated by the noise-source equipment. The microphone registers the noise levels at the same position after the new absorber is applied. A XL2, class 1, certified acoustic analyzer produced by NTi company is used. All the results are collected according to the standards and legislations concerning calibration and testing procedures, as well as measurement error evaluation.

The numerical model of the above described experimental setup is investigated using the Comsol® software. Noise levels in decibels (dB) are measured in both procedures.

The noise reduction level from experimental tests ΔL_{EiI} for i-th frequency are defined as the difference between the measured noise level without absorber $L_{E,i}$ and measured noise level without absorber $L_{EA,i}$:

$$\Delta L_{EiI} = L_{E,i} - L_{E,I}. \quad (1)$$

The noise reduction level from numerical tests ΔL_{NiI} for i-th frequency are defined as the difference between the measured noise level without absorber $L_{N,i}$ and measured noise level without absorber $L_{NA,i}$:

$$\Delta L_{NiI} = L_{N,i} - L_{N,I}. \quad (2)$$

The results about noise reduction with both procedures are given in Table 1.

Table 1. Noise level reduction for test sound barrier – experimental and numerical results

Hz	ΔL_{EiI}, dB	ΔL_{NiI}, dB
125	4,7	4,9
250	8,2	9,7
500	15,6	16,9
1000	10,4	12,3
2000	21,6	23,3

There is a very small difference for each frequency between the two procedures. On the basis of these results, it could be stated that the used numerical procedure is appropriate for the planned investigation.

4 Noise reduction efficiency of the new barrier

The main studied parameter of the newly designed noise insulating barrier is the degree of noise reduction.

First, using the advantages of the advanced digital technologies, a numerical model of the barrier was created. For meshing, defining of the boundary conditions and other parameters, software guidelines and good practices were used.

Next, a numerical simulation of the noise radiation and reduction of the model was done. Commercial software for CAD and for Acoustic analysis were used.

Parameters of the investigated model are:

- Channel material – high density PVC /HDPVC/;
- Width of the channel – 20 mm;
- Diameter of the channel – 600 mm;
- Height of the barrier – 4,2 m;
- Height of the noise source – 1 m;
- Type of the noise waves – spherical;
- Horizontal distance between the source and front of the barrier – 7,5 m;
- Height of the "microphone" position – 1,5 m;
- Horizontal distance between the "microphone" position and the back of the barrier – 5 m.

To compare the results of this study with similar ones, most of the parameters are chosen to replay the standard requirements or to correspond the data from other research works.

The environment around the barrier is air of density of 1,19 kg/m^3 and speed of sound of 343 m/s.

The barrier material is HDPE with 970 kg/m^3 density, 2310 m/s speed of sound, 0,40-0,45 Poisson's ratio, and 1,035 N/mm^2 Young's modulus.

Noise reduction is measured for the central frequencies of the 1/3 octave bands: 125 Hz, 250 Hz, 500 Hz, 1000 Hz, 2000 Hz. Formula 2 is used for evaluation of the degree of noise reduction. A graphical representation of the results for 250 Hz and 1000 Hz is given in Fig. 10 and Fig. 11. Results from the numerical simulations are presented in Table 2.

The maximal level of noise reduction for the studied frequencies is 16,1 dB at 2000 Hz.

On the basis of the obtained results about the level of noise reduction, with the new barrier, it could be stated that:

These results show very good noise reduction of the new noise insulating barrier in general.

There is a big difference in noise level reduction for different frequencies for the studied barrier. It could be the result of resonant modes of the barrier in some frequencies, as well as of the different "noise shadow" at different frequencies.

The minimal level of noise reduction for the studied frequencies is 6 dB at 500 Hz.

 DOI: 10.1051/matecconf/201713306001

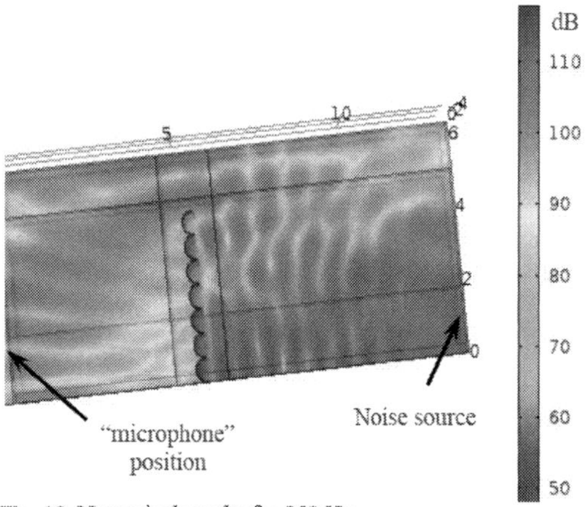

Fig. 10. Numerical results for 250 Hz.

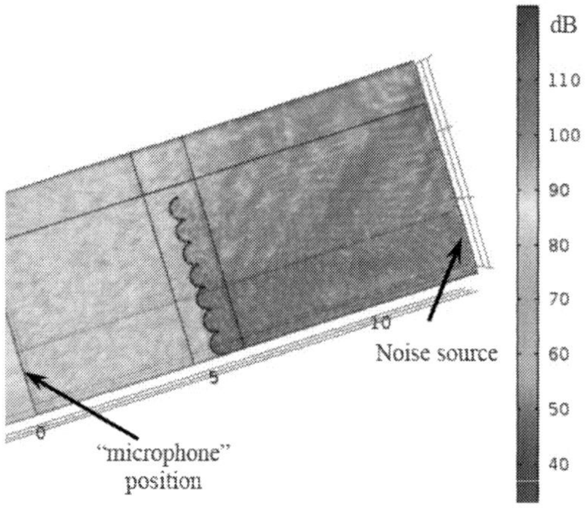

Fig. 11. Numerical results for 1000 Hz.

Table 2. Noise level reduction with the new sound barrier – numerical results

Hz	ΔL_{NI}, dB
125	10,5
250	13,0
500	6,0
1000	16,0
2000	16,1

The next step in the investigation will be to find the influence of some geometrical parameters upon the noise reduction efficiency of the barrier.

5 Conclusion

A new solution for insulation of transportation and industrial noise by means of reflective noise barriers has been proposed and investigated in this study. It combines the main advantages of currently used barriers and it also has additional ones. Numerical simulation shows high degree of noise reduction. To improve the efficiency of noise reduction a further investigation of the influence of the barrier parameters upon the frequency absorption response has to be made.

References

1. S. Banov, I. Kralov, *Noise in Transportation Technics*, TU-Sofia, Sofia, (in Bulgarian), (2003)
2. Nikolov, N., D. Benov, I. Shubin, *Acoustic Design of Transport Noise Insulating Barriers*, ACMO Academic Press, Sofia, (in Bulgarian), (2014)
3. Patent № UD 96/16230 – 30.05.1996.
4. Patent № US № 6305492 B1 / 23.10.2001.
5. Patent № US 7789193 B2 / 07.09.2010.
6. Patent № AT 513615 A4 2014-06-15.
7. Patent № 89203035.4 / 29.11.1989.
8. Patent № DE 42 20 547 A1 / 07.01.1993.
9. Patent № 10 2014 2017 767.7 / 10.03.2016.
10. I. Kralov, S. Terzieva, I. Ignatov, Analysis of methods and MEMS for Acoustic Energy Harvesting with Application in Railway Noise Reduction, *Proceedings of the MECAHITECH'11*, Bucharest, (2011).
11. I. Kralov, P. Sinapov, K. Nedelchev, I. Ignatov, Friction Induced Rail Vibrations, AIP, **1497**, 19-25, (2012)
12. J. Trevor, P. D'Antonio, *Acoustic Absorbers and Diffusers*, Taylor & Francis, (2009)
13. D. J. Thompson, *Railway Noise and Vibration*, Elsevier, (2009)

Energy consumption of auxiliary systems of electric cars

Ivan Evtimov[1,*], *Rosen* Ivanov[1], and *Milen* Sapundjiev[2]

[1]University of Ruse, Department of Engines and Vehicles, Rousse, Bulgaria
[2]University of Ruse, Branch Silistra, Bulgaria

Abstract. The paper analyzes the power demand of the auxiliary systems of electric cars. On the basis of existing electric cars an analysis of energy consumption of different auxiliary systems is done. As a result possibilities for rational use of these systems have been proposed, which can increase the mileage per one charge of the battery.

1 Introduction

An important characteristic of energy performance of the electric car is the distance covered for one charge of the battery [1-3]. Usually, in the technical specification of electric cars, producers give an operational range which is not precisely detailed about the conditions of motion (in city or inter-city traffic, air temperature, use of the auxiliary systems, etc.). For the owners it is very important to know as best as possible the remaining travel distance and impact of auxiliary systems on energy consumption and distance [4-7]. That knowledge will ensure a calm and comfortable trip regardless of the limited energy autonomy of electric cars.

The goal of the paper is to analyze the impact of different auxiliary systems of electric cars on the travel distance in different running conditions and comfort (such as temperature in the car, using lights, audio system, etc.).

2 Exposition

2.1. Fundamentals of electric car energy consumption

The main purpose of the energy accumulated in the battery is to supply the electric motor and to ensure the electric car motion in different running conditions. In addition, the battery also has to supply the auxiliary systems, which guarantee safety (lights, horn, window cleaner etc.) and comfort (climate control, media etc.). During the trip, the value of the specific energy consumption (in Wh/km or kWh/100km) can be different. Depending on the skills and needs of the driver, energy consumption can be 2 times higher than the one indicated in the technical specification of the electric car.

In the general case, the specific energy consumption can be determined theoretically by following the equation

$$E_{100} = \frac{100}{3,6\eta_M \eta_E}\left[\left(f_o + 5.10^{-7}V^2\right)G + k_B S\frac{V^2}{13}\right] + E_{AS_{100}}, \frac{\text{kWh}}{100\,\text{km}}, \quad (1)$$

where f_o is the rolling resistance coefficient at low speed; V – car speed, km/h; G –car weight, kN; k_B – coefficient of aerodynamic resistance, kNs^2/m^4; S – the front area of the car, m^2; η_M – the efficiency coefficient of transmission; η_E – the efficiency coefficient of the electric motor and power electronics; E_{AS100} – the specific energy consumption of auxiliary systems, kWh/100km.

Coefficient k_B is calculated as

$$k_B = 0,5.10^{-3}\rho c_x, \text{kNs}^2/\text{m}^4, \quad (2)$$

where ρ is air density, kg/m^3; c_x – drag coefficient.

The change of the air temperature t from +40 to –20 °C causes a change in its density from 1,127 to 1,395 kg/m^3 [8] and at high speed it can increase the energy consumption by over 10%. The value of the air density can be evaluated with good accuracy (deviation of not more than 0,5% at low temperature) using the equation

$$\rho = 2.10^{-5}t^2 - 0,0048t + 1,2926, \text{kg/m}^3. \quad (3)$$

Mechanical losses in the transmission vary within wide limits and depend on the electric motor load. The efficiency coefficient η_M can be evaluated with good accuracy using the approach, proposed in [1]. The losses in the electric motor and power electronics η_E also depend on thhe working conditions and load. The product of both coefficients varies between 90 – 95%, but can decrease by 50% under some running conditions [9]. It is necessary to have the characteristics of the elements of electric drive, not only at nominal load (whose value is given in the technical specifications), but also at particular load. Some of the researchers assign these two types of losses to the so called drive train losses [3].

* Corresponding author: ievtimov@uni-ruse.bg

 DOI: 10.1051/matecconf/201713306002

Auxiliary systems have great impact on energy consumption– the second part E_{AS100} of equation (1). The approach for their assessment has to be very accurate, especially when the maximal power of these devices is in use, to ensure exact determination of travel distance.

The power supply of auxiliary systems is realized by the second (operational) battery at a voltage of 12V. It can be recharged from the traction battery through DC/DC convertor. The losses during this transformation have to be taken into account by introducing a coefficient marked as η_{DC}. Finally, the specific consumption of the auxiliary systems can be represented as

$$E_{AS_{100}} = \frac{1}{\eta_{DC}}\left(E_{CC} + E_L + E_{WCS} + E_{OS}\right), \frac{\text{kWh}}{100\text{km}}, \quad (4)$$

where η_{DC} is the efficiency coefficient of the convertor between two batteries; E_{CC} – the specific energy consumption of climate control system; E_L – the specific energy consumption of lights and horn; E_{WCS} – the specific energy consumption of windows cleaning system; E_{OS} – the specific energy consumption of other systems as SRS, ABS, TC ESP, electric windows open system etc.

2.2. Distribution of the total energy consumption of the electric car

An analysis of the distribution of the total energy used can be done on the basis of an existing example. Fig. 1 presents a real picture of the energy consumption of the Tesla Roadster electric car with different values of the speed. The climate control does not work [3].

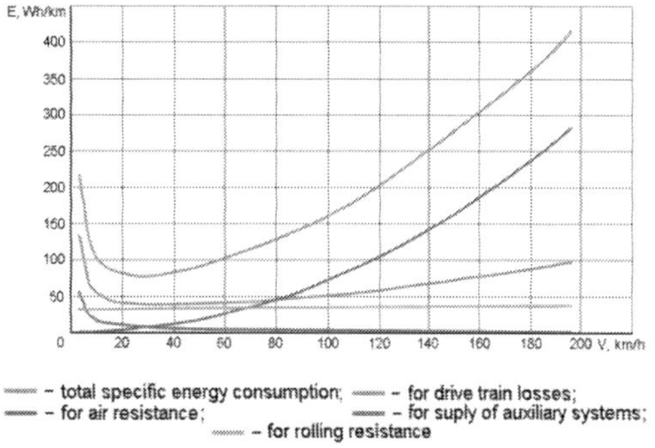

Fig. 1. Distribution the total specific energy consumption of Tesla Roadster vs car speed

The ratio between the different parts of the total specific energy consumption changes with the increase of car speed. At low speed most significant is the part of energy consumption for drive train losses and supply of the auxiliary systems. Higher energy consumption in slow motion is caused by the low values of the efficiency coefficients η_M and η_E. At high speed, the energy spent for air resistance becomes the largest.

The energy spent for rolling resistance is changed within short limits, because of the small impact of the speed on the coefficient *f*.

In fact, in motion the most variable parts of the energy can be those spent for air resistance and supply of auxiliary systems. The last part depends on the atmospheric conditions such as rain, snow, wind etc.

The curves shown in Fig.1 are well represented by the following regression models:

- total specific energy consumption

$$E = 4.10^{-10}V^6 - 3.10^{-7}V^5 + 7.10^{-5}V^4 - \\ - 0{,}009V^3 + 0{,}5715V^2 - 16{,}313V + 234{,}92; \quad (5)$$

- specific energy consumption for drive train losses

$$E = 3.10^{-10}V^6 - 2.10^{-7}V^5 + 5.10^{-5}V^4 - \\ - 0{,}0057V^3 + 0{,}358V^2 - 10{,}26V + 139{,}27; \quad (6)$$

- specific energy consumption for rolling resistance

$$E = 0{,}0297V + 32{,}278; \quad (7)$$

- specific energy consumption for air resistance

$$E = 1.10^{-6}V^3 + 0{,}007V^2 + 0{,}0035V + 86{,}22; \quad (8)$$

- specific energy consumption for supply of auxiliary systems

$$E = 121{,}1\ V^{-0{,}794}. \quad (9)$$

2.3. Energy consumption of auxiliary systems

Approximately, the energy consumption of the auxiliary systems presented as a percentage of the energy accumulated in the main (traction) battery is shown in Table 1 [5, 7].

Table 1. Energy consumption of some auxiliary systems

Auxiliary systems	Part of traction battery energy, %
Climate control:	
– cooling;	Up to 30%
– heating.	Up to 35%
Power steering	Up to 5%
Braking system	Up to 5%
Other (lights, media, locks etc.)	Up to 5%

The information presented in Table 1 is more general and does not include all operational conditions of electric cars. This is the reason to make a review of the impact different factors on energy consumption of each auxiliary system.

Climate control system

The normal internal temperature of the air in the compartment has to be 20-23 °C. To maintain these limits, the energy consumption of climate control depends on the temperature difference in and out of the car. Table 2 presents an example of the needed power of control system at different internal temperatures and high external temperature [5].

 DOI: 10.1051/matecconf/201713306002

Table 2. Needed power for supply of the climate control system as a function of internal temperature in the compartment

External air temperature, °C	Internal temperature, °C	Needed power, kW
43	21	1,5 – 2
43	25	1
43	29	0,5

The maximum value of the power supply of climate control can achieve 3 – 5 kW for some models of car. As heat device they use an electric heater or a thermo pump. In Fig. 2 the impact of power consumption of 2 kW (working climate control) on the travel distance is illustrated for Tesla Roadster electric car [3].

At a speed of 25 km/h, the travel distance per one charge of the battery decreases approximately 2 times when the climate control of 2 kW works. The curves are well represented by the following regression models:

• travel distance without working climate control

$$L = -9.10^{-10}V^6 + 6.10^{-7}V^5 - 0,0002V^4 + \\ + 0,0228V^3 - 1,6088V^2 + 50,131V + 116,1; \quad (10)$$

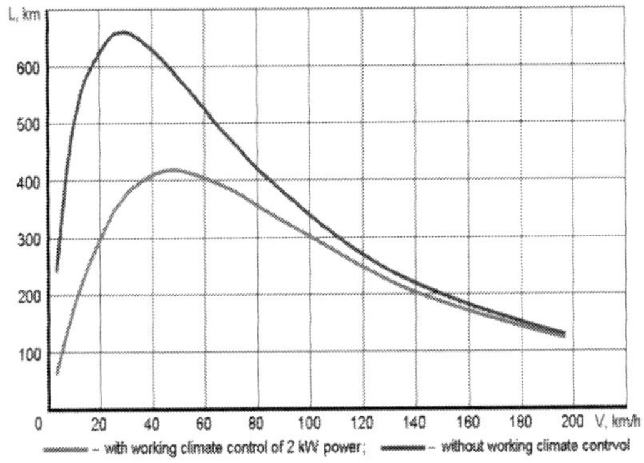

Fig 2. Impact of climate control power consumption of 2 kW on the distance per one charge of battery for Tesla Roadster electric car.

• travel distance with working climate control of 2 kW power

$$L = -4.10^{-6}V^4 + 0,0018V^3 - 0,3157V^2 + \\ + 19,881V + 10,837. \quad (11)$$

There is not much research on the impact of the external air temperature on energy consumption. In [10] a Canadian company, on the basis of over 7000 travels in the whole of North America, has made a generalization of the average energy consumption of Nissan Leaf electric car (Fig. 3).

The curve from Fig. 3 is well represented by the regression model

$$E = 8.10^{-9}t^6 - 3.10^{-6}t^5 + 0,0001t^4 + 0,0028t^3 - \\ - 0,0546t^2 - 2,7979t + 206,22. \quad (12)$$

The same data is shown in Fig.4 as impact on the travel distance L [11]. The respective regression model is

$$L = 6.10^{-8}t^6 - 4.10^{-7}t^5 - 0,0001t^4 - 0,0004t^3 + \\ + 0,0544t^2 + 1,3326t + 99,995. \quad (13)$$

It is obvious that the external air temperature has a significant impact on the energy consumption of an electric car. The explanation for this is connected with the energy for heating or cooling but also with the efficiency of the battery at different temperatures. Taking into account these two factors, one can see an optimal external air temperature at which the energy consumption is minimal and travel distance is maximal (Fig. 3, 4). This optimal value is approximately 20 °C.

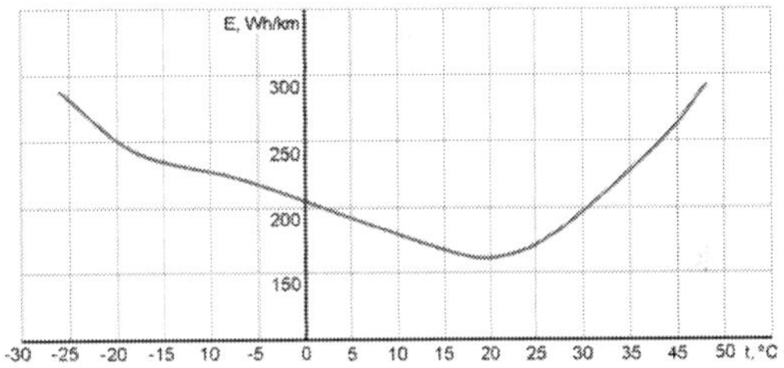

Fig. 3. Impact of the external air temperature on the specific energy consumption of a Nissan Leaf electric car.

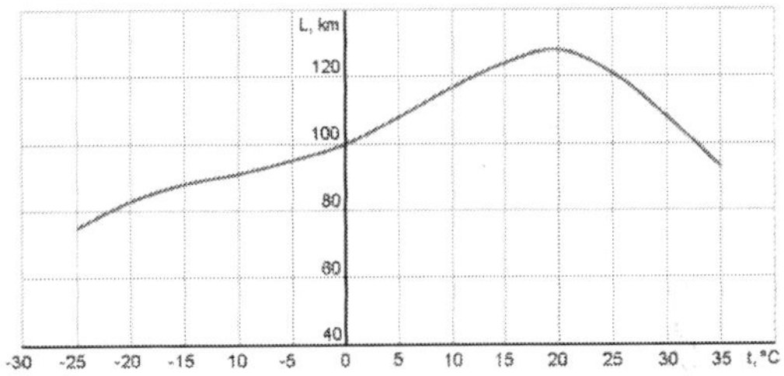

Fig. 4. Impact of the external air temperature on the travel distance of a Nissan Leaf electric car.

Light system, light signalization and horn

The energy consumption of the light system and signalization depend on a twenty-four-hour period – if the trip is realized in the day or at night. That is particularly important for long and short front lights. The usage of elements of light system and signalization, during 100 km travel, are presented in Table 3. The data from different sources [4, 6] was processed and summarized.

The calculations show that the maximal energy consumption of a light system using conventional lamps in night travel is about 150 Wh/100km. The usage of LED-lamps decreases the consumption 2,2 – 3,8 times [2, 4, 6, 12, 13].

In the specialized literature there is no information concerning the time of use and energy consumption of the horn. Probably, this is because the value of the energy used is insignificant.

 DOI: 10.1051/matecconf/201713306002

Table 3. Statistical data for the use of the elements of the light system and signalization.

Elements	Working time, min/100 km	Power consumption for car with conventional lamps, W	Power consumption for electric car with LED-lamps, W
Daily lights	116,5	40	8
Long lights	9,8*	60	34,4
Short lights	97,6*	55	54
Left blinker	5,8	21	6,9
Right blinker	4,6	21	6,9
Stop-lights	18,9	21	5,6
Stop-lights (central position)	18,9	21	3
Rear-lights	107,4	5	1,7
Registration table lights	107,4	5	0,5
Reverse motion light	0,9	21	5,2
* - night time driving only			

Audio system

Energy consumption depends on the power characteristic and the time the system is used. Usually, built-in systems have a power supply of about 200 W. The time during which the audio system is used varies within wide limits and corresponds to the driver and passenger(s) needs.

The actual energy consumption also depends on the sound level. Some authors [9], in simulation models, give an average power supply of 20 W for the audio system and use a ratio of approximately 75% of travel time.

Windows cleaning system and seat heating

This system uses electric motors with maximum power of 30-50 W. The time of use strongly depends on the weather (if there is rainfall or snowfall).

The average consumption of the seat heating system is 30 W and the mean ratio of use – 5% of the time [9].

Other systems

The main systems included in this group are: the system of passive and active safety – SRS; Anti-lock Braking System – ABS; Traction Control System – TC; Dynamic Stability System – ESP; systems for opening and closing of door windows and roof. The biggest consumers in this group are the systems for active safety, but the value of energy depends on driving style.

Internal losses in traction battery

Depending on the battery type, during idle time (no traction) the additional losses can present for the maintenance of the working temperature. For example, natrium-nikelhidrid batteries work at a temperature of approximately 300 °C and permanent consumption power of 60 – 80 W for temperature maintenance. If the capacity of the battery is 18 kWh after 10 days of idle time it will be fully discharged.

Internal losses of the Lithium-ion batteries depend on the number of the connected cells and Battery Management System – BMS.

Every battery has a limited period of exploitation. To extend that period the power electronics controls charge/discharge the process. This means that only part of the battery capacity can be used – full charge and discharge are unavailable. This is done to provide for the possibility of accumulation of the regenerative braking energy.

2.4. Impact of regeneration on travel distance

Regeneration of electric energy is possible during the braking process. Depending on the running conditions and route characteristics, the maximuml value of regenerative energy varies from 10 to 25% in city conditions [14, 15]. The experimental results [14, 15] show that braking deceleration within the limits of 2 – 3 m/s^2 can ensure efficiency of regenerative braking of up to 90% and minimal transformation of kinetic energy to heat and friction in mechanical braking system (Fig. 5).

At bigger decelerations, the battery cannot receive regenerative energy, the mechanical braking system is switched on and the two systems work together to provide the required deceleration (Fig. 6).

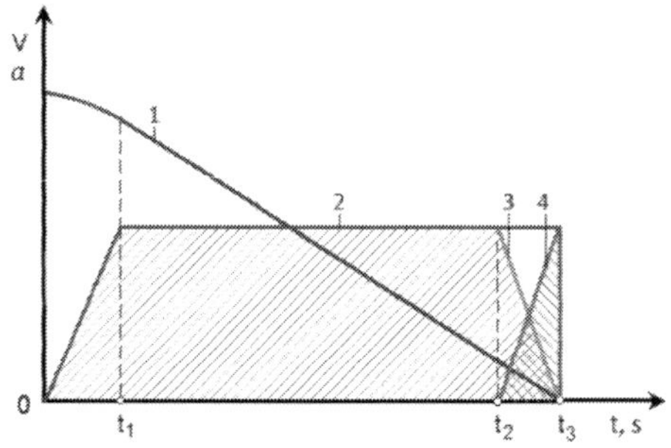

Fig. 5. Example of regenerative braking:
1 – car speed; 2, 3 – deceleration, realized only by regenerative braking; 4 – deceleration, realized only by mechanical braking system.

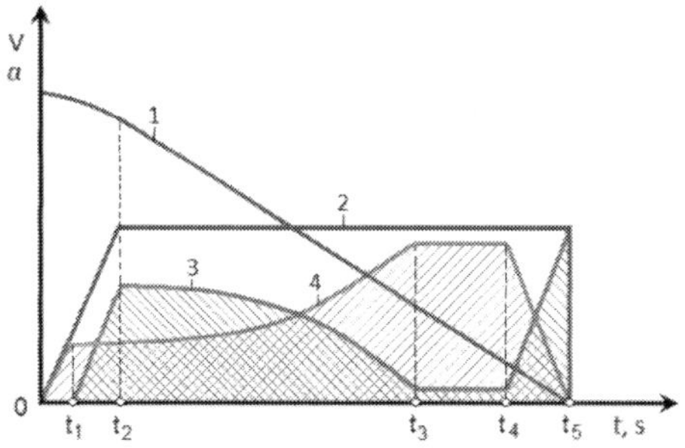

Fig. 6. Interaction of the two braking systems during formation of constant deceleration: 1 – car speed; 2 – total deceleration; 3 – deceleration, realized by mechanical braking system; 4 – deceleration, realized by regenerative braking.

 DOI: 10.1051/matecconf/201713306002

To improve the usage of regenerative energy traction system super capacitors (especially in buses) are often build-in.

3 Conclusions

On the basis of the analysis of the impact of the running conditions and auxiliary systems on the energy consumption of an electric car, the following conclusions can be formulated:

1. The minimal energy consumption of electric cars is realized at lower speed – up to 40 km/h. These values are significantly lower than the respective ones for conventional cars - approximately 65 km/h.

2. At low speed, for example 5 km/h (heavy traffic and jams), the energy consumption can be equal to that one at 100 *km/h*. The reason for this are the low efficiency of the drive train and energy consumption for supply of the auxiliary systems at low speed motion.

3. At high speed – over 50 km/h – the impact of the auxiliary systems in total energy consumption decrease, the energy consumption spent for air resistance becomes dominant.

4. At some values of speed and weather conditions, the energy consumption for the supply of auxiliary systems can decrease twice travel distance of the car.

5. The minimal energy consumption of auxiliary systems is realized at an external air temperature of 20 °C, at which the biggest travel distance is achieved.

6. The light system and signalization consume about 1% of total energy consumption of an electric car.

The authors would like to thank the Scientific Research Fund of the University of Ruse, Project 2017-FT-01 for the financial support.

References

1. R. Ivanov, I. Evtimov, M. Sapundjiev, A model for investigation of energy characteristic of an electric car, *Electric vehicles EM'15*, 131-137, (in Bulgarian), (2015)
2. E. Schaltz. Electrical Vehicle Design and Modeling, *Electric Vehicles – Modelling and Simulations*, 496, (2011)
3. EV Energy Consumption, www.solarjourneyusa.com/EVdistanceAnalysis5.php
4. B. Schoettle, M. Sivak, Y. Fujiyama. Leds and power consumption of exterior automotive lighting: implications for gasoline and electric vehicles. Report No. UMTRI-2008-48, 18 (2008)
5. EV Auxiliary Systems Impacts, http://avt.inl.gov/sites/default/files/pdf/fsev/auxiliary.pdf
6. M. Vražić, O. Barić, P. Virtič. Auxiliary systems consumption in electric vehicle. Przegląd elektrotechniczny, **12**, 172-175 (2014)
7. A. Santiangeli, C. Fiori, F. Zuccari, A. Dell'Era, F. Orecchini, A. D'Orazio. Experimental analysis of the auxiliaries consumption in the energy balance of a pre-series plug-in hybrid-electric vehicle. *Energy Procedia*, **45**, 779-788 (2014).
8. Air - Density and Specific Weight, http://www.engineeringtoolbox.com/air-density-specific-weight-d_600.html
9. A. Duce, P. Egede, G. Öhlschläger, T. Dettmer, H. Althaus, T. Bütler, E. Szczechowicz, *Guidelines for the LCA of electric vehicles*, 158, http://www.elcar-project.eu/fileadmin/dokumente/Guideline_versions/eLCAr_guidelines.pdf (2013)
10. Real-World Nissan LEAF Fleet Data Reveals, http://insideevs.com/real-world-nissan-leaf-fleet-data-reveals
11. Do electric cars work in cold weather? Get the Facts, http://blog.ucsusa.org/dave-reichmuth/electric-cars-cold-weather-temperatures
12. P. Mashkov, B. Gyoch. Thermal loading investigation of led bulbs for automotive headlights. *SR RU&SU*, **55**(4), 66-80, (in Bulgarian), (2016)
13. P. Mashkov, B. Gyoch, R. Ivanov, An investigation on characteristics of led bulbs for car headlights, *BulTrans-2016*, 118-123, (in Bulgarian), (2016)
14. I. Evtimov, R. Ivanov, G. Kadikyanov, Investigation of electric bicycle's regenerative braking, *BulTrans-2012*, 217-221, (in Bulgarian) (2012)
15. D. Mammosser, M. Boisvert, P. Micheau, Designing regenerative braking strategies for electric vehicles with an efficiency map, *21ème Congres Francais de Mecanique*, 6 (2013)

Modelling the impact of cyber attacks on the traffic control centre of an urban automobile transport system by means of enhanced cybersecurity

Yoana Ivanova[1,*]

[1]Bulgarian Academy of Sciences, Institute of ICT, "Information Technologies for Security" Department, 1000 Sofia, Bulgaria

Abstract. This paper aims to show the major role means of protection play for strengthening the cybersecurity of critical transport infrastructure by using the advanced method of simulation modelling. The simulation model of a Traffic Control Centre (TTC) of an urban Automobile Transport System (ATS) is created by the author in the Riverbed Modeler Academic Edition 17.5 computer networks simulation system and is exposed to the impact of a Denial-of-Service attack. In addition, logical conclusions have been made on the basis of the experimental results obtained and evaluated by comparative analysis with results from analogous previous studies.

1 Introduction

The focus of the present research is on the measures of strengthening and maintaining a secure, operational and sustainable critical transport infrastructure and in particular an urban Automobile Transport System (ATS) by building a reliable cyber protection of its Traffic Control Centre (TCC).

For this purpose, the author has studied and analyzed various advanced simulation environments based on agent-based modelling such as NetLogo [1], Aimsun 8.0, Cisco Packet Tracer and Riverbed Modeler. The capabilities of Riverbed Modeler Academic Edition 17.5 for modelling of computer networks by using a rich palette of realistic network devices and components, which could be precisely configured, make it the preferable software for assessing the impact of a Denial-of-Service attack on the TCC.

It is logical that the choice of an impenetrable protection is preceded by a detection of the vulnerabilities in a system. An example of an adaptive conceptual architecture for critical infrastructure cybersecurity is shown in Fig. 1. As it can be seen, the Topological Vulnerability Analysis (TVA) and the strategies for reducing the vulnerability are carried out at the prevention stage of the overall management, monitoring and analysis process, while the reinforcement of security by technical means is recommended to be done at the last stage.

The advantages of complex systems modelling by using professional simulation software are defined on the basis of a comparison with the other publicly known method of "penetration tests" for assessing the vulnerability of the critical infrastructure against cyber attacks.

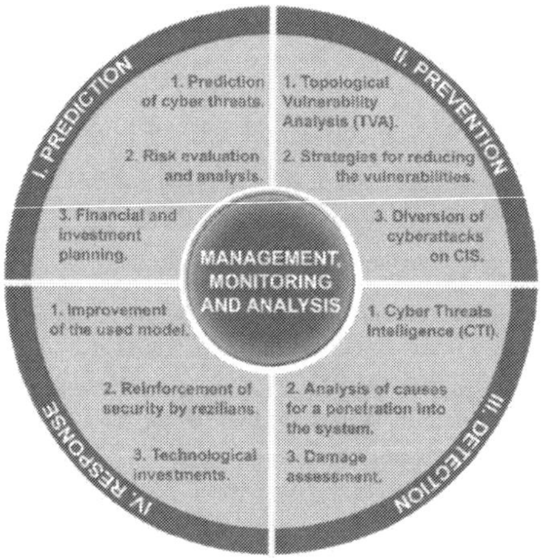

Fig. 1. Adaptive conceptual architecture for critical infrastructure cybersecurity.

2 Advantages of simulation modelling for assessing the vulnerability of critical infrastructure to cyber attacks

Simulation modelling is the main tool of TVA that is expressed in monitoring the state of network assets and maintaining models of network vulnerabilities and residual risk. It combines these to produce models that show the impact of individual and combined vulnerabilities on overall security posture [2].

In Riverbed Modeler Academic Edition 17.5 there is a possibility for specifying the settings of the DoS-attack profile and the vulnerability called "probability of

* Corresponding author: y.ivanova@bas.bg

DOI: 10.1051/matecconf/201713307001

infection" in percentages, as it can be seen from the screenshot shown in Fig. 2. The study could begin with absence of vulnerability (None) and continue to set the maximum vulnerability of 100 %.

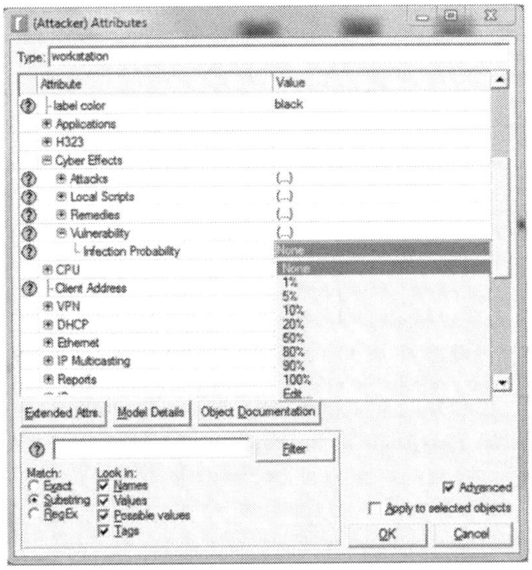

Fig. 2. System vulnerability settings related to a DoS-attack.

Penetration tests are used to verify security and, in particular to detect vulnerabilities in security systems by attacking them in the same way as a potential hacker would. One of the main differences between the two methods is that penetration tests must be implemented in "virtual machines" with vulnerable applications in order to safely test different techniques that target a system breakthrough, using publicly available tools [3]. This can be considered as an advantage of simulation modelling as compared to penetration tests, because the direct installation of simulation software on a personal computer does not pose a threat of infection.

Indeed, both methods in their essence are intended to detect vulnerabilities in the system but differ in the purpose of taking this action. In simulation modelling vulnerability assessment precedes simulation execution, while in penetration tests the attacker wastes time looking for weak points in the system. Practically, every complex system of great importance such as a TCC is protected in some way and it takes the attacker a relatively long time to overcome these means of protection. In this case, the advantage of simulation modelling is the possibility for the impact of a cyber attack to be seen instantly by relevant settings for ignoring the protection.

Actually, the full process of penetration testing is executed in seven successive phases as follows: ***Pre-engagement, Information gathering, Threat modelling, Vulnerability analysis, Exploitation, Post exploitation, Reporting [3].*** An analogous sequence of five phases in modelling a DoS-attack on a TCC in Riverbed Modeler Academic Edition 17.5 can be presented as a comparison: ***Gathering information about the DoS-attack, Attack modelling, Vulnerability assessment, Simulation execution, Summary evaluation of the results.*** By using the method of simulation modelling the phases become fewer as there is no direct and continuous communication with the client who has commissioned the penetration testing. This can be considered as another advantage of simulation modelling, because the client is not constantly involved in the process of performing a penetration test, but gets a finished end product. Additionally, penetration testing is usually conducted on site at the institution or organization that has commissioned it and this causes a disruption of its normal functioning during this process.

3 Means of protection against cyber attacks based on the potential vulnerabilities

The first step in defining an optimal protection strategy is to determine precisely the vulnerability of the specific management system which could be related to lapses in:

- ***network configuration*** – not using flow management methods; unencrypted passwords; bad device configuration in terms of security.
- ***network hardware*** – unsafe physical ports and unreliable physical protection.
- ***means of protection*** – misconfigured firewalls, undefined network security perimeter.
- ***software*** – buffer overflow, DoS-attacks, lack of Intrusion Detection System (IDS), failure to maintain Logs and monitor them in real time.
- ***hardware*** – insufficient testing, unreliable physical protection, unauthorized access, insecure remote access, Electromagnetic Interference (EMI).
- ***communication system*** – unsuitable Logs of firewalls and routers, lack of monitoring, authentication and verification of data integrity.
- ***security policies and procedures [4].***

On the basis of the classified main vulnerabilities the author recommends the following means for overall improvement of cybersecurity:

- ***antivirus software*** – constant update of the software is required to prevent penetration into the system.
- ***router*** – this autonomous device working with IP addresses can be defined as a smart device compared to the switch, because except for distributing the traffic it is also able to reduce the vulnerability of the system to cyber threats from Internet.
- ***Virtual Private Networks (VPN)*** – they ensure remote access, preserving the high quality and integrity of the output information. For example, IP VPN is a service designed to meet the needs of corporate business, where high quality, security and rich capabilities for integrated data, voice, video and multimedia transmission should be guaranteed.
- ***Firewall*** – this is an advanced approach to cybersecurity that is implemented as a strong authentication instead of static passwords through software and hardware complexes, routers, and other security measures. The interpretation of rules is implemented consistently by filters that allow for or prohibit the transmission of data (packets) to the next filter or protocol level.
- ***Honeypots and Honeynets*** – honeypots represent fake computer systems, setup as a "decoy", that are used to

collect data on intruders. This "decoy" appears to contain operating system vulnerabilities that make it an attractive target for hackers. While it appears vulnerable to attack, it actually prevents access to valuable data, administrative controls and other computers. Administrators can collect data on the identity, access, and compromise methods used by the intruder. Honeynets are networks, where all inbound and outbound data is analyzed and collected [5].

- ***Unified Threat Management (UTM)*** – complex solutions for cyber threat protection including all necessary modules.

4 Modelling the impact of a DoS-attack on the Traffic Control Centre of an urban Automobile Transport System through a firewall insertion

Being a professional simulation software Riverbed Modeler is widely used in security and defense for modelling all networks types and technologies (VoIP, TCP, OSPFv3, MPLS, IPv6) and analysing networks. It is suitable for testing and demonstrating technology designs before production; increasing network R&D productivity; developing proprietary wireless protocols and technologies; and evaluating enhancements to standards-based protocols [6].

For a comparison NetLogo is an open source multi-agent programmable modelling environment with a wide range of applications in different areas that makes it very useful for students. But it is not specialized in computer networks modelling and analysis like Riverbed Modeler and Cisco Packet Tracer.

The similarity between Riverbed Modeler and Cisco Packet Tracer is the possibility to simulate the complete network by connecting real types network devices and components. The main advantage of Riverbed Modeler compared to Cisco Packet Tracer is the possibility for directly simulating the impact of cyberattacks on different computer networks. For this reason the author has preferred Riverbed Modeler Academic Edition 17.5 to make the part of this research related to modelling the TCC and a cyberattack on it.

The research can be continued using the professional traffic modelling software Aimsun 8.0 after a logical assumption that the servers or services stop responding to client requests as a result of the simulated DoS-attack and this affects the Traffic Signal Control System. In this paper the focus is on strengthening the protection through a firewall insertion in order to reduce the probability of infection and respectively traffic changes under the impact of a DoS-atack to be prevented. Disturbances in normal signalling of traffic lights are observed after the server has stopped working due to the communication between the server and the signal controllers of the traffic lights [7]. Using reliable protections in TCC should help to support the normal traffic flows without increasing time delay and mean queues observed under the impact of successful cyberattacks.

The research of Prof. J. Alex Halderman from the University of Michigan in the field of computer security can be used in support of the claims about the course and consequences of cyberattacks on an urban ATS. In this study, the team led by Prof. Halderman concludes that the vulnerabilities they discovered in the infrastructure are not a fault of any one device or design choice, but rather show a systemic lack of security consciousness, which is expressed in: unencrypted radio signals; use usernames and passwords by default; a debug port that is easy to be attacked; using an older version of the installed software [8].

The author's reference model (M_{Ref}) of TCC has a typical network configuration including 3 workstations, 3 servers and a switch connected in a linear bus network. It has been developed in Riverbed Modeler Academic Edition 17.5 using built-in protections simulated by specific settings. The model is subjected to the impact of a DoS-attack that causes such an intensive traffic (flooding) that the processing of the requests is impeded [9].

The author has chosen to work with a firewall type *ethernet2_slip8_firewall_adv,* which is a specially programmed router. A screenshot of the model is shown in Fig. 3.

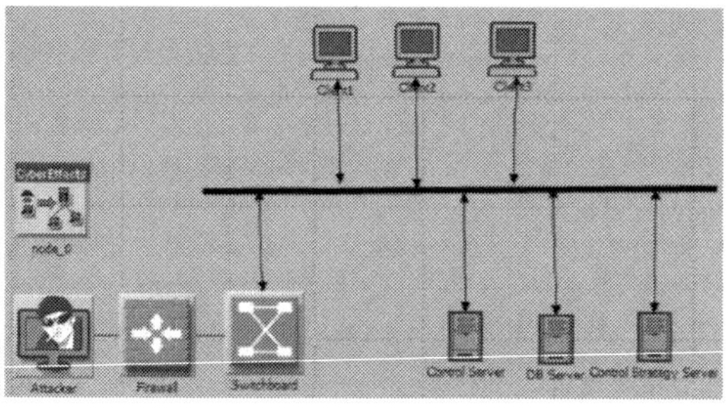

Fig. 3. The model of TCC with a firewall insertion under the impact of a DoS-attack.

The duration of this simulation is 30 s and is divided into six equal intervals of 5 s. Table 1 contains simulation results from 10 scenarios.

In the current research, the method used for verification of the model is based on the reaction time that represents a time interval between the beginning of an action and beginning of the corresponding response. The research is made for 10 consecutive scenarios. This method has been preferred by the author, because it demonstrates one of the great advantages of simulation modelling. It provides an opportunity the research to be done entirely in a virtual environment in order to save financial resources for realization in a physical environment using a hardware prototype. Therefore, the simulation results can be accepted as sufficiently reliable and at this stage it is not required to be compared with results of other analogous studies.

The comparative diagrams in Figures 4, 5 and 6 show in which second of running the simulation are registered peak levels of T_S and T_R, respectively with built-in protections and a firewall. The dark blue chart shows T_R without a firewall, while the red one shows T_R with a firewall. The green chart shows T_S without a firewall, while the light blue one shows T_S with a firewall.

Table 1. Peak levels of "traffic sent" ($T_{S,\,max}$) and "traffic received" ($T_{R,\,max}$), depending on the inter arrival time (T) for the model under the impact of a DoS-attack respectively without and with a firewall.

	Without a firewall		With a firewall	
T s	$T_{S,\,max}$ pack/s	$T_{R,\,max}$ pack/s	$T_{S,\,max}$ pack/s	$T_{R,\,max}$ pack/s
2	10	6,7	6,7	3,4
1	10	6,7	6,7	3,4
0,5	10	13,5	13,5	3,4
0,25	16,5	13,5	13,5	6,7
0,2	16,8	13,5	20	10
0,15	27	13,5	16,4	6,4
0,1	33,5	20	23	13
0,05	44	37	43	20
0,025	80	50	74,5	34,5
0,02	84	58	84	40

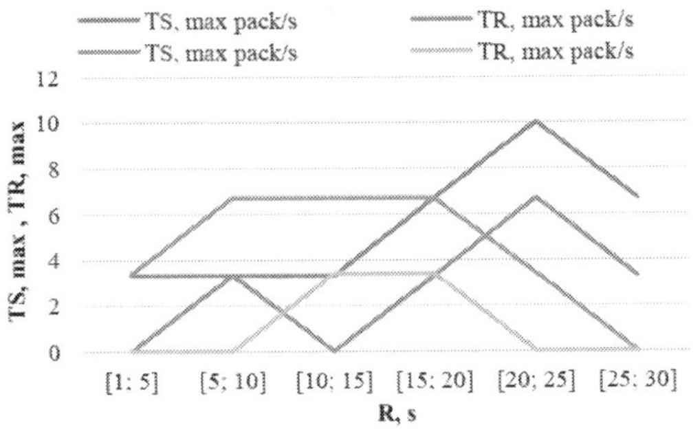

Fig. 4. $T_S = f(T)$ and $T_R = f(T)$ at $T = 2$ s.

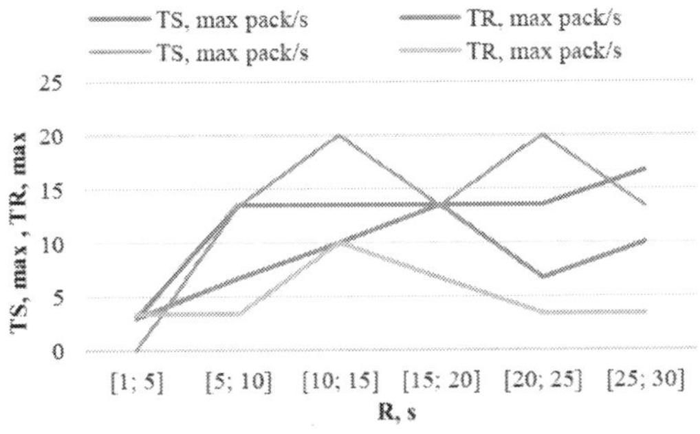

Fig. 5. $T_S = f(T)$ and $T_R = f(T)$ at $T = 0{,}2$ s.

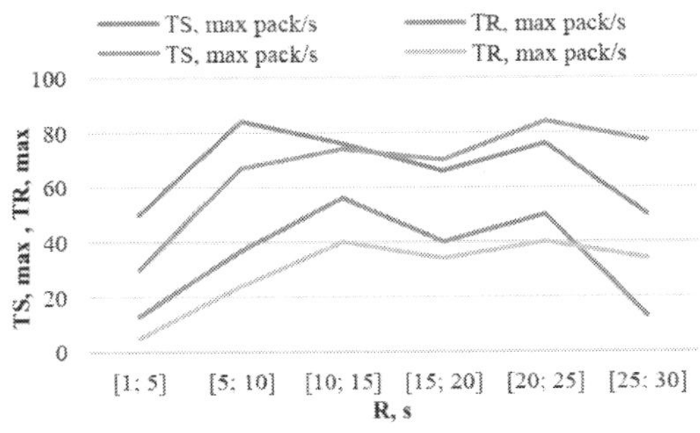

Fig. 6. $T_S = f(T)$ and $T_R = f(T)$ at $T = 0{,}02$ s.

Table 2, 3 and 4 contain the summary results of all three charts and show in which time intervals R in seconds are registered peak levels of T_S and T_R as functions of the selected three values of the inter arrival time T, respectively without and with a firewall.

Table 2. Peak levels of "traffic sent" ($T_{S,\,max}$) and "traffic received" ($T_{R,\,max}$) at $T = 2$ s.

	Without a firewall		With a firewall	
R s	$T_{S,\,max}$ pack/s	$T_{R,\,max}$ pack/s	$T_{S,\,max}$ pack/s	$T_{R,\,max}$ pack/s
[1; 5]	3,3	0	3,4	0
[5; 10]	3,3	3,3	6,7	0
[10; 15]	3,3	0	6,7	3,4
[15; 20]	6,7	3,3	6,7	3,4
[20; 25]	10	6,7	3,4	0
[25; 30]	6,7	3,3	0	0

Table 3. Peak levels of "traffic sent" ($T_{S,\,max}$) and "traffic received" ($T_{R,\,max}$) at $T = 0{,}2$ s.

	Without a firewall		With a firewall	
R s	$T_{S,\,max}$ pack/s	$T_{R,\,max}$ pack/s	$T_{S,\,max}$ pack/s	$T_{R,\,max}$ pack/s
[1; 5]	3	3	0	3,4
[5; 10]	13,5	6,7	13,4	3,4
[10; 15]	13,5	10	20	10
[15; 20]	13,5	13,5	13,4	6,7
[20; 25]	13,5	6,7	20	3,4
[25; 30]	16,7	10	13,4	3,4

Table 4. Peak levels of "traffic sent" ($T_{S,\,max}$) and "traffic received" ($T_{R,\,max}$) at $T = 0{,}02$ s.

	Without a firewall		With a firewall	
R s	$T_{S,\,max}$ pack/s	$T_{R,\,max}$ pack/s	$T_{S,\,max}$ pack/s	$T_{R,\,max}$ pack/s
[1; 5]	50	13	30	5
[5; 10]	84	37	67	24
[10; 15]	76	56	74	40
[15; 20]	66	40	70	34
[20; 25]	76	50	84	40
[25; 30]	50	13	77	34

Summary evaluation

On the basis of the tabular and graphical results the author has come to the following conclusions. When a firewall is used the number of sent packets $T_{S,\,max}$ is generally lower than the number of packets sent when built-in protections by vulnerability settings are used.

Before the time interval [25; 30], when the number of sent packets is less than 80, a better filtration of the packets sent is observed due to the firewall insertion as compared with the case of using only built-in device protections. In the last time interval, when the number of sent packets exceeds 80, there is a "saturation" which may be due to a limitation of the used type of a firewall.

It is also necessary to establish whether the packets themselves represent a threat or not. In case that certain packets threaten the system is necessary to improve the selected firewall model by replacing it with another model, or by integrating it with other means of protection.

In addition, when a firewall is used the cases of flooding are reduced to only one in the last time interval from all ten scenarios, while without the inserted means of enhanced protection the flooding starts in the time interval [20; 25] and it deepens in the last time interval [25; 30]. This means that a system denial can be observed in at least two of all 10 scenarios. If this should be presented with the probability of an adverse event, then when a firewall is used, this probability is equal to 10 %, but it is at least twice as much if there is not a firewall insertion.

Recommendation to enhance the protection used

One of the reliable solutions for analyzing the received packets are Intrusion Detection Systems (IDS), which evaluate each packet and assess whether it is a hazard or not. If a packet is defined as a source of a cyber threat, the system decides whether to ignore it completely, log it to be analyzed by the administrator or immediately alerts that such a packet has infiltrated the network.

The IDS can be placed in different locations in the network, as well as in the firewall itself. The advantage of placing it in the firewall is that many of the suspicious packets will be blocked at the entrance and potential cyber threats will be prevented. Besides, IDS control the firewall. An IP address of the network controller that monitors the traffic should not be set when configuring IDS to prevent a potential attacker from detecting the device. The aim is for the IDS sensor to be protected from network scans and attacks that it is trying to find. In this case the disadvantage is that the removal of the IP address causes problems for administrators to manage the process. The solution is to use two network interfaces, one of which is configured without IP and works as a sensor, and the other one is connected to a separate local area network that collects information and manages IDS devices [10].

5 Conclusion

This research is an example of applying simulation modelling for solving problems related to cybersecurity of critical transport infrastructure. There are possibilities for the study to be extended in order to improve the methods and means of protection using the simulation results.

It has been demonstrated that the negative impact of a cyber attack on the TCC can be reduced by using appropriate protection, but the wide variety of existing cyber threats requires that the means of protection are constantly tested, updated and improved. The demonstrated method proves that the use of a simulation environment is a very effective way to achieve that aim.

One of the directions in which the research can be expanded is an additional verification based on the margin of error by comparing the simulation results with measurements in a physical environment. In the present case, such study is not intended because the aim is to emphasize the advantages of simulation modelling itself.

References

1. Y. Ivanova, Military Journal, Sofia, 4, 113-126 (2015).
2. S. Jajodia, S. Noel, B. O'Berry, *Managing Cyber Threats: Issues, Approaches and Challenges*, Kluwer Academic Publisher, 248-250 (2005).
3. G. Weidman, No Starch Press Inc., San Francisco, 2-6 (2014).
4. E. Stoilov, *Department of Informatics at NBU*, Sofia, 21-28 (2010).
5. SANS Institute. *Honey Pots and Honey Nets - Security through Deception*, 1-5 (2003).
6. Riverbed Technology, *Riverbed Modeler: A Suite of Protocols and Technologies with a Sophisticated Development Environment*, San Francisco (2017).
7. Y. Ivanova, IJITS, issue 3, vol. 9, 117-142 (2017).
8. Y. Ivanova, IJITS, issue 2, vol. 9, 83-95 (2017).
9. B. Ghena, W. Beyer, A. Hillaker, J. Pevarnek. & A. J. Halderman, *Green Lights Forever: Analyzing the Security of Traffic Infrastructure*, Proceedings of the 8th USENIX Workshop on Offensive Technologies (WOOT '14) (2014).
10. E. Stoilov, *Department of Informatics at NBU*, Sofia, 32-36 (2011).

Low-shrink airfield cement concrete with respect to thermal resistance

Małgorzata Linek[1,*]

[1]Kielce University of Technology, al. Tysiąclecia Państwa Polskiego 7, 25-314 Kielce, Poland

Abstract. The paper presents theoretical background to the occurrence and propagation of imposed thermal load deep inside the structure of airfield pavement. The standard composition of low-shrink cement concrete intended for airfield pavements was presented. The influence of recurring temperature changes on the extent of shrinkage deformations was assessed. The obtained lab test results, combined with observations and analysis of changes of the hardened concrete microstructure, allowed the authors to draw conclusions. It was proven that the suggested concrete mix composition makes it possible to obtain the concrete type of better developed internal microstructure. More micro air voids and reduced distance between the voids were proven, which provides increased frost resistance of concrete. The change of size, structure and quantity of the hydration products in the cement matrix and better developed contact sections resulted in the improvement of the mechanical parameters of hardened concrete. Low-shrink concrete in all analysed cases proved to have increased resistance to the variable environmental conditions. Increased concrete resistance is identified through reduced registered shrinkage deformations and growth of mechanical parameters of concrete. Low-shrink concrete used for airfield structure guarantees extended time of reliable pavement operation.

1 Introduction

Thermal resistance of the airfield structure is defined as the ability to transfer diversified types of thermal load by pavement layers (especially pavement surfacing). Thermal loads of three types are generated onto airfield pavements. The first type includes loads generated as a result of the thermal processes occurring in the hardening concrete mix. The second type includes natural thermal loads occurring in concrete as a result of ambient temperature variations. The third type of thermal loads is the imposed loads. The factor which causes the occurrence of the imposed thermal loads on the airfield pavements are aircrafts in operation. Jet engines [1] suck the air, which combusts rapidly after being compressed and mixed with the fuel, thus producing a gas with high temperature and pressure. The gas gets out of the engine. Placing the engines under the aircraft wings contributes to directing the exhaust gas stream straight onto the pavements. The diameters of exhaust nozzles range between 0,45 m and 1,58m [2]. The velocity of the gas stream emitted from aircraft engines is irregular, turbulent and it depends mainly on engine performance [3, 4]. The width of stream of one engine is from 12 m to 20 m. Parameters of exhaust gas stream at nozzle outlet reach the streamflow velocity of 500-600 m/s and gas temperature of 530-900 °C. In the case of turbine jet engines without afterburner, in external channel, the velocity of air stream is 300-400 m/s and its temperature is 30-230 °C [3].

Hot exhaust gas stream, after leaving the engine nozzle on contact with airfield pavement, spreads in ellipsoidal shape forming the so called "stream core". Core characteristics depend on the individual structure solutions of the aircraft power unit and geometry thereof. The core width usually ranges from 5 to 10 m and it spreads from 70 to 90 m behind the nozzle edge [3]. The local heating of airfield exceeds 200 °C [4] within this area. The highest temperatures of hot gas stream occur along the extension of the engine axis. The intensity of the impact of such temperatures increases, together with the increase of inclination angle of the longitudinal axis of the aircraft, with respect to the pavement vertical alignment. It ranges between approx. 140 °C at 0° and over 520 °C at 16° [3].

Imposed thermal loads are multiple repeated gas stream impact of high kinetic energy and high temperature onto the pavement [3]. The turbine slope of the common aircraft structures is at a maximum of 2 to 3° and the minimum nozzle height from the airfield surface – 1,3 m. Taking into consideration such an assumption, the stream temperature at slab surface can reach even 250 °C. Concrete pavement can be resistant to such temperature for a short period of time of 5-10 s, however, multiple repeated deformations cause structural concrete changes. Stream dynamics can cause pavement defects and its intensity lowers with the distance from the nozzle and within immediate vicinity thereof. During wintertime, at low ambient temperatures, the degree of gas stream impact on the surface changes. Repeated

* Corresponding author: linekm@tu.kielce.pl

influence of high temperatures in alternating freezing and defrosting of the surface pavement layer increases the temperature difference and the extent of deformations reduces pavement resistance and durability [2].

Fig. 1. Change of concrete colour within the area of the so called stream core on the airfield pavement.

Within the area of the pavement under operation, within the location where the exhaust gas meets the pavement, there are areas of changed colour [5]. Visually analysing the colour of concrete exposed to the influence of high temperatures, subject to the indirect method, the temperatures to which concrete was heated, were estimated. The colour of the analysed concrete within such areas changed from natural to dark grey, dark yellow and even yellow (Fig. 1).

Such colour proves the influence of temperatures ranging from 600 °C to 900 °C on the pavement. Independent of the process length, heating usually affects 0,003-0,005 m of the concrete slab external layer. Surface layer with a thickness of approximately 0,0010-0,0015 m (Fig. 2) is particularly exposed to the intensive influence of temperature. Even under short stream impact, the dynamics of the heat flow can cause pavement vibrations. Vibrations of undefined characteristics can be the reason for the intensity of the damages.

Due to the fact that the degree of exposure of the airfield pavement to the impact of hot exhaust gas depends, among other things, on the properties of the component materials of the concrete mix, it is advisable to modify the mix composition intended for airfield pavements. During the varying heating and cooling, the range of concrete stress changes. As a result of the immediate temperature and humidity changes, shrink deformations can occur, contributing to the occurrence of micro cracks of concrete. Heat inside the concrete will move from the surface layer – with a higher temperature, towards cooler areas – located lower. Water contained in concrete evaporation during heating and its volume increase can form additional stress areas in concrete structure. Interruption of concrete structure continuity during heating and cooling causes additional hydration, dehydration and reactions in solid reaction in concrete components. Insufficient consideration of temperature changes, caused by recurrent heating can result in concrete damage. Such damage, including the loss of concrete continuity, will be represented by the occurrence of cracks.

Fig. 2. Change of concrete colour.

2 Purpose and scope of the research

The purpose of this research is to define the influence of the applied modification of the concrete mix composition on the change of the parameters of hardened concrete exposed to the influence of imposed thermal loads. The analysis included determination of changes of basic mechanical parameters of concrete and determination of the extent of rheological deformations.

The research included two series of paving-grade concrete. Types of concrete of C40/50 class intended for the airfield pavements, complying with the requirements of [6, 7], were analysed. The first research series included concrete whose composition contained the materials recommended by the requirements of [7]. This concrete was defined as CC-1. The results obtained in the case of CC-1 concrete were taken as reference values. The second research series was low-shrinkage concrete of modified composition (determination: CC-2). In CC-2 concrete composition, apart from materials consistent with the requirements [8-11], an optimum quantity (determined on the basis of pilot studies) of ceramic modifier in the form of dust was included. The following preliminary tests for both concrete types were conducted: compressive strength f_{cm}, tensile strength in bending f_{cf} and flexural strength f were determined. The tests were conducted, in accordance with the guidelines of the applicable standards [12-14] after the assumed curing period (28 days). The influence of the applied ceramic dust on the changes in compressive strength of concretes exposed to the influence of regular heating and cooling was determined (Δf_{cm}). Prepared samples intended for laboratory tests were cured in water of temperature 20 °C ± 2 °C for the first 28 days [15, 16]. Subsequent laboratory tests assumed the diversified number of test cycles. A single test cycle included the process of alternating heating and cooling - Table 1. Recurrent influence of temperature stimulated the impact of aircraft on airfield pavement. The assumed heating and cooling time corresponded to the time during which the aircraft affected the most frequently used airfield pavement within the area of Poland. The velocity of gas stream emitted during the test was 30 m/s.

Table 1. Conditions assumed during the research.

Concrete	Number of cycles	Curing conditions
CC_{50}	50	Heating: time: 60 s, temperature: 250 °C ± 2 °C
CC_{100}	100	Cooling: time: 900 s, temperature: 20 °C ± 2 °C

In order to assess the impact of the used ceramic dust on the changes of their internal structure, concretes were subject to observations in scanning electron microscope (SEM). Preparation of samples and interpretation of the obtained results were in compliance with those described in literature [17, 18]. Fresh fractures were prepared using concrete CC-1 and CC-2. The preparation surface was observed by means of scanning electron microscope SEM and each time it was more than 1 cm^2. The magnification ranged from 200x to 100000x.

During the basic tests, the values of shrinkage of concrete cured for the first 28 days and concrete exposed to the influence of variable thermal loads were defined.

3 Research materials

The assumed characteristics of aggregate composition (Table 2) determine the physical and mechanical parameters of hardened concrete and its resistance to the occurrence of shrinkage deformations. Granite grit was used in the total quantity of 1400 kg/m^3, out of which 37% was 2/8 mm fraction, 9% was 8/16 mm fraction and 32% was 16/32 mm fraction. In CC-1 mix, fine aggregate in the quantity of 415 kg/m^3 was included, while in the CC-2 mix it was included in the quantity of 370 kg/m^3. The change of aggregate quantity applied to the mixes resulted from the modifier contents. Aggregate compositions of the designed mixes (Table 2) were considered as good grain size distribution of airfield pavements. Mixes CC-I and CC-II included cement CEM I 42,5 N to the amount of 377 kg/m^3. Admixtures in the form of air entrainment agent (1,7 kg/m^3) and plasticizing admixture (0,7 kg/m^3) were used in these mixes.

Table 2. Aggregate compositions of the mixes.

Size of the reference sieve mesh [mm]	Passes through the sieve [%]		Mixtures curve [%]	
	CC-1	CC-2	lower	upper
16,00	100,0	100,0	100	100
11,20	83,6	80,4	79	85
8,00	69,9	67,8	65	75
5,60	58,3	60,9	49	61
4,00	47,1	41,2	39	52
2,00	33,6	27,5	25	40
1,00	20,6	23,9	16	28
0,50	11,6	13,4	9	18
0,25	3,0	2,7	2	2

4 Testing methods

For research purposes, shrinkage was defined as the difference in sample length. A basic measurement was the sample dimension measured directly after unforming thereof. This length referred to sample length in the course of its drying.

The extent of the observed shrinkage deformations is variable and depends closely on climatic conditions. It was assumed that defining the influence of the applied dust on the change in shrinkage deformations will take place in the case of five series. Each time, 6 samples of concrete CC-1 and CC-2 stored in the same environmental conditions were subject to the analysis. The first (A) series, reference series, included CC-1 and CC-2 concrete curing for the period of 28 days completely immersed in water with a temperature of 20 °C. Series B and D included CC-1 and CC-2 concrete stored from the 28th day of curing in air-dry conditions. Series C and E included CC-1 and CC-2 concrete exposed to the influence of thermal cycles. The comparison of conditions during the test has been presented in Table 3.

Table 3. Test conditions for the research series (A-E).

series	Time	Temperature		Humidity	
	[s]	[°C]		[%]	
A	2419200	20 ± 2		100	
B	48000	20 ± 2		46-64	
C	48000	25 ± 3	20 ± 2	42-50	56-65
D	96000	20 ± 2		46-64	
E	96000	28 ± 2	20 ± 2	40-47	56-65

In order to determine single shrinkage of CC-1 and CC-2 concrete lots A, B, C, D and E 6 samples were used, in accordance with the guidelines [18], in the form of beams with a cross section of 150×150×600 mm. Changes in sample length in the course of their drying were measured by means of a measuring instrument of reading capacity of more than 0,005 mm. Directly after removing each sample from a mould, the locating distance l_0 was determined. Socket extensometer and benchmarks stuck with a quick-drying glue to the sample surface were used.

The assumed frequency of measurement series A performance was determined every 2 days in the period between the 1st and the 10th day, every 4 days for the period between the 7th and 28th day. The assumed frequency of measurement series B, C, D and E performance was determined every 2 days for the entire period.

The adopted research procedure complies with the instructions [18]. In the case of each analysed sample, after specified time, shrinkage deformation was determined ε_s, according to formula (1), in which d_0 is initial-basic distance (directly after unforming) between measurement points and d_t is the distance determined after the expected drying time:

$$\varepsilon_s = (d_0 - d_t) / d_0. \quad (1)$$

Selected samples of CC-1 and CC-2 concrete subject to 50 and 100 thermal cycles were observed in a scanning electron microscope. During observation, special attention was paid to the structure of the contact areas between aggregate grains and cement matrix, porosity characteristics and changes of cement matrix crystallization.

5 Results and discussions

On the basis of the analysis of the obtained destructive test results, it was proven that CC-2 series concrete, regardless of the analysed feature, is distinguished by more favourable mechanical parameters (Table 4). In the case of compressive strength, after 28 days of curing in standard conditions, CC-2 concrete reached slightly higher compressive strength values. In the case of tensile strength, there is a significant growth in the tested feature of 18%. With reference to the airfield pavements, the flexural strength is identified with operational durability of pavement. And its higher value refers to higher fatigue limit of concrete. Also, in the case of flexural strength, higher parameters of CC-2 concrete were observed.

Table 4. Mechanical parameters ($\overline{X}$ - average value [MPa], σ - standard deviation).

Concrete	f_{cm}		f		f_{cf}	
	$\overline{X}$	σ	$\overline{X}$	σ	$\overline{X}$	σ
CC-1	57,0	0,04	6,1	0,05	4,76	0,05
CC-2	59,4	0,03	7,3	0,03	5,59	0,04

According to the obtained results from the shrinkage deformation analysis it was proven that dust additive used in the composition of CC-2 concrete has significant impact on the extent of reduced shrinkage (Table 5). Regardless of environmental conditions and the time of thermal cycles influence, CC-2 concrete is distinguished by reduced deformations. It was proven that CC-1^{A} concrete cured for 28 days in water with a temperature of 20 °C is distinguished by a deformation of 0,0417‰, while in the case of CC-2^{A} concrete this value does not exceed 0,0150‰ (Table 5 – A series). It was proven that change in the curing conditions (series B and C and series D and E) also influences the phenomenon of increased shrinkage deformations. Samples of CC-1 and CC-2 concrete stored for 50 and 100 cycles in air-dry conditions (average temperature of 20 °C and humidity range of 46-64%) are distinguished by average deformation lower than in case of thermal cycles. In each of these cases, deformations of CC-2 concrete are significantly lower than in CC-1 concrete. Deformations of CC-1^{B} concrete amount to 0,000128‰ and they are greater by order of magnitude than deformation of CC-2^{B} concrete. Similar tendency is also observed after extending the period with subsequent 50 thermal cycles. The average deformation in CC-1^{D} concrete is 0,000135‰, while in CC-2^{D} concrete it is only 0,000071‰. A favourable phenomenon is also the fact that with CC-2 concrete, the growth of deformations after 100 cycles is lower than after 50 cycles. It is evidenced by the fact of the favourable influence of the applied dust on the reduction of deformations within a longer period of time. This phenomenon was also proven by the conducted compressive strength tests. The tested feature of CC-2 concrete after 56 days of curing was 63,7 MPa and it was higher by 4,5 MPa than the resistance of CC-1 concrete. It was proven that CC-2 concrete exposed to alternating thermal cycles is distinguished by significantly less shrinkage deformation. It should be emphasized that deformations observed in CC-2 concrete, in each of the discussed cases, reached significantly lower values than those of CC-1 concrete (Fig. 3).

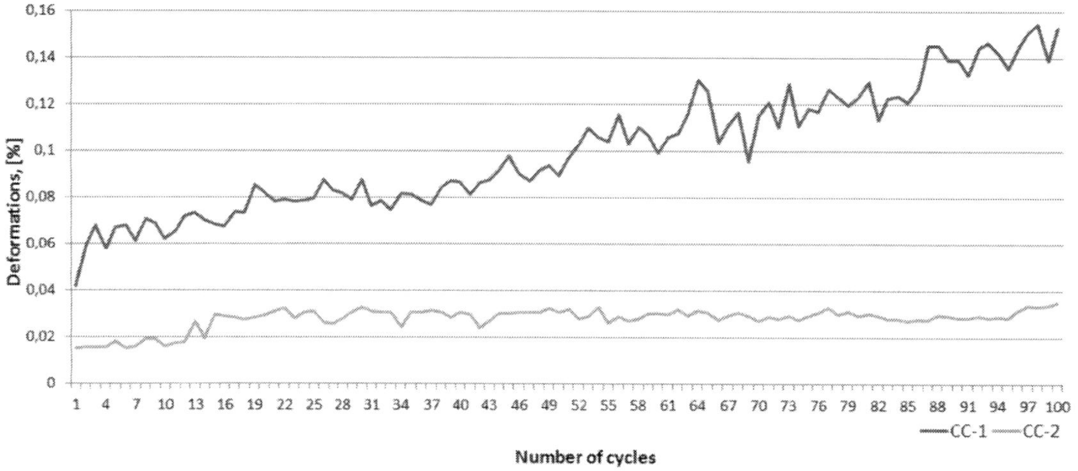

Fig. 3. Deformation of concrete CC-1 and CC-2.

Table 5. Increasing shrinkage deformations of CC-1 and CC-2 concrete in various conditions (where: T indicates days, ε_p indicates initial deformation [‰], and ε_k indicates final deformation [‰]).

series	T	Concrete CC-1			Concrete CC-2		
	[days]	$\varepsilon_p \div \varepsilon_k$	$\varepsilon_k - \varepsilon_p$	$\frac{\varepsilon_k - \varepsilon_p}{t}$	$\varepsilon_p \div \varepsilon_k$	$\varepsilon_k - \varepsilon_p$	$\frac{\varepsilon_k - \varepsilon_p}{t}$
A	28	0,0012÷0,0417	0,0405	$1,45\times10^{-3}$	0,0004÷0,0150	0,0146	$5,21\times10^{-4}$
B	13,3	0,0417÷0,0434	0,0017	$1,28\times10^{-4}$	0,0150÷0,0161	0,0011	$8,27\times10^{-5}$
C	13,3	0,0417÷0,0977	0,0056	$4,21\times10^{-4}$	0,0150÷0,0327	0,0018	$1,35\times10^{-4}$
D	26,6	0,0417÷0,0453	0,0036	$1,35\times10^{-4}$	0,0150÷0,0169	0,0019	$7,14\times10^{-5}$
E	26,6	0,0417÷0,1547	0,1094	$4,11\times10^{-3}$	0,0150÷0,0352	0,0202	$7,59\times10^{-4}$

Physical and mechanical parameters (among others, average compressive strength and flexural strength) are internal concrete structure derivatives. The following factors influenced, among others, the concrete structure of series I and II: air voids characteristics, occurrence of aggregate grains (fine, coarse and ceramic dust) of various mineral composition and diversified shapes and dimensions, structure of cement matrix and contact area between matrix and grains. Modification of concrete mix composition of series II with ceramic dust, contributed to the provision of concrete of better developed internal structure. According to the observation made of CC-1 and CC-2 series concrete by means of SEM, after 28 days of curing and exposure to thermal cycles it was proven that the influence of the alternating heating and cooling cycles contributes to significantly unfavourable changes in the microstructure of the hardened concrete.

The microstructure of CC-1^{50} and CC-1^{100} concrete, in comparison to CC-1 concrete, showed a change in the upper, surface layer of a sample exposed to the influence of exhaust gas. This layer is distinguished by the increased concentration of carbonates and increased porosity and width of micro scratches. In matrix, crystallization of fine-grained hydrated calcium silicates of type C-S-H prevails. There are also cracked aggregate grains (Fig.4).

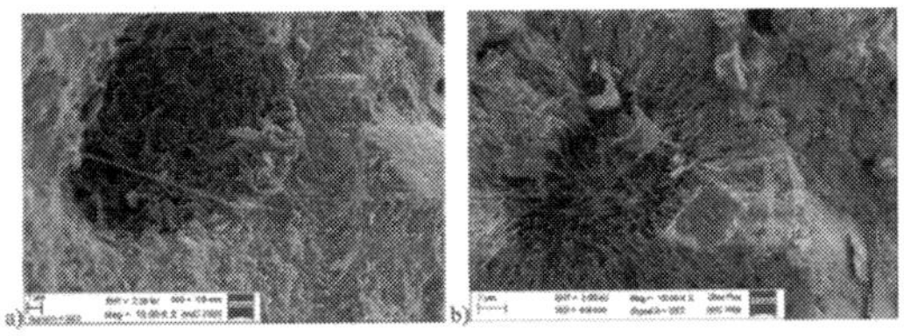

Fig. 4. Microstructure diversification: a) concrete CC-1 and b) concrete CC-1^{50}.

The microstructure of CC-2^{50} and CC-2^{100} concrete as compared to CC-2 is distinguished by a change within the surface area, accumulation of carbonates, increased porosity and a few scratches. Ettringite crystals occur inside air voids. Contact areas of cement matrix with aggregate grains remain uninterrupted (Fig. 5).

The microstructure of CC-2^{50} and CC-2^{100} concrete when compared with CC-1^{50} and CC-1^{100} concrete demonstrated a lower influence of temperature within the surface layer. Scratches in the modified concrete occur less often than in the reference concrete, at the same time they are distinguished by a smaller width. In the case of voids in both types of concrete, crystallization of ettringites prevails, however crystals in CC-1 concrete are predominantly bigger than in CC-2 concrete. In the matrix of CC-2 concrete prevail fine-grained hydrated calcium silicates of type C-S-H, whereas contact areas with aggregate grains remain uninterrupted (Fig. 6).

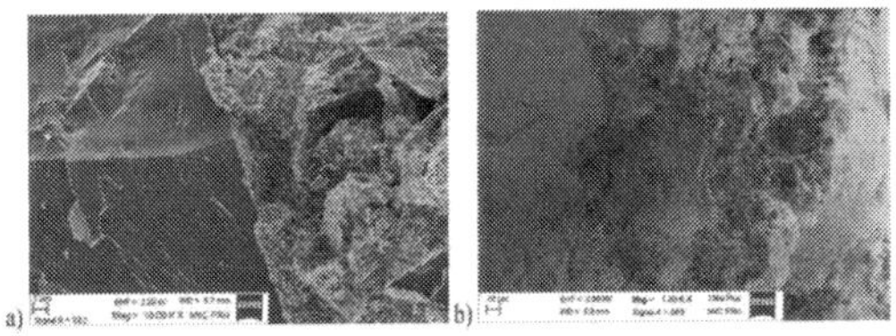

Fig. 5. Microstructure diversification: a) concrete CC-2 and b) concrete CC-2^{50}.

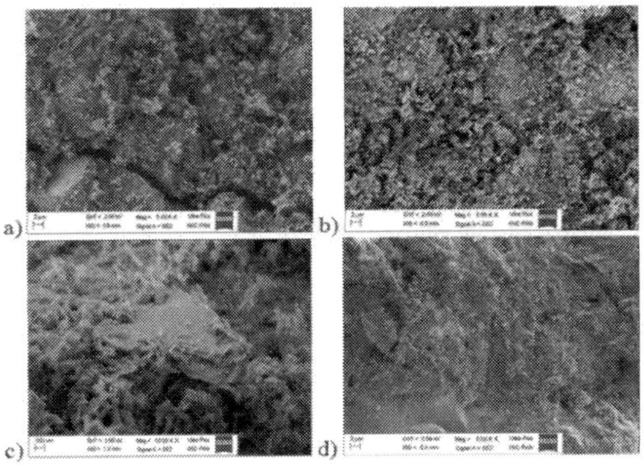

Fig. 6. Microstructure diversification of cement matrix: a) and c) concrete CC-1^{50}, b) and d) concrete CC-2^{50}.

6 Conclusions

On the basis of the tests, which have been conducted, the following conclusions were drawn:

a) Application of dust in the mix favourably influences the changes in concrete micro structure. In the case of CC-2 concrete, the content of distinguished components and their mutual ratios changes with respect to CC-1 concrete. In CC-2 concrete, stored in standard conditions and subjected to heating cycles, there were uninterrupted contact areas between aggregate grains and cement matrix. Discontinuities in the cement matrix

DOI: 10.1051/matecconf/201713307002

occurred rarely and their width did not exceed 6μm. The microstructure of CC-1 concrete was distinguished by significant number of micro cracks occurring both in cement matrix and contact areas with aggregate grains. Micro cracks width was varied and strictly dependant on curing conditions and areas of occurrence. The process of micro cracks formation in the case of CC-1 concrete clearly increased heating cycles. In the case of CC-2 concrete, less developed ettringite crystallization occurred (the length of single crystals did not exceed 10 μm) and different, clearly fine-grained C-S-H phase than in CC-1 concrete.

b) Application of dust in the mix favourably influences the changes occurring in concrete porosity characteristics. In the case of CC-1 concrete, air voids were located within greater distances than in CC-2 concrete. Moreover, the diameters of these voids were much bigger than in those in CC-2 concrete. General air voids content in CC-1 concrete was 2,2%, and 3,7%in CC-2 concrete.

c) The application of dust in the mix significantly influences the reduction of observed rheorogical deformations of hardened concrete. With concrete cured at a temperature of 20 °C, deformations in CC-2 concrete are almost three times less than in CC-1 concrete. When the curing period is extended, CC-2 concrete provides higher resistance to the occurrence of shrinkage deformations. The observed average deformation for this type of concrete is 0,0000827‰ after 13,3 days and 0,0000714‰ after 26,6 days and they are lower in magnitude than in CC-1 concrete. The most significant influence of dust on the reduction of deformation was seen in concretes exposed to the influence of alternating heating and cooling cycles. Deformation in CC-1 concrete exposed to 50 thermal cycles is 0,000421‰, and in CC-2 concrete it is only 0,000135‰. A much more significant reduction in deformation was observed after 100 thermal cycles. In the case of CC-1 concrete, deformation amounted to 0,00411‰ and in the case of CC-2 concrete it was 0,000759‰.

d) The application of dust in the mix favourably influences the changes in porosity characteristics and internal structure of CC-2 concrete, which results in obtaining low-shrinkage concrete.

References

1. A. Jankowski, Theses Air Force Institute of Technology, 28, (2010)
2. M. Linek, Surface concrete with improved parameters of physical and mechanical on the loads caused by forced temperature, PhD Thesis, Kielce, (2013)
3. P. Nita, Concrete surface airport. *Theory and structural dimensioning*, Publishing Air Force Institute of Technology, (2005)
4. B. Trigoni, *Struinnaja erozja aerodromow*, Transport, (1981)
5. M. Linek, P. Nita, Problems of research and operation of aviation technology, **9**, 85-100, (2016)
6. PN-EN 206-1 Concrete-Part 1: Specification, performance, production and conformity
7. NO 17-A204:2015 Airfield concrete pavements - Requirements and test methods for cement concrete pavements (2015)
8. M. Linek, P. Nita, The impact of the ceramic modifier on selected characteristics of the physical and mechanical properties of concrete pavement airport, *VII Conference Dni Betonu*, 83-92, (2012)
9. M. Linek, P. Nita, Influence of ceramic additive on the size of rheological deformations of airfield pavement concrete, *IX Conference Dni Betonu*, 131-142, (2016)
10. M. Linek, P. Nita, Journal of KONES, **23 1** (2016)
11. M. Linek, Low Shrinkage Cement Concrete Intended For Airfield Pavements, in print, (2017)
12. PN-EN 12390-3:2011, Testing hardened concrete - Part 3: Compressive strength of test specimens
13. PN-EN 12390-5:2011 Testing hardened concrete - Part 5: Flexural strength of test specimens
14. PN-EN 12390-6:2011 Testing hardened concrete - Part 6: Tensile splitting strength of test specimens
15. PN-EN 12390-1:2013 Testing hardened concrete - Part 1: Shape, dimensions and other requirements for specimens and moulds
16. PN-EN 12390-2:2011 Testing hardened concrete - Part 2: Making and curing specimens for strength tests
17. M. A. Glinicki, *Durability of concrete in road surfaces. Effect of microstructure, materials design, diagnostics*, Road and Bridge Research Institute, (2011)
18. ITB Instruction No 194/98, *Study of mechanical properties of concrete on samples taken in the forms*, Building Research Institute, (1998)

Methodology for testing a system for remote monitoring and control on auxiliary machines in electric vehicles

Vasil Dimitrov[1,*], and *Petko* Kostadinov[1]

[1]Todor Kableshkov University of Transport – Sofia, Department of Power Engineering and Electrical Equipment in Transport, Faculty of Telecommunications and Electrical Equipment in Transport, 158 Geo Milev St., 1574 Sofia, Bulgaria

Abstract. A laboratory system for remote monitoring and control of an asynchronous motor controlled by a soft starter and contemporary measuring and control devices has been developed and built. This laboratory system is used for research and in teaching. A study of the principles of operation, setting up and examination of intelligent energy meters, soft starters and PLC has been made as knowledge of the relevant software products is necessary. This is of great importance because systems for remote monitoring and control of energy consumption, efficiency and proper operation of the controlled objects are very often used in different spheres of industry, in building automation, transport, electricity distribution network, etc. Their implementation in electric vehicles for remote monitoring and control on auxiliary machines is also possible and very useful. In this paper, a methodology of tests is developed and some experiments are presented. Thus, an experimental verification of the developed methodology is made.

1 Introduction

Speed regulation of auxiliary machines in electric vehicles (pumps, fans, air coolers, compressors, etc.) is often unnecessary. These drives need to be started and stopped smoothly but their speed can be constant through the operating modes. Asynchronous motors controlled by soft starters can be used in those cases – their implementation enhances the starting performance of the drives by allowing the motor to start gradually and smoothly in a controlled manner [1-6]. Contemporary soft starters can be set up to meet the requirements of the individual application and the motor is adjusted to the machine's load behaviour [4, 5].

Therefore, the study and examination of soft starters is very important. A laboratory system has been designed and built and remote monitoring and control by an intelligent energy meter and PLC has been provided [7]. In this paper, a methodology of tests has been developed and verified.

2 Possibilities of testing and research

The existence of PLC with touch panel and intelligent energy meter Sentron PAC3200 connected to a PC in the developed laboratory system allows for simultaneous drive control, monitoring and parameters' tracing and capturing through the relevant software product Powerconfig V3.7 installed on the PC [7-9]. Thus this system offers various possibilities of tests and research, for example:

- Training the programming skills for setting-up the soft starters, intelligent measuring devices, PLCs and touch screens, as well as communication between these devices and the PC – knowledge of the relevant software products is necessary;
- Remote monitoring of the drive operation and analysis of the electrical system's current consumption and power flows;
- Capturing the dynamic characteristics when starting and stopping the drive at different settings and loads;
- Determination of the correlation between the soft starter settings and the drive performance and energy parameters (current, power, voltage, power factor, efficiency, THD, etc.);
- Study of the impact of start-up and ramp-down time on drive performance;
- Synthesis of algorithms for optimal control of the drive systems;
- Evaluation of the PLC's possibilities of using it in the drive systems for efficiency optimization;
- Examination of the asynchronous drive at variable load, etc.

2.1 Remote monitoring and analysis of the drive's parameters

The Sentron PAC3200 device must be connected to the PC and the Powerconfig V3.7 software must be run [8, 9]. This software package is a combined service and commissioning tool for Sentron measuring devices. It offers the possibility for remote parameterization of the device, as well as monitoring and documentation of the system parameters. The measured values can be observed in real time in numerical and graphic form - the moment values are written and the graphics can be plotted (Fig. 1).

* Corresponding author: ydimitroff@abv.bg

 DOI: 10.1051/matecconf/201713308001

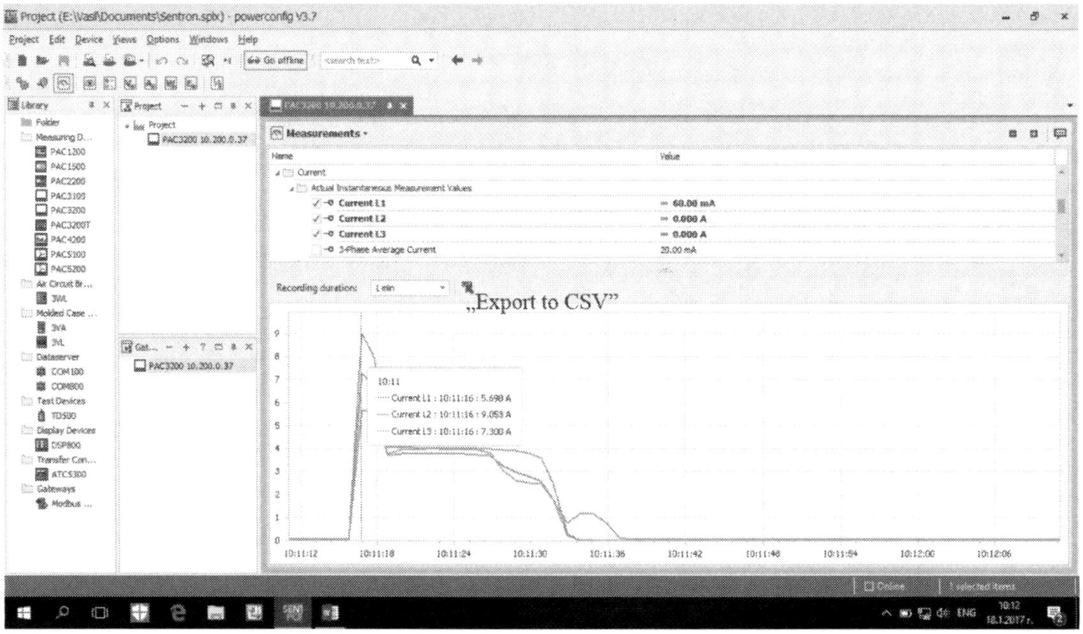

Fig. 1. Software Powerconfig V3.7 – current monitoring.

All the relevant system parameters can be shown on the PC display selected through the menu: voltage (phase-neutral and phase-phase), currents, active, reactive and apparent power (per phase and total), power factor (per phase and total), frequency, total harmonic distortion (THD) for voltage and current, phase unbalance (voltage), etc. The date, time and value of the selected parameters are displayed in real time. The refresh time could be changed (the minimum value is 330 ms) and the differences could be observed. Read-out and saving of message lists, load profiles, and histories is also possible.

The variation of the motor currents during different regimes is shown in the Fig. 1. The big values of the starting currents are clearly visible. The currents are constant during the steady-state operation and decrease slowly after the Stop command. The difference between the currents is caused by the unbalanced three-phase voltage and the increased value (~1A) of the current I1 after a Stop command is generated by affecting the electrodynamic brake. Thus these graphics show a proper operation of the drive.

2.2 Capturing the dynamic characteristics

The Powerconfig V3.7 software provides the opportunity to export data to Excel tables using the "Export to CSV" button (see Fig. 1). The duration of recording has to be defined in advance. The data can be processed in Excel and many graphics can be plotted and analyzed.

The methodology of capturing the dynamic characteristics $I = f(t)$ when starting and stopping the drive are as follows:

1) The laboratory system must be switched on by the breaker. Three phase voltage is displayed on the Sentron PAC3200. Many other parameters could be visualized by pressing F2 or F3 function keys [8]. These parameters can also be seen on the PC monitor.

2) Direction of the motor rotation has to be selected (by *F* or *R* buttons).

3) The acceleration time t_{acc} has to be set by the potentiometer. The deceleration time t_{dec} and the initial voltage U_{in} can be also set by other potentiometers. The Boost–function could be also activated by the switch on the Touch Panel.

4) Sentron PAC3200 has to be set to display the currents. Min. / max. values stored in the memory can be checked using function key F1 and they have to be cleared by pressing key F3.

5) The drive starts by pressing the button on the Touch Panel. The values of the currents have to be observed. If there are voltmeters connected between the soft starter and drive, the values of the three phase motor voltage could also be observed. Acceleration time can be measured by a chronometer. The yellow indicator of the soft starter lights up after 10s – the motor has received maximum power supply.

6) The maximum currents reached during starting can be displayed using function key F1. They may be compared to the values shown in the graphics.

7) The drive can be stopped by pressing the button on the Touch Panel. After the soft starter is switched off the drive could be stopped by means of the *Br* button if it is necessary.

8) The time t_{acc} or the initial voltage U_{in} must be changed and tests 3) – 7) have to be performed again.

After the end of the recording the data are exported in an Excel table and the graphs can be plotted. The three-phase current is shown in Fig. 2.

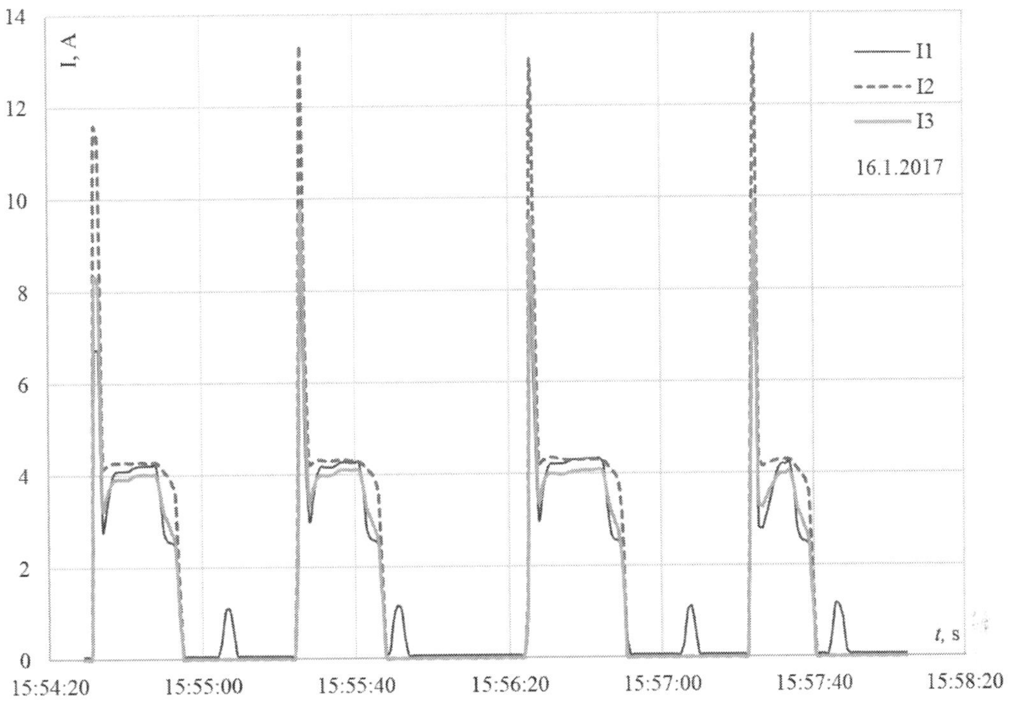

Fig. 2. Captured dynamic characteristics $I = f(t)$ at different settings of the acceleration time t_{acc} and initial voltage U_{in}.

All the relevant system parameters changes (P, Q, cosφ, THD, etc.) can be captured by similar methods.

It is clearly visible in Fig. 2 that the big values of the starting currents are different with changes in the acceleration time t_{acc} and initial voltage U_{in}. The currents are independent of the soft starter settings during the steady-state operation and decrease slowly after the Stop command.

2.3 Determination of the correlation between the soft starter settings and the drive performance

The developed methodology of tests allows to determine the correlation between the parameters of the drive (starting current I_{st}, power factor cosφ, efficiency η, THD-I, THD-U, power – active P, reactive Q, apparent S, etc.) and the settings of the soft starter:

$I_{st} = f(t_{acc};t_{dcc};U_{in})$, $P = f(t_{acc};t_{dcc};U_{in})$,
$Q = f(t_{acc};t_{dcc};U_{in})$, $S = f(t_{acc};t_{dcc};U_{in})$,
$\cos\varphi = f(t_{acc};t_{dcc};U_{in})$, $\eta = f(t_{acc};t_{dcc};U_{in})$,
$THD_I = f(t_{acc};t_{dcc};U_{in})$, $THD_U = f(t_{acc};t_{dcc};U_{in})$.

The data obtained during the appropriate tests must be previously exported to an Excel table and the graphs have to be plotted. In this paper, the correlation between the acceleration time and starting current at different settings of the initial voltage is determined. Data and graphs analogue to these shown in the Fig. 2 have been used. The methodology of determining this correlation is as follows:

1) The average values of the starting current at different settings must be calculated:

$$I_{st} = \frac{I_{1st}+I_{2st}+I_{3st}}{3}. \qquad (1)$$

2) The graph representing the correlation between the starting current and acceleration time at constant initial voltage has to be plotted.

3) A family of characteristics can be obtained at various settings of the initial voltage as a percentage of the nominal voltage (Fig. 3).

It is obvious that the change of the acceleration time setting, respectively the rate of the supply voltage increasing, results in a different maximum reached value of the starting current. When acceleration time is increased, the starting current is smaller, and when the initial value of the supply voltage is increased, the starting current also increases.

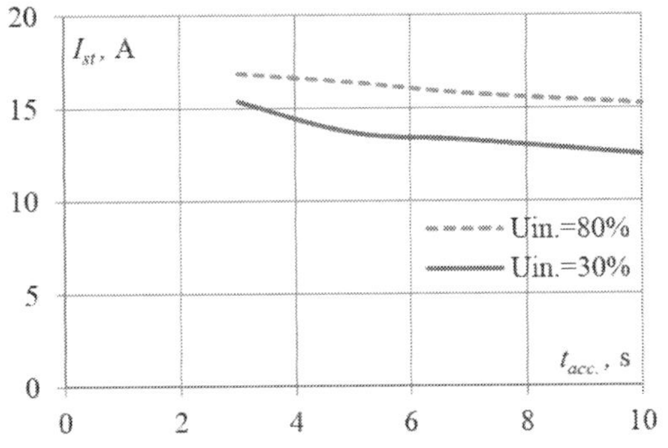

Fig. 3. Impact of the acceleration time t_{acc} on the starting current I_{st} at different settings of the initial voltage U_{in}.

2.4 Study of the impact of start-up time on the drive performance

Using the described methodology, the correlations between the acceleration time and the power factor *PF* (Fig. 4) and the coefficient THD-I (Fig. 5) have been plotted on graphs. It can be noted that the load of the drive is constant during the tests because the load of the auxiliary machines in electric vehicles does not change during their operation.

It can be seen that the power factor improves when reducing the acceleration time. This interdependence can be explained by the bigger starting torque developed by the asynchronous motor for a short time, and hence the greater value of the active power.

The THD-I value deteriorates as the acceleration time increases. This interdependence can be explained by the operation of the thyristor voltage regulator. Longer acceleration time is provided by increasing the unlocking angle α of the thyristors, which leads to a distortion of the sinusoidal form of the current [1, 2].

Therefore, changing the settings has a different impact on the drive parameters. The optimal solution for each particular case could be achieved but a complex research and testing have to be done.

Similar tests can be conducted to determine the effect of the Boost function and the deceleration time on the drive performance and energy efficiency. For example, the energy consumption during transient regimes of the auxiliary machines could be optimized: the initial voltage can be reduced when a centrifugal mechanism is driven (fan, pump, rotational compressor, etc.) because its starting torque is very low. However, if a piston compressor is driven, a big value of the initial voltage has to be chosen and even the boost function may be activated.

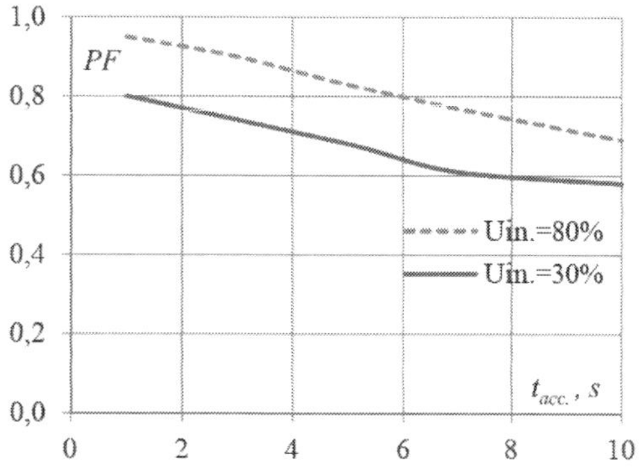

Fig. 4. Impact of the acceleration time t_{acc} on the power factor *PF* at different settings of the initial voltage U_{in}.

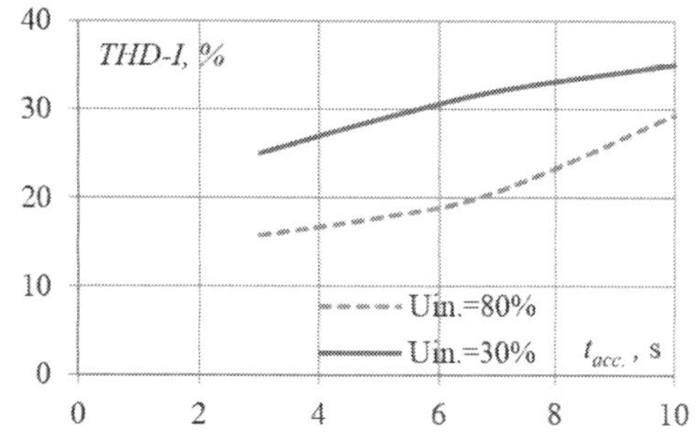

Fig. 5. Impact of the acceleration time t_{acc} on the coefficient THD-I at different settings of the initial voltage U_{in}.

3 Conclusion

A laboratory system has been designed and built using contemporary devices. An asynchronous motor is controlled by a soft starter and opportunities for remote monitoring (by an intelligent energy meter) and control (by a PLC and Touch panel) have been provided.

This laboratory system offers various possibilities for research and implementation into practice. Methodologies for different tests are developed. The impact of soft starter settings on drive performance and energetic parameters amendment could be examined. The interdependencies obtained during the experiments are similar to the characteristics known in the technical literature [1-3]. Thus an experimental verification of the described methods has been carried out.

If such a system is involved in electric vehicles for remote monitoring and control on auxiliary machines, energy efficiency and reliability will increase. The starting current will be reduced and the power factor and THD could be optimized in comparison with the direct connection of the motor to the grid. The optimal setting of the soft starter for each particular case must be done in advance. As a result of real time monitoring of drive performance any failure or damage can be discovered immediately and the motor can be instantly switched off from a distance by the operator in the control center.

References

1. A. Hughes, Electric Motors and Drives – Fundamentals, Types and Applications. Third edition, Elsevier Ltd (2006)
2. U. Bakshi, M. Bakshi, Electrical Drives and Control, Technical Publications Pune, India (2009)
3. I. Boldea, A. Nasar, Electrical Drives, CRC/Taylor & Francis Group, NY (2006)
4. The essential guide of Automation & Control (Schneider Electric, 2012)
5. Soft Starts for Single-Phase and Three-Phase Asynchronous Motors, Catalog, Schneider El (2011)
6. E. Dimitrova, Building automation and control systems, "MTC", **14 3/2**, art. ID: 1395 (2016)
7. V. Dimitrov, Design and setting up of a system for remote monitoring and control on auxiliary machines in electric vehicles, BulTrans Proceedings (2017)
8. SENTRON PAC3200, Technical specification, Siemens (2011)
9. http://w3.siemens.com/powerdistribution/global/en/lv/portfolio/pages/powerconfig.aspx

 DOI: 10.1051/matecconf/201713308002

Design and setting up of a system for remote monitoring and control on auxiliary machines in electric vehicles

Vasil Dimitrov[1,*], and *Petko* Kostadinov[1]

[1]Todor Kableshkov University of Transport – Sofia, Department of Power Engineering and Electrical Equipment in Transport, Faculty of Telecommunications and Electrical Equipment in Transport, 158 Geo Milev St., 1574 Sofia, Bulgaria

Abstract. Systems for remote monitoring and control of the proper operation, energy consumption, and efficiency of the controlled objects are very often used in different spheres of industry, in the electricity distribution network, etc. Various types of intelligent energy meters, PLCs and other control devices are involved in such systems. Proper operation of the auxiliary machines in electric vehicles is of great importance and implementation of a system for their remote monitoring and control is useful and ensures reliability and increased efficiency. A system has been designed and built using contemporary devices. An asynchronous motor is controlled by a soft starter and opportunities for remote monitoring (by an intelligent energy meter) and control (by a PLC and Touch panel) have been provided. Soft starters are widely used in industry for control on asynchronous drives when speed regulation is not a mandatory requirement. They are cheaper than inverters and frequency converters and allow for temporal reduction of the torque and current surge during start-up, as well as smooth deceleration. Therefore they can also be used in electric vehicles to control auxiliary machines (pumps, fans, air coolers, compressors, etc.). The present paper presents a methodology for their design and setting up.

1 Introduction

Proper operation of the auxiliary machines in electric vehicles is very important. They have to provide reliable cooling of the traction equipment, as well as compressed air with a certain pressure required for the systems controlling the doors, the brakes, and some pneumo-electrical devices. On the other hand, increased energy costs have resulted in greater emphasis on the power consumption of drive systems. It is extremely important to utilize the full potential for minimizing energy consumption using systems for control on motors with improved efficiency and performance. Different types of low-voltage asynchronous (induction) motors are very often used for driving auxiliary machines. These motors have minimized dimensions and weight, higher efficiency and enhanced reliability as opposed to DC motors.

In this paper, a laboratory system for control of an asynchronous motor has been designed and built with contemporary devices. Opportunities for remote monitoring and control have been provided. The developed methodology for its design and setting up can be used when building systems for control of auxiliary machines in different spheres of industry and transport.

2 Selecting the motors

Auxiliary machines in electric vehicles are mostly pumps, fans, air coolers, compressors. The driving motor has to be selected according to the rated power and speed of the machine [1]. The following correlation can be used:

- calculation power P_{calc} of the motor driving a pump:

$$P_{calc} = \frac{k_s.Q.H.\gamma}{\eta_p.\eta_t}, W, \quad (1)$$

where k_s= 1,1-1,3 is a stocking factor taking into account the possibility of overloading;
Q – pump flow rate, m^3/s;
H – computational feed height (head, liquid column), m;
γ – volume weight of the liquid, N/m^3;
η_p – efficiency of the pump;
η_t – efficiency of the gear (transmission used).

- calculation power P_{calc} of the motor driving a fan:

$$P_{calc} = \frac{k_s.Q.p}{\eta_f.\eta_t}, W, \quad (2)$$

where k_s = 1,25-2 at motor power under 5 kW and k_s = 1,1-1,25 at motor power above 5 kW;
Q – fan volume flow rate, m^3/s;
p – computational fan pressure, N/m^2;
η_f – efficiency of the fan.

- calculation power P_{calc} of the motor driving a piston compressor:

$$P_{calc} = \frac{Q.A_s}{\eta_c.\eta_t}, W, \quad (3)$$

where Q is the compressor performance, m^3/s;
A_s – specific work required to compress 1 m^3 of air to the appropriate pressure, J/m^3 (determined in Table 1);
η_c – efficiency of the compressor.

* Corresponding author: vdimitroff@abv.bg

 DOI: 10.1051/matecconf/201713308002

Table 1. Determining the specific work A_s required to compress 1 m³ of air to the appropriate pressure p.

Pressure: p, Pa	Specific work: A_s, J/m³
$1,5.10^5$	42.10^3
2.10^5	73.10^3
3.10^5	12.10^4
4.10^5	16.10^4
5.10^5	18.10^4
6.10^5	21.10^4
8.10^5	25.10^4
9.10^5	26.10^4
12.10^5	30.10^4

- calculation power P_{calc} of the motor driving a rotary compressor:

$$P_{calc} = \frac{168.Q.(p_2 - p_1)}{\eta_c.\eta_t}, W, \quad (4)$$

where p_1 and p_2 are the pressures of suction and exhausted air, Pa.

If the rated motor speed corresponds to the machine speed, a gear will not be necessary – a coupler should be used. As can be seen from the above-mentioned interdependencies, the motor power varies widely, but in most cases a motor with low or middle power is used. This is why a 3 kW motor has been selected and included in the developed laboratory system. An electromagnetic brake (EMB) has been built and coupled to the motor. It makes it possible to change the load of the motor [2].

3 Design of a control system

Asynchronous motors in contemporary drive systems can be controlled mostly by frequency converters (inverters) or soft starters. The speed regulation of the auxiliary machines in most cases is not a mandatory requirement. These drives need to be started and stopped smoothly but their speed can be constant through the operating modes. Intelligent devices called "Soft starters" can be used in those cases. They are cheaper than inverters and frequency converters and allow for temporal reduction of the torque and current surge during start-up, as well as smooth deceleration. Altistart 01 soft start / soft stop units ATS01N2 developed by Schneider Electric Company has been selected [3, 4]. This unit controls the motor power supply to limit the starting current and for deceleration using thyristor regulator. It also features an internal bypass relay switched on during steady state operation. This device is equipped with 3 potentiometers for setting the start and deceleration time and for adjusting the starting voltage threshold according to the motor load. Two logic inputs provide Run/Stop commands and a third logic input can be used for the Boost function (+24V positive logic). One logic output indicates the end of starting and one relay output indicates that the voltage is switched off at the end of the deceleration stage. A green LED indicates that the unit is powered and a yellow LED indicates that the motor is powered at nominal voltage. The appropriate model is selected according to the technical data of the asynchronous motor. Altistart ATS01N2**QN is available for nominal voltage of 380V. The ATS01N209QN model for nominal current 9A is appropriate for motors with power of 3-4 kW [5].

The necessary protection should be ensured and an appropriate 3 poles thermal magnetic motor circuit breaker MCB GV2-ME14 has been selected (rated current 10A and thermal protection adjustment range 6...10 A depending on the load) [6].

The developed scheme and a photo of the laboratory system are shown in Fig. 1. The possibility of reverse has been provided by two buttons F (forward) and R (reverse) and circuit breakers K1 and K2. A stop button S and the necessary locks have been also included. An opportunity for electrodynamic brake has been provided by the transformer Trf, rectifier, button Br and circuit breaker K. A 220 V/ 36 V transformer has been selected and the DC voltage is 32 V at current 1 A (acceptable for the rectifier diodes). Circuit breakers and buttons for operation in 220V circuits have been selected.

The logical inputs and outputs of the soft starter have been connected to the PLC Simatic S7-200. TP177micro Touch Panel has been configured for Start / Stop and Boost commands that are transmitted through the PLC to the soft starter. Two-wire control has been realized – the run and stop commands are controlled by a switch and single logic input LI2 of the soft starter. State "1" controls the run process and state "0" controls the stop process. Another switch has been configured for the Voltage Boost function via a third logic input of the soft starter – activating this input enables the function for supplying a starting overtorque capable of overcoming any mechanical friction. When the input is in state "1", the function is active and the starter supplies the rated voltage to the motor for a limited period of time (200 ms) before starting. The digital inputs of the PLC are used for receiving information about the drive condition. A bar has been configured on the Touch Panel and shows the motor speed indicated by the logic output LO1 of the soft starter (the bypass relay has been switched on and the motor has reached nominal speed). An alarm has also been configured on the Touch Panel and indicates a fault: the relay contact R1A-R1C of the Fault relay (built into the soft starter) closes with the Run command and opens when the motor voltage is around 0 with a decelerated stop or stop immediately on a fault.

Measuring equipment with high accuracy has been used. The power monitoring device Sentron PAC3200 developed by Siemens provides an analysis of the electrical system's current consumption and power flows [7-9]. It has been connected directly to the voltage system. Three 50/5 A current transformers CT are used for measuring the current (the maximum current at starting is 50 A). This device precisely and reliably detects the power values of electrical consumers and provides important measured values for assessing the system state and power quality. It has a range of useful monitoring, diagnostics and service functions, a two-tariff active and reactive energy counters (imported and exported energy), a universal counter, and a working hours counter for monitoring the running time of loads.

Four function keys can be used for a simple configuration and quicker selection of the desired display menu. The integral Ethernet interface is provided for easy

 DOI: 10.1051/matecconf/201713308002

integration into the remote monitoring system. In the laboratory system, it is used for communication between the device and the computer (PC). Modbus TCP Ethernet protocol is used (10 Mbps) and the relevant software product Powerconfig V3.7 has been installed on the PC. Sentron PAC3200 transmits the measured values to the supervisory systems, where the data can be further processed for display and control (there is an opportunity for exporting it to Excel tables). All the relevant system parameters can be shown on the graphical display or the monitor through the communication interface and the Powerconfig software:

- Voltage (Phase-neutral – $U_{1\text{-}n}$ / $U_{2\text{-}n}$ / $U_{3\text{-}n}$ and Phase-phase $U_{1\text{-}2}$ / $U_{2\text{-}3}$ / $U_{3\text{-}1}$);
- Currents (per phase – I_1 / I_2 / I_3);
- Active power (per phase – $\pm P_1$ / $\pm P_2$ / $\pm P_3$ and total $\pm P_{total}$);
- Reactive power (per phase – $\pm Q_1$ / $\pm Q_2$ / $\pm Q_3$ and total $\pm Q_{total}$);
- Apparent power (per phase – S_1 / S_2 / S_3 and total S_{total});
- Power factor (per phase – $|PF_1|$ / $|PF_2|$ / $|PF_3|$ and total $|PF_{total}|$);
- Frequency f;

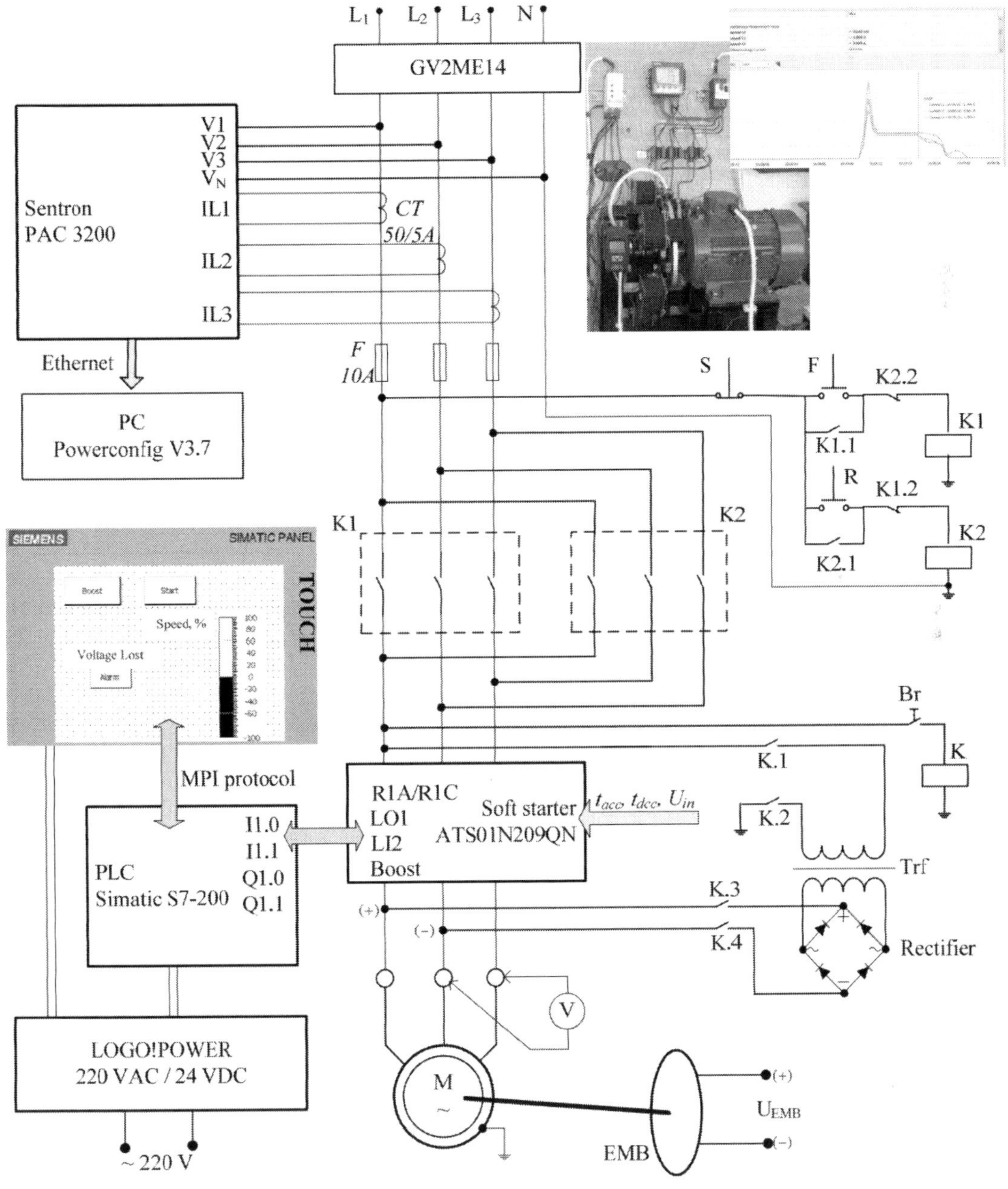

Fig. 1. Laboratory system – electrical scheme, exterior view of the drive and monitoring of its parameters on the PC.

- Total harmonic distortion (THD) for voltage and current (per phase – THD-U_1 / THD-U_2 / THD-U_3 / THD-I_1 / THD-I_2 / THD-I_3);
- Phase unbalance (Voltage);
- Min. / max. values (Voltage / Current / Power / Power factor / THD / Frequency / Three phase average voltage and current / Three phase average rating for active and reactive power), a possibility of reset is provided.

An example of remote monitoring of the drive performance on the PC (by the software Powerconfig) is shown in the Fig. 1. In addition, Sentron PAC3200 has a multifunctional digital input and output (simple logic functions for alarming can be set up).

Finally, it should be noted that most of the selected devices (or their analogs) may also be used in single-phase networks.

4 Setting up the devices

The presence of a PLC with a touch panel and a PC connected to the Sentron PAC3200 allows for simultaneous drive control, monitoring and parameters tracing and capturing. An appropriate setting up of devices to be used has to be made in advance.

Setting up the acceleration time (t_{acc}) using one of the soft starter potentiometers means that the time of the voltage ramp applied to the motor can be adjusted and gradual starting time can be achieved, which is dependent on the motor load. The initial voltage applied to the motor when switching on the soft starter (U_{in}) should also be set up in advance using the second potentiometer. It can be adjusted within the limits of 30-80 % U_n according to the load. Therefore, the starting torque is 30...80% of the one of a motor directly connected to the supply line.

The high values of the starting current and power can be reduced by increasing the acceleration time or decreasing the initial voltage. The correlation between the torque developed by the asynchronous motor and the supply voltage should be taken into account during the set up. The torque varies with the square of the voltage at a fixed frequency – for ex. at $U_{in} = ½\ U_n$ the starting torque will be reduced ¼ $M_{st\text{-}n}$. Therefore, if the voltage is very low, the drive could not start at high load and some devices could be damaged. Then the Boost function must be activated. On the other hand, the gradual increase in the voltage prevents the instantaneous current peak on power-up.

Setting up the deceleration time (t_{dcc}) can be done in a similar way using third potentiometer. A short time could lead to overvoltage that is not desirable. A smooth deceleration prevents the voltage drop on the line and damage of other devices in the network. This is why t_{dcc} should also be adjusted according to the motor load.

Initial setup of the measuring device Sentron PAC 3200 includes specifying the operating parameters in the device settings. The connection type (3P4W – 3 phases, 4 conductors, unbalanced load in the developed system), voltage (in our case direct measurement on the system) and measuring input voltage (400 V), current (primary 50 A and secondary 5 A) must be properly set up. Parameterization of the language, regional settings, and password protection could also be made. The universal counter and the limit values could be defined.

The next step is setting up communication between Sentron PAC3200 and PC. The following parameters have to be defined in both devices: IP address, Subnet mask, Gateway address and Protocol used (Modbus TCP). The correct setting ensures that the two devices can work properly together.

The configuration of the parameters can be done either directly from the display with the function keys or through the communication interface using the PC.

5 Conclusions

The developed system ensures the proper operation of the controlled machines. The existence of a PLC with a touch panel and a PC connected to Sentron PAC3200 allows for simultaneous drive monitoring and control. Therefore, any failure can be discovered immediately and the reliability of the drive system is increased. The energy consumption can be optimized through the proper setup of the soft starter for each particular application. Many other drives can be also connected to the PLC and a common control system would be realized. The methodology for its design and set up can be used in developing systems for remote monitoring and control of different electrical drives.

References

1. G. Pavlov, V. Dimitrov, *Design guide for electrical equipment* (Propeller, Sofia, 2013)
2. V. Dimitrov, *Development of laboratory system for electrical drives examinations*, "MTC", **13 3/3**, art. ID: 1246 (2015)
3. *Soft starters Altistart 01 for asynchronous motors*, Catalog (Schneider Electric, 2014)
4. *Soft Starts for Single-Phase and Three-Phase Asynchronous Motors*,Catalog (Schneider El., 2011)
5. *ATS01N209QN Soft Starter for Asynchronous Motors, Product data sheet* (Schneider El., 2011)
6. *TeSys GV2-Circuit breaker – thermal - magnetic – 6...10 A, Product data sheet* (Schneider El., 2016)
7. E. Dimitrova, *Conceptual description of the processes of remote control and monitoring of mobile objects*, "MTC", **14 3/2**, art. ID: 1394 (2016)
8. E. Dimitrova, *Building automation and control systems*, "MTC", **14 3/2**, art. ID: 1395 (2016)
9. *SENTRON PAC3200, Technical specification* (Siemens, 2011)
10. http://w3.siemens.com/powerdistribution/global/en/lv/portfolio/pages/powerconfig.aspx

AUTHOR INDEX